AF718681

Structural systems: behaviour and design

Volume 2: Spatial structural systems, foundations and dynamics

Structural systems: behaviour and design

Volume 2: Spatial structural systems, foundations and dynamics

L. T. Stavridis

thomas telford

Published by Thomas Telford Limited, 40 Marsh Wall, London E14 9TP.
www.thomastelford.com

Distributors for Thomas Telford books are
USA: Publishers Storage and Shipping Corp., 46 Development Road, Fitchburg, MA 01420
Australia: DA Books and Journals, 648 Whitehorse Road, Mitcham 3132, Victoria

First published 2010

Also available from Thomas Telford Limited
Structural systems: behaviour and design. Stavridis L.T. ISBN: 978-0-7277-4105-9
Structural systems: behaviour and design. Volume 1: Plane structural systems. Stavridis L.T. ISBN: 978-0-7277-4106-6
Structural analysis with finite elements. Rugarli P. ISBN: 978-0-7277-4093-9
Steel–concrete composite buildings: Designing with eurocodes. Collings D. ISBN: 978-0-7277-4089-2
Designers guide to eurocode 1: Actions on bridges. EN 1991-2, EN 1991-1-1, -1-3 to -1-7 and EN 1990 Annex A2. Calgaro J.-A., Tschumi M., Gulvanessian H. ISBN: 978-0-7277-3158-6

www.icevirtuallibrary.com

A catalogue record for this book is available from the British Library

ISBN: 978-0-7277-4107-3

© Thomas Telford Limited 2010

All rights, including translation, reserved. Except as permitted by the Copyright, Designs and Patents Act 1988, no part of this publication may be reproduced, stored in a retrieval system or transmitted in any form or by any means, electronic, mechanical, photocopying or otherwise, without the prior written permission of the Publisher, Thomas Telford Limited, 40 Marsh Wall, London E14 9TP.

This book is published on the understanding that the author is solely responsible for the statements made and opinions expressed in it and that its publication does not necessarily imply that such statements and/or opinions are or reflect the views or opinions of the publishers. While every effort has been made to ensure that the statements made and the opinions expressed in this publication provide a safe and accurate guide, no liability or responsibility can be accepted in this respect by the authors or publishers.

FSC
www.fsc.org
MIX
Paper from responsible sources
FSC® C013604

Typeset by Academic + Technical, Bristol
Index created by Indexing Specialists (UK) Ltd, Hove, East Sussex
Printed and bound in Great Britain by CPI Antony Rowe, Chippenham and Eastbourne

'To the memory of my parents'

Contents

"Πεπαιδευμένον γαρ εστιν
επι τοσουτω το ακριβές επιζητειν
όσον η του πράγματος φύσις επιδέχεται"
Αριστοτέλης

"Because it is the essence of education
to seek as much accuracy
as the nature of things allows"
Aristoteles

Preface

What a technically educated person (engineer, architect or constructor) understands today by the term 'structural design' is practically the same as what his fellow man meant 500 – or even more – years ago, namely a procedure based on the application of a particular item of knowledge, and because of which a structure will 'stand up' and not 'fall down', under whatever actions it may be exposed to during its life span.

However, what changed over the years and took the so-called 'structural theory' out of the realm of empiricism and intuition was the introduction of analysis in the assessment of structural behaviour and its elevation to an applied scientific field. Of course, the evolution of computational methods together with the wide availability of computer facilities played decisive roles in this.

Structural mechanics is now a highly demanding subject, not only from the point of view of its analytical treatment regarding structural behaviour but also with regard to its evaluation and practical application in structural design. Both of these directions have quite distinct characteristics.

The analytical approach always poses the question: *Given a structural configuration and a certain loading, what is the response and its corresponding deformations?* This is a problem governed by strict analytical conditions and requirements, establishing in this way the scientific character of the subject matter, but sometimes creating the illusion that the process of analysis is an end in itself. Of course, the solution to this problem is nowadays ensured, due to powerful numerical methods and the wide availability of personal computers.

On the other hand, the practical aspects focus on the application of load-carrying behaviour in the conceptual and working design of structures, by posing the essential question: *Given the physical environment and the prevailing service requirements, what structural system made of the appropriate materials will meet the necessary load-bearing requirements in an economical and aesthetically satisfactory way?* This problem does indeed constitute an end in itself.

It is the first approach which has become established, over the years, in education for many reasons, all of them hinging on the fact that it is this approach which ensures the appropriate 'scientific' profile. The second approach, although more realistic, seldom attracts the attention it deserves in the educational civil engineering curricula, a fact that the student unfortunately becomes aware of at a rather later stage, to his or her disappointment.

Indeed, the almost exclusive concentration on the computational aspect of structural mechanics dramatically deprives the young and inexperienced engineer from the physical perception of the load-carrying characteristics of a structural system, something which is not due solely to his or her lack of experience. The student concerned, being used to the study of complicated computational scenarios, tends to lack the mental clarity that permits the direct structural perception which is so often required either by the collaborating architect or the constructor on the building site.

In this book, comprised of two volumes, an approach to the understanding of load-carrying mechanisms and the behaviour of a wide range of structural systems is presented, with subsequent application to relevant design decisions, which, relying in principle on physical comprehension, is carried out through simple analytical reasoning. However, despite the prevailing non-computer-oriented philosophy, it is recognised throughout this book that the computational procedure in the design office practice needs in every case to be underpinned by the appropriate computer software. A necessary condition, though, for the successful use of such software is the ability of the user to derive, even approximately, results based on a well-cultivated structural perception regarding the load-carrying action and the subsequent preliminary design, which should in any case precede any computer modelling and the corresponding computation: this is exactly what the present book aims at.

Thus, through the study of basic load-bearing actions and the behaviour of typical systems, such as simply supported and continuous beams, frames, arches, cable structures of any type (volume 1), grids, plates, shells, rectilinear and curved thin-walled beams, and multi-storey systems (this volume), an attempt has been made to bring insights into these load-carrying characteristics that are necessary for their design and are usually overshadowed by a strictly analytical examination. In this respect, particular attention has been given to the use of reinforced and prestressed concrete, as well as to composite structures, in addition to the 'traditional' consideration of steel as the most 'convenient' material. Due attention has also been given to the plastic analysis and design of skeletal structures as well as to second-order theory and stability effects, with an emphasis on their practical and efficient use.

However, it has been deemed an absolute necessity to consider in advance and in some detail the handling of plane skeletal statically determinate and indeterminate structures in two respective chapters. Moreover, the dynamic behaviour of discrete-mass structures has been examined in the penultimate chapter, where not only the response of multi-storey systems to earthquake but also that of beams and plates to human-induced dynamic actions as well as to machine operation is discussed.

Of course, complete consideration of structural design must also include the foundation problem. Thus, the final chapter is dedicated to this issue, dealing primarily with the behaviour and design of shallow foundations, but also with soil–structure interaction and, to a lesser extent, the pile foundations.

It is hoped that, through the systematic examination of the above subjects, the basic elements of structural perception are emphasised which permit the safe preliminary design and dimensioning of any structure, such as a bridge, building, roofing of a large space etc. These elements constitute the basis of not only appropriate computer modelling but also of the deliberate acceptance – or not – of the numerical results produced by software, which is a very important issue.

This book follows a strictly progressive path in the presentation of various subjects. Thus, later chapter build only on previously examined concepts, and some basic knowledge of elementary mechanics is considered a prerequisite.

As is well known – since the first century BC, from the Roman architect Vitruvius – a successful structural concept requires the satisfaction of four characteristic properties, namely *technical safety*, *functionality*, *economy* and *aesthetic quality*. Technical safety means that the available strength should be greater than the resulting response; functionality means, structurally, the control of different deformational characteristics, annoying vibrations included; economy means the successful selection of the structural and foundation system, as well as of the appropriate construction method; and aesthetic quality means the achievement of structural elegance.

While the first two criteria are the subject of a knowledgeable technical analysis, successfully meeting the last two requires, on the part of the engineer, ingenuity, creativity and aesthetic judgment – properties not acquired by studying the prevailing design characteristics of the structural systems but are nevertheless prerequisites. Thus, although the successful design of a structure is based on the technical and aesthetic talent of the engineer, the knowledge of the structural principles put forward in this book are an essential basis for any such endeavour.

This book is aimed at everyone engaged in the study of structural analysis and design, either as a student, a practising engineer, or even as an architect seeking for a more profound structural understanding of bridge or building engineering. I hope that this books proves useful and achieves its goals.

Finally, I would like to express my gratitude to Thomas Telford Ltd for undertaking the publication of this edition of this book and the excellent collaboration with the editorial team, whose authoritative and knowledgeable direction under Matthew Lane has led to an entirely satisfactory result.

L.T. Stavridis
Athens, June 2010

Foreword to volumes 1 and 2

In the process of designing, engineers have to consider the different alternatives for structures, aiming generally towards an optimum of functional performance, resistance, durability and also towards a minimum of costs. In the dialogue with architects, the overall design as well as the suitable structural elements are looked for. Hence, the engineer has to master the art of structural design, has to understand the behaviour and the functions of beams, frames, plates, shells etc. Engineering knowledge is more than the ability to analyse a structure. It enables engineers to propose structural systems and also to determine a first estimation of the overall and sectional dimensions needed.

A computer aided numerical analysis can be applied only after the design is chosen. It is a complementary tool, especially helpful to obtain results for dimensioning the structure and for parameter variations. For checking and validating computer results it is needed to apply engineering judgement, plausibility considerations and e.g. simple equilibrium checks.

An engineer is a good engineer when he or she – even on vacation – does not only admire structures like bridges, towers, buildings, but when he or she is visualising the flow of forces, the structural elements for tension and those for compression, perhaps even imaging moment diagrams. This is rather easy where 'form follows function' (L.H. Sullivan) as for instance in suspension bridges, yet difficult for more complex structures such as thin shell structures.

The book of Professor Stavridis *Structural systems: behaviour and design* follows the above issues very consistently, making a remarkable contribution to this rather rarely encountered book genre. It is an excellent textbook, especially for Civil Engineering students and young professionals. It covers the principal approaches to determine states of equilibrium and to calculate the deformations. For statically indeterminate structures, alternative methods are demonstrated using either forces or deformations as unknowns. In explaining the structural behaviour of beams, frames, plates, shells, thin-walled elements, pre-stressed concrete, foundation elements and dynamically incited systems, basic concepts and directly applicable approaches are presented that allow an estimation of crucial quantities for the design of a very wide range of structures, including bridges, buildings, roofing of larger spaces and foundations. Many graphic figures help to visualise the underlying mechanics. Thus, the main objective is very well achieved: to promote structural understanding as a complementary tool to computer analyses, to make students and readers fit for the art of designing.

Heinz Duddeck
Professor emeritus Dr.Ing., Dr.Ing. E.h.
Institute of Structural Analysis, Technical University of Braunschweig

Foreword to volumes 1 and 2

Methods of structural analysis have experienced an explosive growth during the last 40 years. But it was the advent of powerful personal computers, along with the evolution of numerical tools (based mainly on the finite element method) and the parallel development of numerous reliable, comprehensive, commercially available computer software, that have enabled the engineer to tackle very complex structural systems. As a consequence, in today's design offices, analysis of even some rather simple systems is performed (especially by the younger generation of engineers) with the use of such computer codes. Classical as well modern methods of structural analysis (based on the principles of virtual work, compatibility of deformations, matrix analysis) are rather rarely invoked in everyday practice. Yet, these theoretical tools often constitute the major (if not the only) part of the curriculum in civil engineering schools.

Several problems may arise from this state of affairs. First, the danger of the '*black-box syndrome*': when a sophisticated code is used without the analyst having the ability to check if the results are indeed reasonable and to spot any errors in the physical meaning of his/her implicit assumptions and on how these assumptions are materialised in the model. Second, there is little if any training to help the young engineer develop a deeper understanding of how structural systems behave, let alone to sharpen his/her physical intuition; such understanding and intuition are necessary especially in the conception and preliminary design stages. Indeed, conceptual clarity and physical insight are rarely mentioned as key objectives of structural analysis courses.

The book by Professor Leonidas Stavridis offers a much needed addition to classical computational structural analysis: a physical approach is developed in which a structural system is decomposed into elements whose behaviour to the applied loads is easily computed '*from the basics*'. Starting in the first chapters with fundamental concepts and applications, the step-by-step exposition becomes progressively more advanced. Structural analysis blends naturally with mechanics of materials – the latter include reinforced and pre-stressed concrete, steel and composites. The in-depth analysis of standard structural systems (such as simply and multi-supported beams, frames, arches, cabled beams) is followed by the exposition of some more advanced topics such as buckling, plates and shells, thin-walled and box girders, grids and curved beams, laterally-loaded multi-story frames and shear walls.

It is amazing how the analysis of such complex systems is made so simple, clearly understandable even to a non-specialist civil engineer, as the present writer. This is accomplished to a large extent thanks to the numerous illustrative figures (sketches) which go far beyond the usual 'formalistic' figures of most available textbooks: they are imaginative, vivid, self-explanatory. What a difference they make when trying to comprehend difficult topics! For instance, the chapter on 'Shells' contains 51 elaborate figures, most of which comprise several sketches while a few of them are a whole page long. The 3D nature of cylindrical, spherical, paraboloid, and conical shells is elucidated with the help of ingeniously-selected isometric views and numerous cross-sections, so that the reader feels that this is a rather simple subject.

As an engineer with special interest in soil–foundation–structure interaction I was particularly happy with the comprehensive treatment of foundations. Viewed mainly from a structural engineer's viewpoint, the pertinent chapter deals not only with some classical deformation-settlement and stress-distribution problems, but also with the interplay between foundation stiffness and structure distress.

I believe this book will prove invaluable to both students and active engineers in helping them not only to absorb a huge volume of material but (more significantly) to cultivate '*engineering intuition*' and develop insight into the physics of structural analysis. For students, in particular, all this will offer the motivation for further study and the desire to later apply in real-life projects both the material and the methodology developed in the book.

George Gazetas
Professor of Soil Mechanics and Foundation Engineering
National Technical University of Athens

10

Grids

10.1 Overview

A grid is a plane formation of beams connected to each other with joints (basically crossing each other). The aim is to receive loads acting perpendicular to the grid plane and to transfer them in two or more directions. Despite the fact that the grid nodes are almost always formed as rigid, it is useful to consider the case where the cross-beams are simply supported by each other, thus transferring only an internal vertical force.

In the case of rigid joints, the grid beams develop three types of sectional force according to Section 2.5.1: bending moments, torsional moments and shear forces. The grid joints develop a vertical displacement and a rotation, represented by a horizontal vector (Figure 10.1). In the case of nodes simply supported by each other, the beams develop only bending moments and shear forces. It should be recalled that the loads acting on any cut-out part of the grid must satisfy three equilibrium conditions, specifically equilibrium of the vertical forces, and the equilibrium of the vector projections of the moments with respect to two arbitrary horizontal axes of the grid plane (see Section 2.5).

It is obvious that the grids constitute statically indeterminate structures with a large degree of redundancy, and their analysis almost always requires the application of suitable computer software. Nevertheless, it is essential to examine some basic characteristics of the load-carrying behaviour of grids.

10.2 Main characteristics of the structural behaviour of grids

In order to examine the main structural action of a grid, i.e. the transfer of vertical loads in two horizontal directions, the simple model shown in Figure 10.2 is employed, where a vertical load P is transferred in two directions via two cross-beams. The symmetry existing in both directions involves the development of only one (vertical) displacement at the node. This displacement is common to both beams. However, the corresponding curvature $1/r$ is different. The shorter beam will develop a greater curvature than the longer beam and, according to the relation $M = EI/r$, it will also develop a higher bending moment, provided that the stiffness EI of both beams is the same.

As has been shown in a previous chapter (see Section 2.3.8), the bending stiffness or 'rigidity' of the simply supported beam with regard to its midpoint displacement is the force required to cause a unit displacement equal to $48 \cdot EI/L^3$. The factor 48

Structural systems: behaviour and design
978-0-7277-4105-9
Copyright Thomas Telford Limited © 2010
All rights reserved

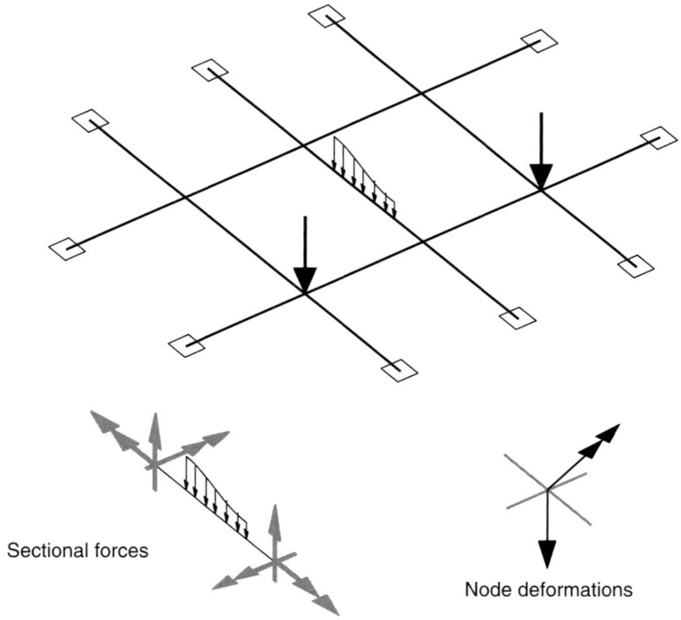

Figure 10.1 Internal forces and displacements in grids

becomes 192 for a beam with fixed ends. It thus appears that, for example, doubling the length of the beam causes an eightfold decrease in the above stiffness, while it is proportional to EI.

Analysing this simple system (using the method of forces or the method of deformations), the important conclusion is found that if P_1 and P_2 are the forces

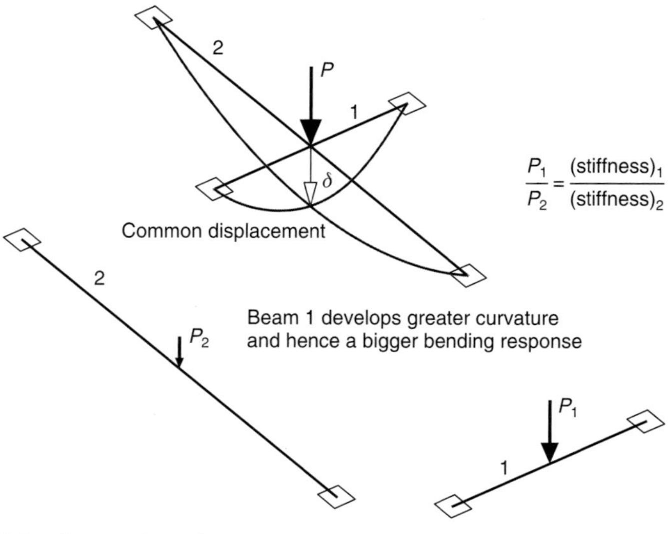

Figure 10.2 Load distribution depending on member stiffness

carried by beams 1 and 2 at their midpoints, respectively, then

$$\frac{P_1}{P_2} = \frac{(\text{stiffness})_1}{(\text{stiffness})_2} \quad \text{(see Section 3.2.10)}$$

It is easily concluded that if beam 2 has double the length of beam 1, then the load carried by beam 2 is one-ninth the applied load P, while the maximum moment of the short beam 1 is four times greater than that of beam 2. If beam 2 is even longer, it is clear that its contribution to carrying the vertical load will be insignificant.

It is, of course, clear that even in the case of beams with different lengths it is possible to achieve a balanced load distribution based on the above equation, by designing the beams in such a way that $(\text{stiffness})_1 = (\text{stiffness})_2$. ■

The conclusions resulting from the examination of the above simple model are also valid for regular grids that cover areas with an orthogonal outline. If the beams have the same cross-sectional characteristics, then those with a shorter length are more stressed, while for an aspect ratio larger than 1.50 the beams in the longitudinal direction are clearly 'underworking' (Figure 10.3).

The above relationship is obviously valid also in the case where the two cross-beams are not at right angles (see Figure 10.3). On the basis of this fact, it can be seen that, to cover oblong areas (regardless of the aspect ratio), the adoption of a skew layout of beams ensures essentially the same stressing for each group of beams, as they have the same length. Thus, the balanced load transfer through the two groups of beams clearly implies smaller bending moments (i.e. smaller beam depths) and also smaller deflections compared with the orthogonal layout.

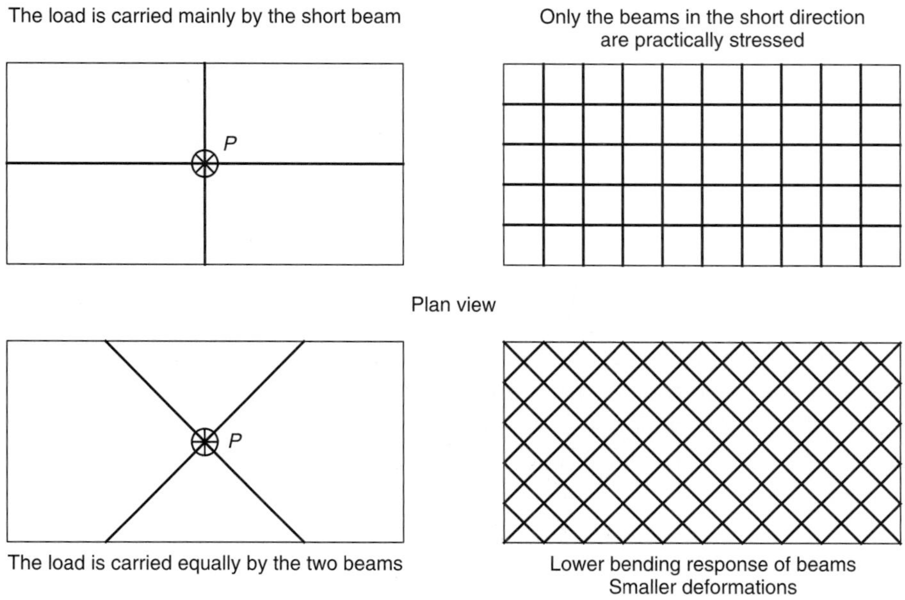

Figure 10.3 Influence of the grid layout on the load-carrying behaviour

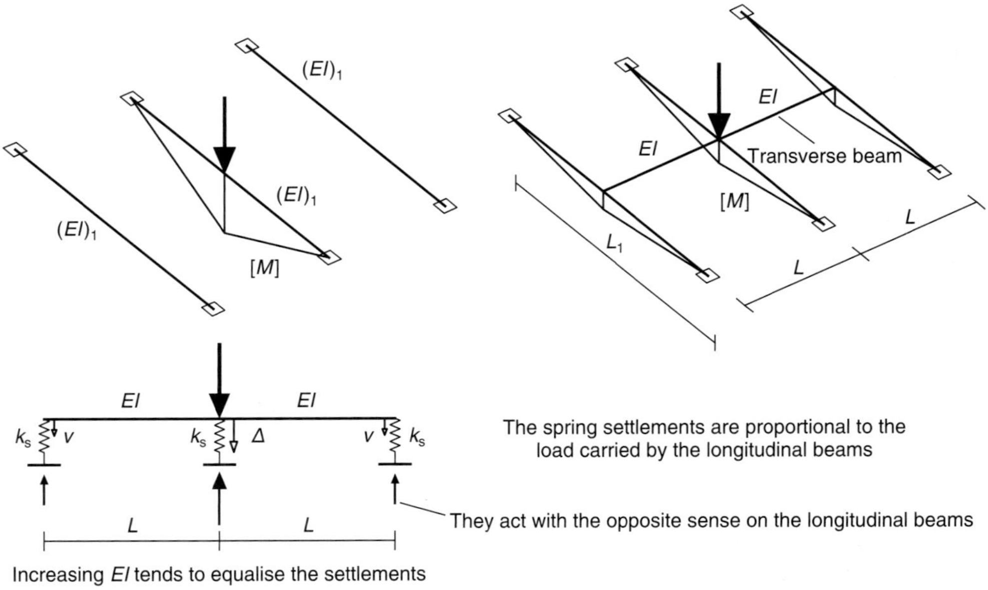

Figure 10.4 The influence of a cross-beam on the transfer of a concentrated load

Moreover, the presence of beams of short lengths and hence of high stiffness near the corners provides fixed conditions for the relevant diagonal beams, which is favourable for both the bending response and the deformation. ■

The model examined above shows the contribution of crossed beams in transferring the load. An additional characteristic of the load-bearing action of grids is the possibility that other neighbouring parallel beams also contribute to the transfer of a load that is applied to the beam under consideration.

More specifically, if a load is applied to the intermediate of three longitudinal beams with a parallel layout, as shown in Figure 10.4, the two extreme beams will obviously remain unloaded, whereas the presence of a transverse beam connected to them transfers the load also to the other two beams, thus relieving the intermediate beam. In order to examine the influence of the transverse beam, this beam may be considered as a continuous beam of two spans resting on three flexible supports with spring stiffness $k_s = 48 \cdot EI_1/L_1^3$, according to Section 2.3.7.

Under the influence of the applied concentrated load, the intermediate longitudinal beam deflects downwards by Δ, imposing the same displacement on the central support of the transverse beam too. At the extreme supports of the transverse beam, upward reactions will develop, which will be applied with opposite sign to the longitudinal beams (see Figure 10.4). From the analysis (see Chapter 3), the downward end displacement v is

$$v = \frac{\Delta}{1 + \dfrac{k_s \cdot L^3}{3 \cdot EI}}$$

The stressing of each longitudinal beam depends exclusively on its midpoint displacement, whereby, the effect of EI according to the above equation becomes clear. A small value of EI causes a small v, whereas with an increase in EI the displacement v tends to approach the value Δ. This means that the load is also transferred to the two extreme beams due to the stiffness of the transverse beams. The utilisation of a very stiff transverse beam ensures that, in practice, the load is equally distributed between the three parallel beams.

In view of this conclusion, it becomes clear that in oblong orthogonal areas, where, as was previously examined, only the beams in the shorter direction participate substantially in the distribution of loads, the longitudinal beams, despite the fact that they are ineffective for the uniformly distributed permanent loads, nevertheless must be placed in order to ensure, by their rigidity, the distribution of any concentrated load to the neighbouring main beams too. ■

In the above considerations it has been assumed that the cross-beams of the grid are simply resting on each other by transferring only vertical support forces. In practice, the connections between beams are not designed this way, being usually rigid (or monolithic), which means the development of additional torsional moments in the beams. This fact does not alter the above conclusions at all, while it contributes to the more favourable structural action of grids with respect to bending stresses and deformations. This favourable contribution of torsion to the bearing capacity of grids is explained immediately below.

The symmetrical model in Figure 10.5 is considered, where the beams are monolithically connected to each other. The cross-beam is simply supported at its ends while the two longitudinal beams are supported so as to be able to take up torsional moments. Two equal forces P are applied to the internal nodes of the model.

It is clear from the above that the longitudinal beams provide the transverse beam with an elastic support at the corresponding nodes with a spring stiffness equal to $k_s = 48 \cdot EI/L_1^3$. It is also clear that, besides the settlement of the support points, a rotation φ will also occur at these points in the transverse beam, which will be freely developed if the longitudinal beams are simply supported, but not if these are connected monolithically. The reason for this is that in the last case the above rotation will be necessarily developed in the longitudinal beam as well, which will develop torsional moments (see Figure 10.5), because the extreme cross-sections cannot rotate, thus applying a bending moment simultaneously to the transverse beam that resists this rotation. It is clear that this moment will decrease the bending stress state and deformation of the transverse beam accordingly (see Figure 10.5).

More specifically, in order for the longitudinal beam to develop a unit rotation at its considered internal node, it requires – or in other words, opposes the member imposing this rotation – a specific moment k_φ, is called the 'torsional stiffness' (see Section 2.5.2). Thus, the imposition of a rotation φ requires the application of a torsional moment $(k_\varphi \cdot \varphi)$, which is obviously also applied with opposite sign to the member imposing this rotation, i.e. to the transverse beam. Therefore, the transverse beam does not only have the translational springs k_s in its two internal nodes but also the torsional

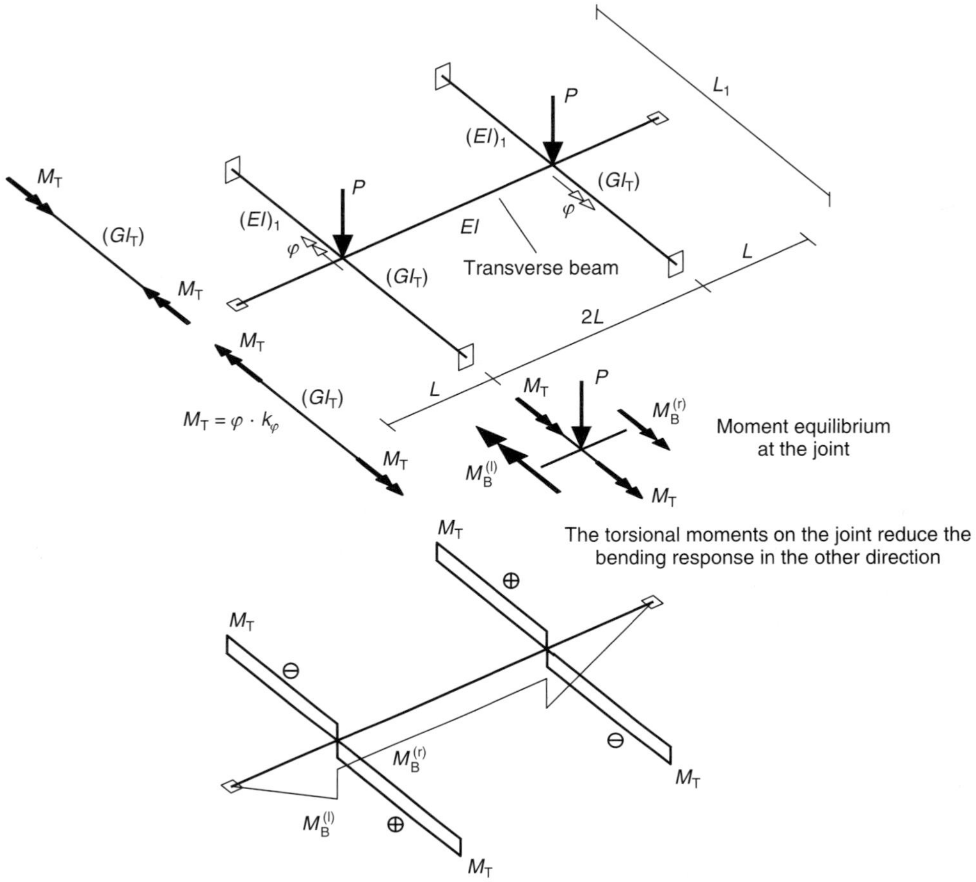

Figure 10.5 Relieving effect of torsion in the bending and deformation states

springs k_φ that oppose the rotation of the nodes (Figure 10.6). According to Section 2.5.3, as well as Section 3.2.10,

$$k_\varphi = \frac{4 \cdot G \cdot I_T}{L_1}$$

For the applied loads P and after direct application of the method of deformations (see Chapter 3) it is found that, for the particular layout shown in Figure 10.6, the developing vertical shift δ results from the relationship

$$\delta = \frac{PL^3}{3 \cdot EI} \cdot \frac{1}{1 - \dfrac{3}{4 + k_\varphi \cdot L/EI} + \dfrac{k_s \cdot L^3}{3 \cdot EI}}$$

From this equation one can draw the conclusion that the existence of torsional rigidity in the longitudinal beams that is expressed by the presence of a rotational spring k_φ causes a reduction in the developing settlement δ of the supports of the transverse beam, in

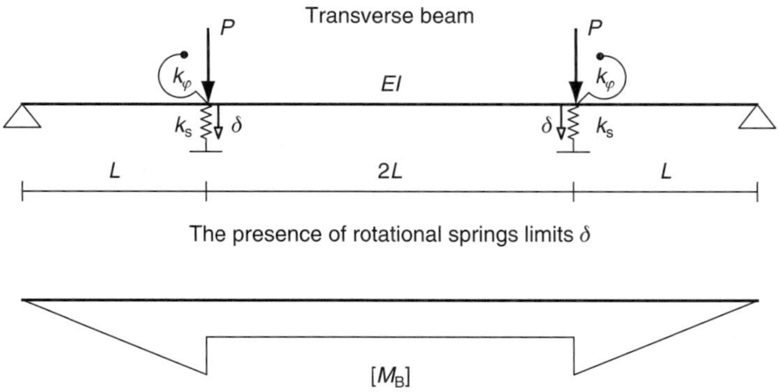

Figure 10.6 Justification of the relieving effect of torsion

contrast to the case where this torsional rigidity does not exist ($k_\varphi = 0$). In other words, the developing torsion of the longitudinal beams decreases the bending stress in both the transverse and the longitudinal beams, given the fact that this reduction in δ also implies decreased corresponding curvatures (see Figure 10.2). The system with monolithic nodes thus acquires higher stiffness. This is, of course, at the expense of torsional moments, which means additional shearing stresses.

It is clear that these conclusions are also applicable to larger grids: the existence of monolithic nodes implies the development of torsional moments, which reduce the bending response of beams and increase the stiffness against vertical loads.

The above 'contribution' offered by torsion – as well as its magnitude – depends directly on the torsional rigidity of the beams, i.e. on the value of the torsional constant I_T. Thus, in grids with beams of I cross-section, the developing torsional moments are much smaller than, for example, in grids of the same dimensions with hollow sections of the same moment of inertia, so that in the latter case the relieving influence of torsion on bending is more intense.

It becomes clear that the structural action of grids actually constitutes an interaction between torsional and bending rigidity, and is generally based on the ratio κ of these values, that is, $\kappa = GI_T/EI$. An increase in the ratio κ means an increased torsional contribution – and a corresponding bending restriction – in carrying the loads. ■

At this point, it should be pointed out that the presence of torsion in the grids is not essential for the equilibrium, as is the case in all the models previously examined. This means that equilibrium in a grid node, as shown in Figure 10.7, can be ensured – besides the requirement for vertical equilibrium – by itself, in the presence of bending moments. The presence of torsional moments acts simply as a helpful contribution for equilibrium to each direction, and is due to the requirements for compatibility of deformations. This has a direct repercussion on the approach to the design of a grid with respect to the ultimate state.

In particular, for beams of reinforced concrete with orthogonal section the imminent cracking will reduce their torsional much more than their bending stiffness, so that the torsional moments are reduced whereas the bending moments are increased.

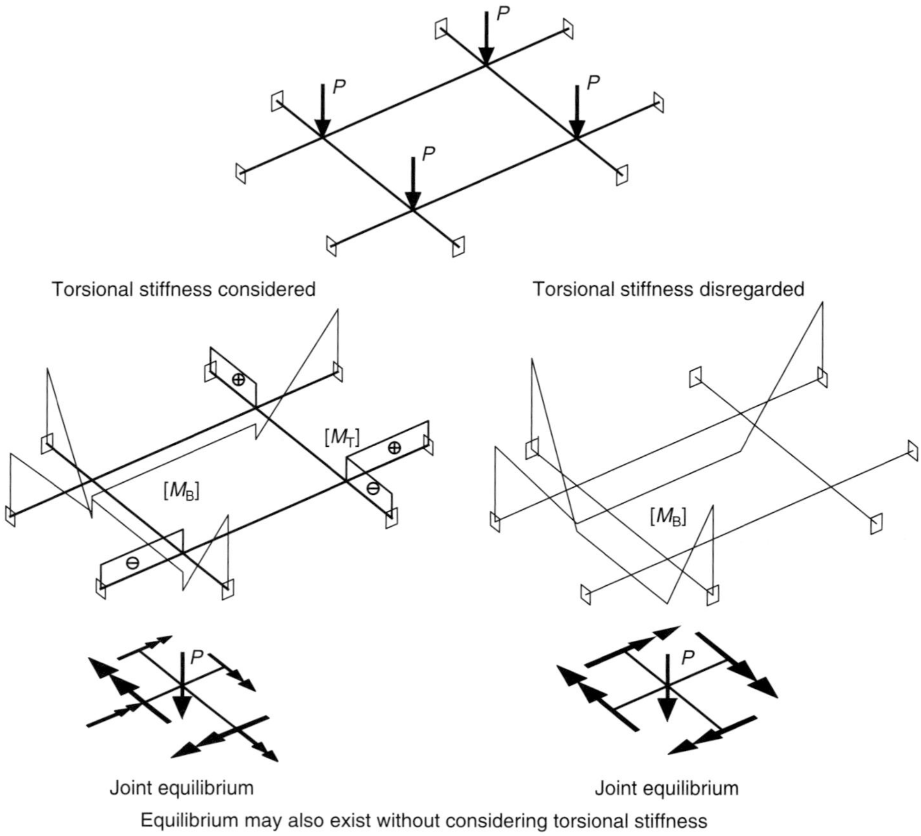

Figure 10.7 Load bearing with or without the development of torsion

As an extreme situation, the torsional rigidity can be totally disregarded, implying that the highest possible bending, can, of course, be considered. In any case, according to the static theorem of plastic analysis (see Section 6.6.2) and provided that the conditions for equilibrium are maintained, the design involving the omission of torsional moments (and a corresponding increase in the bending moments) is, of course, on the safe side, and can be followed, at least for the aims of preliminary design, as the estimate of the effective torsional stiffness due to cracking is rather uncertain.

10.3 Layout and structural action of skew bridges

From the previous section, it can now be understood how beams with relatively large spans and relatively high torsional rigidity can reduce their bending, because of the grid action, through the bending stiffness of the transverse beams connected to them.

This possibility is used in bridge design with the so-called *skew layout*, where the existing transport line cuts sideways the axis of an existing obstacle (river, underpassing road, etc.) and results in an increase in the length of the structure compared with what would be the case if the two axes were crossing each other at a right angle.

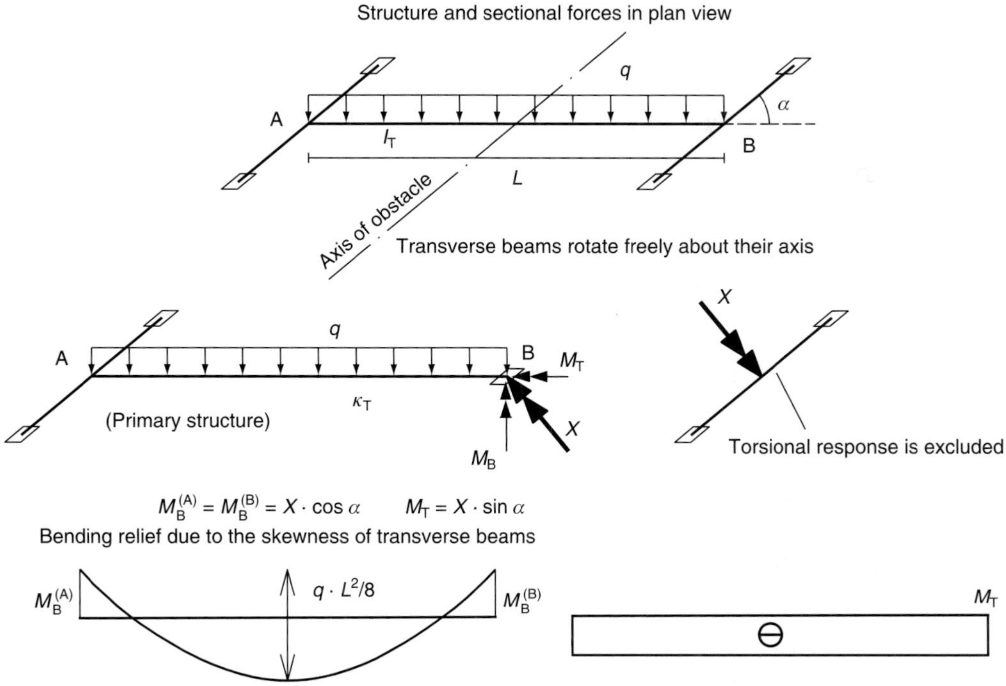

Figure 10.8 Skew bridge layout

The beam to be used to support each of the bridge beam ends should be arranged parallel to the axis of the obstacle, which is sideways with respect to the main beam (Figure 10.8). This support beam, which has a high bending stiffness, is simply supported on bearings that leave free (unhindered) the rotation about their connecting axis, so that no torsional stiffness is offered in practice. For the efficient structural action of this system, the main beam must carry significant torsional moments, and, as mentioned in Section 2.5.3 (see also Section 13.1), only hollow-type beam sections can meet this requirement.

The above-described structural system is shown in Figure 10.8. The main beam with length L is loaded with a uniformly distributed load q. At each node, the vector of the transferred moment can only be perpendicular to the axis of the support beam, as the least deviation from the perpendicular direction would create a torsional loading that the beam will not be able to carry. Considering this moment X as a statically redundant magnitude, it is found from the requirement for compatibility of the developing rotations, by applying the force method (see Chapter 3), that

$$X = \frac{q \cdot L^2}{8} \frac{2}{3 \cdot (1 + t \cdot g^2 \cdot \alpha/\kappa) \cdot \cos\alpha} \quad (\kappa = GI_T/EI)$$

It is clear that the moment X with its transverse component to the longitudinal axis of the main beam provides the relieving bending moment M_B for the span, while with its component M_T on the axis it provides the torsional moment of the beam. This action arises because of the angle of skewness α, and, as long as this angle decreases, the greater is the bending relief in the main beam and the greater is its torsional response. As previously pointed out, the latter increases as the ratio κ increases.

With regard to concrete structures, it is understandable that prestressing, which is used to limit the bending response of the main beams, also contributes substantially to guarantee their full torsional rigidity, which is necessary for the appropriate development of torsional moments.

11

Plates

Plates or *slabs* are flat monolithic structures that extend in two dimensions and have a relatively small, constant thickness. A plate is basically intended to take up distributed loads transversely to its plane; however, it presents a particularly high stiffness when receiving loads within its plane.

The support of the plate – to where naturally its loads are transferred – is set along the whole length or a part of its border, along the length of any line, or even at preselected points on its plane (Figure 11.1).

In this chapter, only plates receiving gravity loads are examined. Such plates are made almost exclusively of concrete, and constitute the largest part of the entire concrete volume used for load-bearing construction in general.

11.1 The plate equation as a consequence of its load-bearing action

The bearing action of plates can primarily be considered as arising from the bearing action of grids, which was examined in Chapter 10. However, the fact that the plate constitutes a continuous medium adds some particular bearing characteristics that are important for both the understanding of the load-carrying mechanism and for its design.

The idea of modelling a plate as a dense grid of beams allows, first, the conclusion that the bearing action of a plate is based on its bending in two directions and on the corresponding torsion. It is reasonable for the considered grid to be oriented in the directions of a system (x, y), which is also common to the plate (Figure 11.2). The natural parallel of the plate with the grid enables the load-bearing action of the plate to be understood in analogous terms to those for beam structures.

With regard to the load resistance mechanisms, the load $p(x, y)$ applied on a cut-out square element of unit dimensions is equivalent to four 'resistance' forces, namely: the forces developed through the bending in the x and y directions, denoted as $p_{b,x}$ and $p_{b,y}$, respectively; and the forces developed through torsion about the axes x and y, denoted by $p_{t,x}$ and $p_{t,y}$ respectively. Thus (Figure 11.3)

$$p = p_{b,x} + p_{b,y} + p_{t,x} + p_{t,y}$$

To understand this better, the orthogonal plate shown in Figure 11.2 and the corresponding grid with its beams placed at unit distances are considered. The deformation picture in the region of a considered point A shows that the deflections along the length in each direction are increasing, whereas the slopes are decreasing towards the

Structural systems: behaviour and design
978-0-7277-4105-9
Copyright Thomas Telford Limited © 2010
All rights reserved

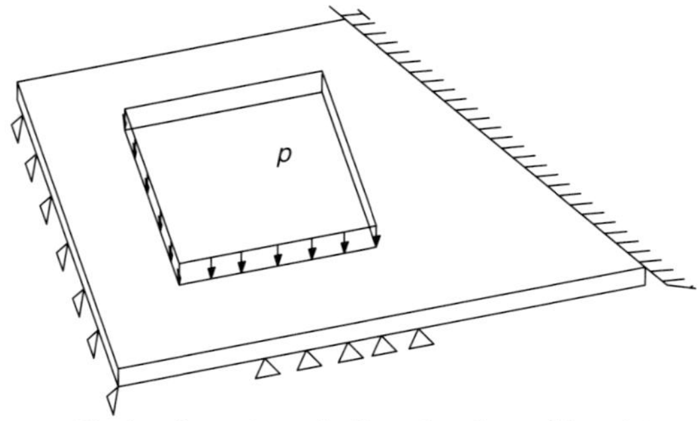

Figure 11.1 A plate as a continuous plane structure, loaded transversely

centre of the plate, with a gradual increase in curvature in both the x and the y directions. In the cut-out square element, the vector bending moments m_x and m_y applied on the corresponding sides are considered, with the moments on the opposite sides determined, according to the law of differential increase, as $m_x + (\partial m_x/\partial x) \cdot 1$ and $m_y + (\partial m_y/\partial y) \cdot 1$, respectively (see Figure 11.3). It is clear from the total picture of deformations that all the above moments are causing tension on the bottom fibres of the plate, this tension increasing toward the centre. This means that in the x and y directions the element is acted on by the resultant moments $(\partial m_x/\partial x) \cdot 1$ and $(\partial m_y/\partial y) \cdot 1$, respectively (see Figure 11.3).

Considering first the x direction, the moment $(\partial m_x/\partial x) \cdot 1$ requires, for equilibrium, the development of shear forces $q_{b,x} = (\partial m_x/\partial x) \cdot 1/1$, which, decreasing toward the centre, offer through their variation along x the corresponding 'resistance force' $p_{b,x}$ to the load $p(x, y)$, i.e. $p_{b,x} = -\partial^2 m_x/\partial x^2$.

Now, if $w(x, y)$ expresses the deflection of the plate, then the change in slope $\partial w/\partial x$ at point A along y is $\partial^2 w/\partial x\, \partial y$. Given, however, that this change in slope is of a torsional nature along the y direction, expressing the relative twisting angle of the unit-length element, a torsional moment m_{xy} will develop in the y direction of the plate, according to relationship

$$m_{xy} = -\frac{\partial^2 w}{\partial x\, \partial y} \cdot GI_T \quad \text{(see Section 2.5.3)}$$

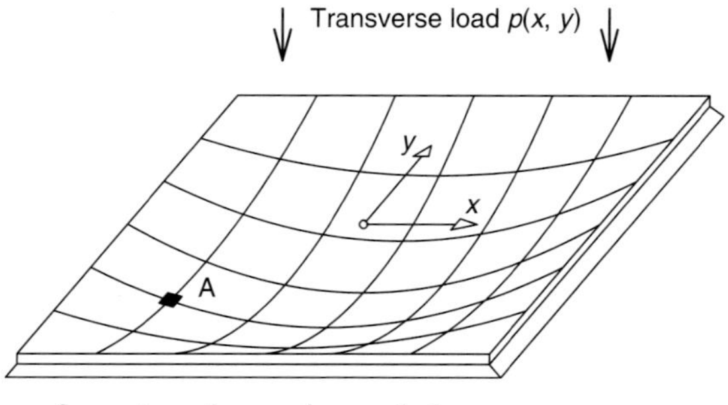

Figure 11.2 Deformed configuration of an orthogonal plate

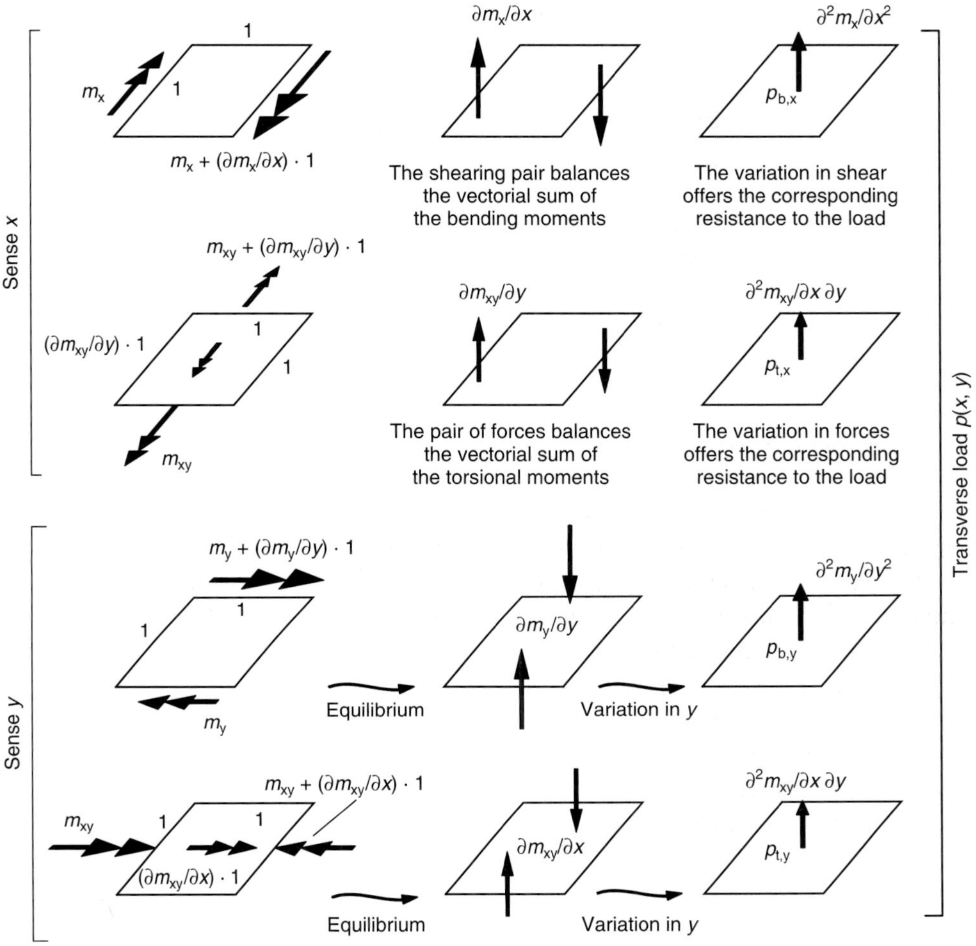

Figure 11.3 Development of resistance forces due to the transverse load on the plate

As the relative twisting angle $\partial^2 w/\partial x\,\partial y$ is decreasing along the y axis, the torsional moment m_{xy} will also be decreasing, and its change $(\partial m_{xy}/\partial y)$ requires for equilibrium the shear forces $(\partial m_{xy}/\partial y)/1$, which, decreasing along the x direction, offer by their variation the corresponding 'resistance force' $p_{t,x}$ to the load $p(x, y)$, which has the same sign as $p_{b,x}$ and is equal to $p_{t,x} = -\partial^2 m_{xy}/\partial x\,\partial y$.

In a similar manner, consideration of the y direction leads to the other two resistance forces (with the same sign as the previous ones) associated with the corresponding parallel sides of the element, i.e. the 'bending' resistance force $p_{b,y} = -\partial^2 m_y/\partial y^2$ and the 'torsional' resistance force $p_{t,y} = -\partial^2 m_{xy}/\partial x\,\partial y$. It is obvious that the torsional moment m_{xy} is the same in both directions.

Thus, the above formulated equilibrium of load p with the four resistance forces can be written on the basis of their corresponding expressions (see Figure 11.3):

$$\frac{\partial^2 m_x}{\partial x^2} + 2 \cdot \frac{\partial^2 m_{xy}}{\partial x\,\partial y} + \frac{\partial^2 m_y}{\partial y^2} = -p$$

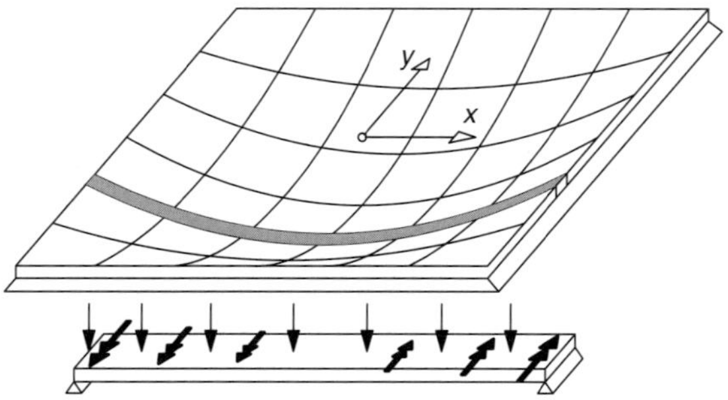

Figure 11.4 Influence of the twisting moments on the bending stress state and deformation

In this way, the relieving role of the twisting moments m_{xy} in the bending stress state of the plate becomes evident, either by the offer of the corresponding resistance forces $p_{t,x}$ and $p_{t,y}$, or through the consideration of a bent strip of unit width (e.g. along the direction x (or y)) receiving along its length the action of the resultant of moments m_{xy} on its two longitudinal edges (Figure 11.4).

It must be emphasised that the moments m_x, m_y and m_{xy} are referred to a unit length. The bending stress state is proportional to the corresponding curvature $1/r$ (see Section 2.3.2), that is

$$m_x = -\frac{\partial^2 w}{\partial x^2} \cdot EI$$

and

$$m_y = -\frac{\partial^2 w}{\partial y^2} \cdot EI$$

The expression for m_{xy} has been determined previously. In the present case, $I = d^3 \cdot 1/12$, $I_T = d^3 \cdot 1/6$ (see Section 2.5.3) and $G = E/2$. (Note that only half the torsional inertia I_T is considered because the shear stresses receiving the torsional moment m_{xy} are activated only through the thickness of the relevant section, and not through its width also, as is the case in a typical cross-section.)

The last expressions for the bending moments are not strictly correct, as they do not contain the influence of the so-called Poisson's ratio, which is related to the transverse deformation that generally accompanies an axial stress state. However, in any case the concrete cracking limits this influence, and so it can be ignored for preliminary design purposes.

The substitution of the expressions for the moments m_x, m_y and m_{xy} in the above equilibrium equation leads to the classical equation for plates with respect to deflection w:

$$\frac{\partial^4 w}{\partial x^4} + 2 \cdot \frac{\partial^4 w}{\partial x^2\, \partial y^2} + \frac{\partial^4 w}{\partial y^4} = \frac{p}{K}$$

where K is the plate bending stiffness, which is practically equal to $Ed^3/12$.

In the past, this differential equation, together with the effective support conditions also formulated in terms of deflection w, was used for the analytical determination of the stress state in plates with orthogonal or circular boundaries. But this always constituted a particularly difficult problem. The rapid spread of computers in recent years has allowed the application of numerical methods that were developed long time ago, and hence, using suitable and accessible programs, the reliable analysis of plates is nowadays possible, even when their geometry is highly complex. The need for a natural interpretation of the load-bearing behaviour of plates remains, nevertheless, essential, as does an appraisal of the characteristic particularities of certain typical constructional layouts and their design. The present chapter will move within this frame. ■

First, it may be pointed out that the reactions on a straight boundary do not coincide with the shear forces q corresponding to the transverse bending mechanism, as occurs in a grid, but are formed by the simultaneous presence of the twisting moments. For example, the twisting moments m_{xy} that appear along the freely supported straight boundary of the orthogonal plate shown in Figure 11.5 can be represented, from a

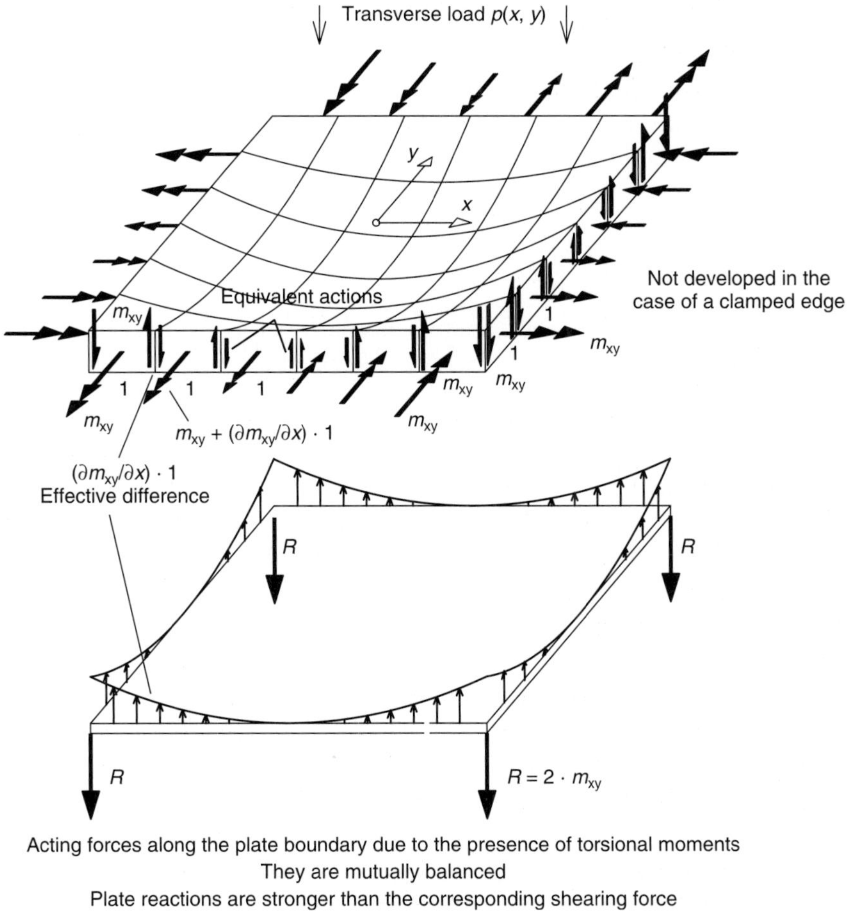

Figure 11.5 Development of reactions at the plate boundaries

natural point of view, only by corresponding pairs of forces with a unit lever arm, being of course equal to m_{xy} (see Figure 11.5). The variation in these forces, for example in the x direction, involves the additional loading of this straight boundary of the plate by the distributed load $\partial m_{xy}/\partial x$ (upwards), with the value of this load increasing towards the ends of the boundary and resulting in a reaction equal to $(q + \partial m_{xy}/\partial x)$. Thus it appears that the sum of the reaction forces is higher than the total load applied to the plate. The reason for this paradox lies in the fact that at the four corners of the plate the above local 'fictitious' forces m_{xy} have the same direction, and are added up in each corner to give a resultant downward force R, which cancels out the distributed load $\partial m_{xy}/\partial x$. Thus the equilibrium between the external load, the distributed reactions $(q + \partial m_{xy}/\partial x)$ and the corner forces R is finally ensured. It thus results that $R = 2 \cdot m_{xy}$. Note that the force R is of the order of roughly 9% of the total uniform load of the plate, and the take up of this force requires a suitable anchorage of each corner of the plate, so that this downward force can actually be offered.

It should be pointed out that, in the case where the boundary of the plate is fixed, the moments m_{xy} are not developed along it. Because the value of the corresponding zero slope is not altered in the other direction, the magnitude $\partial^2 w/\partial x\, \partial y$ is thus equal to zero. Consequently, the reactions remain unaffected (see Figure 11.5). ■

At this point it should be noted that, as the loads on the beams are transferred to the supports via principal stresses (see Section 4.1.1), the loads on the plates are not taken up along the arbitrary directions x and y (involving the development of the bending and twisting moments m_x, m_y and m_{xy}, respectively) but are taken up through the so-called principal moments. These principal moments are bending moments applied to each point of the plate in specific directions perpendicular to each other.

Thus, at each point of a plate there are two mutually perpendicular directions, each of which is acted on exclusively by a moment vector parallel to the cut-out side, without any twisting action along it, which is the case in all other directions (Figure 11.6). This fact has particular importance as these directions reveal the paths that are followed for an exclusively load-bearing bending action. The two perpendicular directions at each

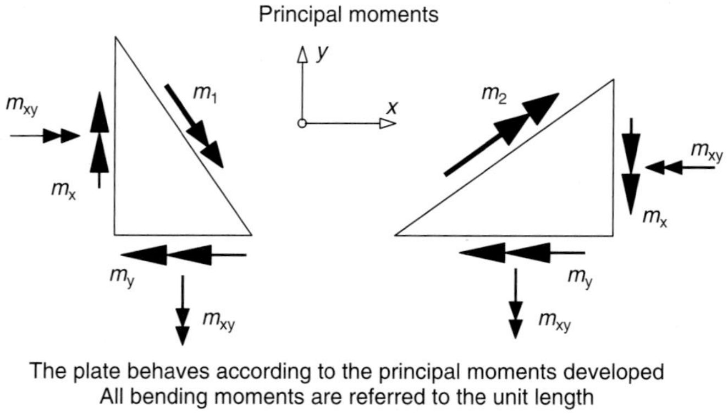

Figure 11.6 The presence of principal moments

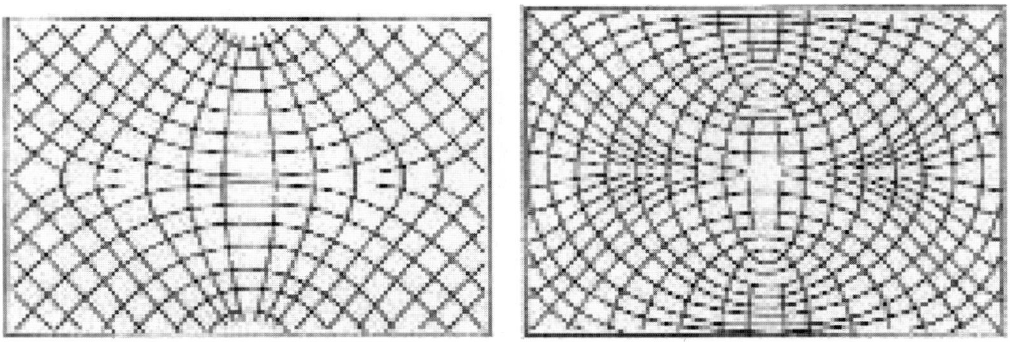

(a) (b)

Figure 11.7 Trajectories of the principal moments under (a) free and (b) fixed support conditions

point, as well as the principal moments m_1 and m_2 themselves, are determined through the values of m_x, m_y and m_{xy} that prevail at the point considered, on the basis of equilibrium of the two triangular cut-out elements (see Figure 11.6).

That this is the 'correct' way of perceiving the structural action of plates is confirmed by the fact that the developing cracks in a concrete plate correspond precisely to the principal moments considered. Thus it is understandable that, regardless of the grid model used to approach the bearing action of a plate (see above), the essential characteristic of its structural operation results from its monolithic continuity, which is not valid for the grid, which exhibits a naturally structural discontinuity.

The fact that a plate transfers its loads only by bending in the directions of the principal moments (Figure 11.7) makes it possible to model it with a grid, the elements of which may be aligned as closely as possible to the principal directions, transferring to each other one vertical force (i.e. bending only, and no torsion). Thus, the obsession with the grid returns again, having once been utilised in an aesthetically very persuasive way by the Italian architect–engineer Pier Luigi Nervi for covering a specific area using beams with negligible torsional stiffness (Figure 11.8) (Nervi, 1956). ■

Considering now, as previously, the equilibrium of the elements (see Figure 11.6), it is evident that the degree of divergence of the principal moment direction from the directions x and y is related to the magnitude of the twisting moment m_{xy}. Thus, for the corner regions of the simply supported plate, where the biggest m_{xy} occur and the bending moments m_x and m_y are vanishing, it is possible to obtain a clear picture of the principal moments m_1 and m_2 developing there (Figure 11.9). Equilibrium of the two corner elements (a) and (b) in the same region, with equal twisting moments m_{xy} acting at the vertical sides, implies that at the 45° cuts only the vector of principal bending moments m_1 and m_2 is acting and, in particular, $m_1 = m_2 = m_{xy}$ (see Figure 11.9).

The resulting sense of m_1 in Figure 11.9(a) shows that the top face is in tension, with tensile stresses acting along the diagonal, and this is confirmed by the developing cracks in the oblique direction. Moreover, equilibrium of element (b) shows that the vector of the moment required for equilibrium on the skew side is also of a purely bending nature,

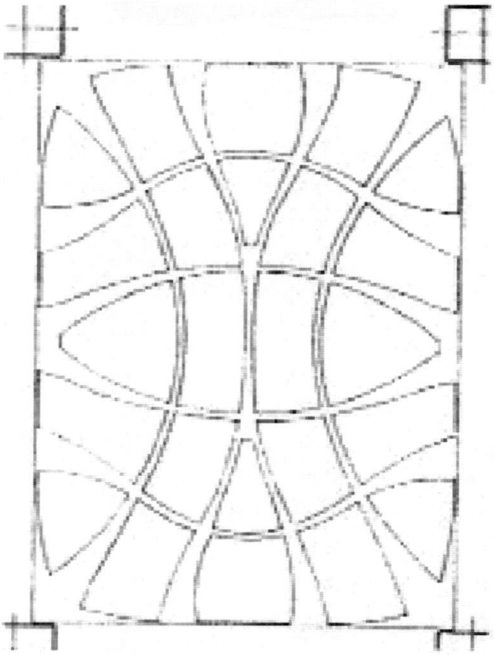

Figure 11.8 Configuration of a grid following the trajectories of the principal moments

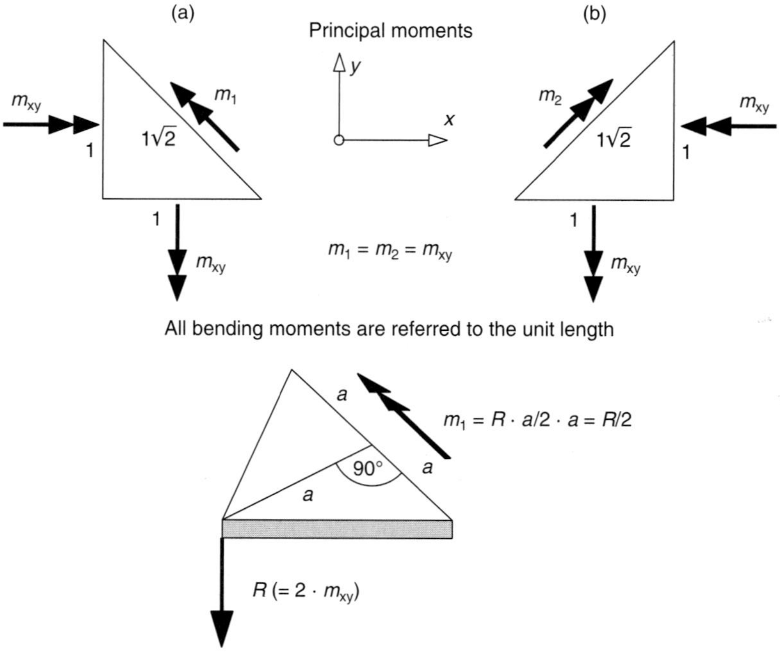

Figure 11.9 Equilibrium in the corner region of a plate

representing the principal moment m_2, causing cracks at the bottom face in the direction of vector m_2.

Note that the principal moment m_1 relates directly to the corner force R, as this causes a moment $(R \cdot a)$ at a distance a from the diagonal, applied over a length $2 \cdot a$. Thus $m_1 = (R \cdot a)/(2 \cdot a) = R/2$, a result that is supported by the above conclusions (see Figure 11.9).

11.2 Orthogonal plates

Orthogonal plates are the most used form of plates in building construction and in bridges. In bridges, because of the concentrated wheel loads of vehicles, a specific load-bearing characteristic should be clarified as early as possible.

The presence of a concentrated load causes curvature in both directions of the plate and, consequently, a proportional bending response. The apparent paradox is that this bending response is, in practice, independent of the dimensions of the plate (Figure 11.10). The reason for is that for a specific concentrated load P the total moment developing in each direction is proportional to the corresponding length, whereas for the corresponding unit width of application the bending moment comes out practically as a constant (see Figure 11.10).

11.2.1 *Two-side supported slabs*

11.2.1.1 Stress state

Two-side supported plates are those that are supported on two opposite sides by either simple or fixed supports, while the other two sides are either free or are supported at such a large distance apart that the transfer of loads in effect takes place within the shorter span distance L.

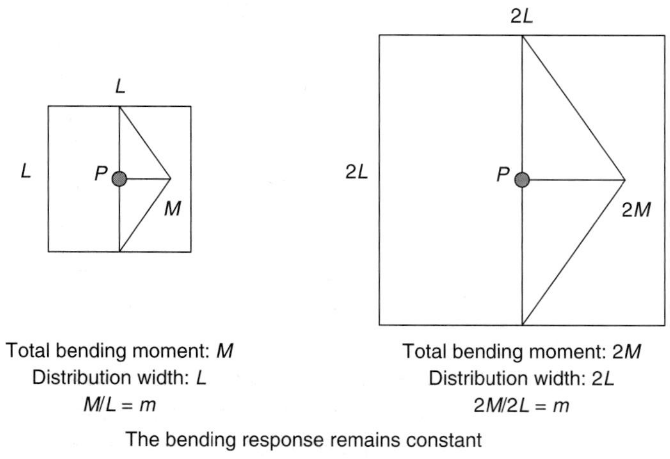

Figure 11.10 The action of a concentrated load on a plate

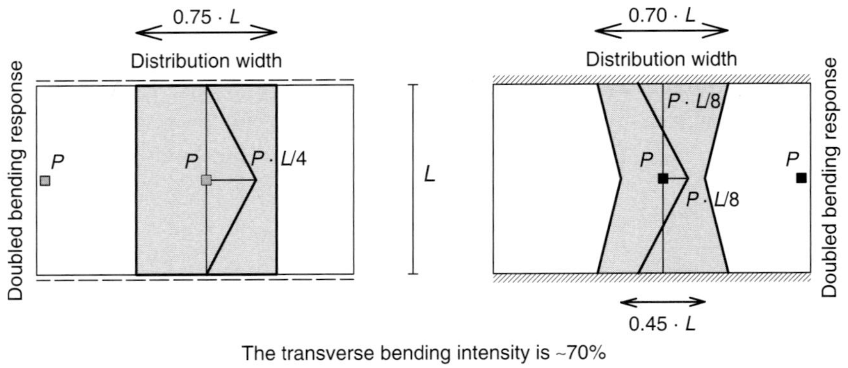

Figure 11.11 The influence of a concentrated load on the stress state of a two-side supported plate

These plates, when loaded with a uniform load p, act as beams and are substantially stressed only in the supporting direction, with maximum span moments of $m = p \cdot L^2/8$ and $m = p \cdot L^2/24$ for the simply supported and fixed cases, respectively.

However, in the case of a concentrated load P (which of course is not referred to a geometrical point, but is considered here as being distributed over a small square of area $0.0025 \cdot L^2$), the fact that the locally developing deflection causes curvature also in the transverse direction implies the development of moments in this direction too (Figure 11.11).

The transfer of the concentrated load to the supports can also be considered similar to the situation for a beam. The transfer is through an effective zone of width b_m that is of the order of 75% of the span L for the simply supported plate and of the order of 45% of the span L for the fixed supported plate, assuming that the load is applied at the midspan. Thus, the statically developed total bending moment M, which at the midspan is $P \cdot L/4$ for the simply supported plate and $P \cdot L/8$ for the fixed supported one, is distributed over an effective width of $0.75 \cdot L$ and $0.45 \cdot L$, respectively. Therefore, the developing moments can be estimated as $m = M/b_m = P/3$ and $m = M/b_m = P/3.60$, respectively. Special attention must be paid in the case where the concentrated load is applied at the midpoint of the free edge. In this situation the effective width is decreased by half, and the bending moments are doubled accordingly (see Figure 11.11). Note that the transverse bending moments developed are about 70% of the longitudinal ones. Moreover, the total fixed support moment $M = P \cdot L/8$ in the fixed plate is distributed to the support over a larger width b_m, which is of the order of 70% of the opening L, and therefore the corresponding bending moment is estimated as $m = M/b_m = P/5.60$. Thus, as noted initially, the bending moments due to concentrated loads are independent of the span length, a fact that is of particular importance in bridge engineering, given the significant loads imposed by vehicle wheels.

Finally, it is pointed out that for a linear load q (kN/m) extended over the span L of a simply supported plate (Figure 11.12), the total developed bending moment $q \cdot L^2/8$ is distributed over a width $b_m = 1.35 \cdot L$, so that the bending moment m is roughly equal to $0.10 \cdot q \cdot L$, which is of course dependent on the length L.

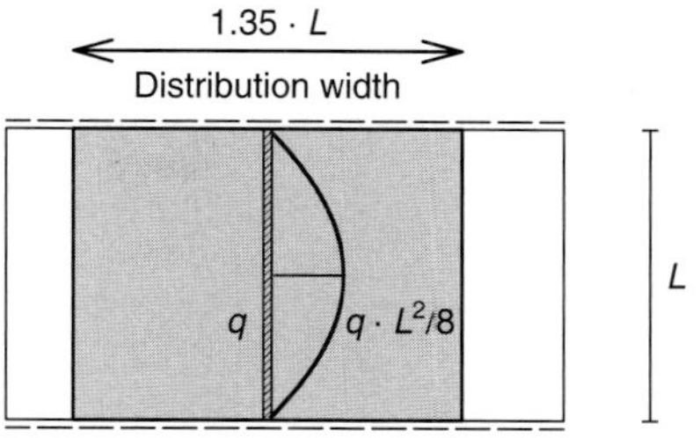

Figure 11.12 The influence of a linear load on the bending state of a two-side supported plate

11.2.1.2 Design

Plates are generally designed and reinforced for bending rather than for shearing. This is because the developing principal tensile stresses in the regions of intense shear and negligible bending are – for the applied design thicknesses – definitely smaller than the resistance 'allowed' in tension. With regard to simply supported plates, reinforcement is placed on the bottom region and is anchored at the ends. Thus, when the load is increased, with the accompanying development of the expected intense cracking, the compact compression zone of the concrete acts as a tied arch that takes up the loads without the need for stirrups (Figure 11.13).

The design of the bending reinforcement a_s (cm^2/m) follows that examined in Section 4.2.2.1, but in the present case the lever arm z of the internal forces is considered as equal to $0.95 \cdot h$. Thus for a design moment m_d, it is obtained that $a_s = \gamma_R \cdot m_d / (f_{sy} \cdot 0.95 \cdot h)$. ■

The design of plates always includes the control of deflections in the service state. For permanent loads, for which one should not consider the uncracked section but perform a more detailed calculation instead, according to Section 4.2.1.1, the creep influence intervenes, always adversely. The creep increases the deformation δ_L due to the loads by $\varphi \cdot \delta_L$. Thus the total deflection may be estimated as $\delta_L \cdot (1 + \varphi)$. This should be smaller than an acceptable ratio of the opening L, usually $L/250$, while in the case where brick walls are supported on the plate this ratio must be smaller than $L/500$. These conditions are usually met for plate thicknesses of the order of $d \sim L/25$.

Prestressing

The possible need for prestressing is treated as developed in Section 4.3. Prestressing is mainly applied in bridges with spans over 15 m. The longest span that is economically acceptable is roughly between 35 m and 40 m. The maximum thickness of a slab bridge should not exceed 1.00–1.10 m.

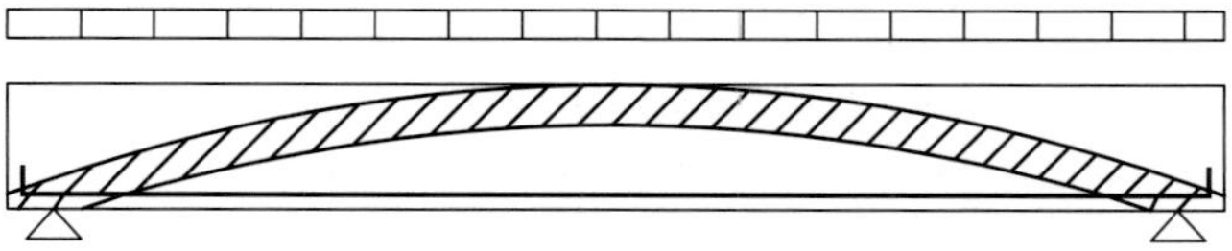

Figure 11.13 The action of a concrete plate in the cracked state

In this respect, attention must be paid to the more intense bending response, as well as the deformation, developed at the free edges, compared with that in the middle region of the slab, due to vehicle loading (see Section 11.2.1.1). The use of special computer software is necessary to determine the stress state and the deformations.

To estimate the prestressing force two basic approaches may be followed, as described in Section 4.3: full prestressing and partial prestressing. In full prestressing, the size of the prestressing force (i.e. the total section of the cables) is determined such that under the permanent and traffic loads the tensile stresses in the concrete are practically eliminated, or at least tolerated if they are smaller than the resistance of the concrete in tension ($2\,\mathrm{N/mm^2}$). However, as explained in Section 4.3, this approach is uneconomical.

In partial prestressing the prestressing force is limited to that size which allows only the permanent loads to be taken up, without developing tension. The traffic loads that will be applied may lead to stresses higher than the tensile resistance of the concrete, and cause cracking (of limited width), which, however, disappears after their removal (see Section 4.3.2.1). Thus the partial prestressing approach is clearly more economical.

Prestressing is usually distributed uniformly over the slab width, i.e. the cables are placed and anchored at equal distances from one other (Figure 11.14). For the full take-up of a permanent load g ($\mathrm{kN/m^2}$) over a span length L, the required prestressing force P_g per metre is (see Section 4.3.1.1)

$$P_g = \frac{g \cdot L^2}{8 \cdot (d/2 - c)}$$

where d is the slab thickness and c is the cable cover.

Under the permanent loads and the prestressing force P_g no bending develops, only a uniform compression across the plate thickness. As a reasonable consequence, this

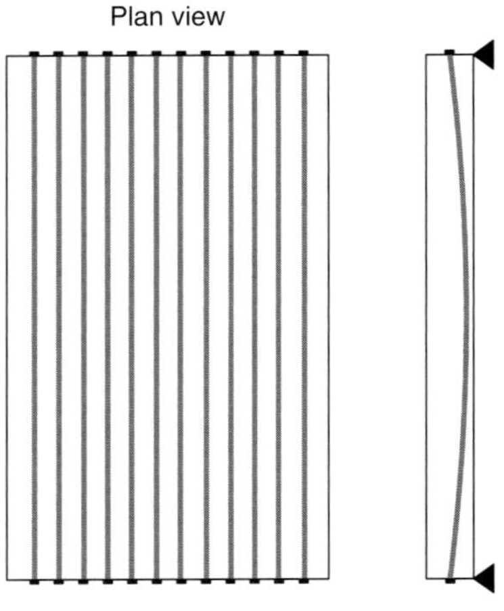

Figure 11.14 Loads are taken up through distributed prestressing

prestressing value usually offers, together with the necessary percentage of steel reinforcement (of the order of 1‰), a bending resistance m_R that is more than sufficient to satisfy the design criterion $\gamma \cdot (m_g + m_p) < m_R$, where $\gamma \cdot m_g$ and $\gamma \cdot m_p$ are the design moments for the permanent and traffic loads, respectively. For this reason, and in order to limit the 'distance' between the two members of inequality, thus reducing the necessary cross-sectional area of cables, a percentage of the prestressing force P_g of the order of 75% is applied.

In the case of a bridge, the bending response state along the free edges due to traffic loads m_p^r is clearly more than that which occurs in the interior of the slab m_p^m, as explained previously. Thus, the above checking of the strength must be performed not only for the interior of the slab under the design moment $S_d^m = \gamma \cdot (m_g + m_p^m)$, but also for the extreme zones with the moment $S_d^r = \gamma \cdot (m_g + m_p^r)$.

At this point it should be noted that the force P_g is transferred as a compression force in the slab only if the slab support allows movement in the direction of the force, so that the corresponding shortening can take place. If this is not the case, the slab cannot receive axial forces (because the prestressing force is absorbed by the immovable supports). Nevertheless, the slab is acted on by the full deviation forces, and thus, from a stress state point of view, the bending response, as previously considered, is still valid (see Section 4.3.1.2). ■

The saving in steel through partial prestressing may be even greater if the cables are concentrated in two bands at the outer edge zones of the slab, over a width of the order of 20% of the total width B of the slab, as shown in Figure 11.15 (Menn, 1990; Stavridis, 2001). In this case, due to prestressing, the plate develops a bending moment m_P that is practically constant over the entire width of the plate, and can be easily calculated on the basis of the deviation forces as follows. If the width of each zone is $(\lambda \cdot B)$ and ω is the applied reduction factor for the prestressing force P_g (i.e. of the order of 50%), then the applied prestressing $\overline{P}$ per unit width of the zone is $\overline{P} = \omega \cdot P_g/2\lambda$ and the applied deviation forces u_P on the zones are

$$u_P = 8 \cdot \overline{P} \cdot (d/2 - c)/L^2 \quad (\mathrm{kN/m^2})$$

Thus the bending due to prestressing, which is practically constant across the entire width of the slab, is

$$m_P = -u_P \cdot (2 \cdot \lambda \cdot B) \cdot L^2/8B = -0.25 \cdot u_P \cdot \lambda \cdot L^2 \quad \text{(tension on the top fibres)}$$

The above bending moment is composed of the so-called 'statically determinate' part m_{OP} (tension on the top fibres), which is limited to the outer zones only and affects, as already stated, the bending resistance only in these regions, and the 'statically redundant' part m_{SP} (see Section 5.4.1). Hence

$$m_{SP}^r = m_P - m_{OP} = -0.25 \cdot u_P \cdot \lambda \cdot L^2 + \overline{P} \cdot (d/2 - c)$$

while

$$m_{SP}^m = -0.25 \cdot u_P \cdot \lambda \cdot L^2$$

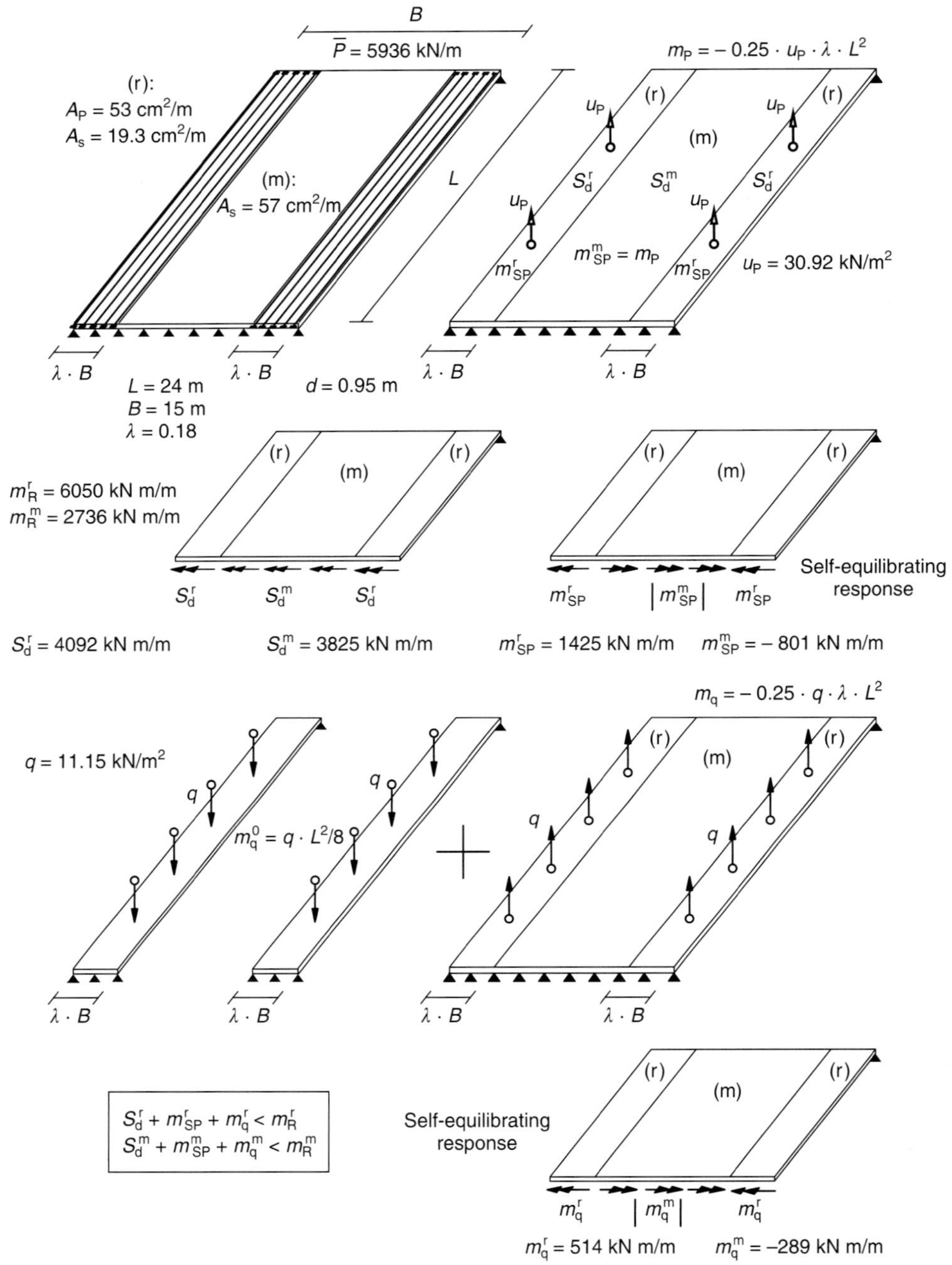

Figure 11.15 Taking up the load through prestressing of the edge zones

As this last part is self-equilibrating, as explained in Section 5.4.1 (check: $m_{SP}^r \cdot 2 \cdot \lambda \cdot B + m_{SP}^m \cdot (B - 2 \cdot \lambda \cdot B) = 0$), it may be superposed on the bending moments due to the external loads in the strength check, according to the equations

$$S_d^r + m_{SP}^r < m_R^r$$

and

$$S_d^m + m_{SP}^m < m_R$$

for the outer and interior regions, respectively (see Figure 11.15).

It may always be asserted that the moment m_{SP}^r burdens the bending state, and the moment m_{SP}^m relieves it. It is also known that the resistance moment m_R^r is formed by both the prestressing and the steel reinforcement, while the resistance moment m_R^m is formed only by the steel reinforcement. Of course, taking into account the favourable influence of the distributed compression force n_P due to prestressing, a good approximation is obtained: $n_P = 2 \cdot \overline{P} \cdot \lambda$.

As the above resistance check for the outer region of the slab is generally satisfied, whereas that for the interior region – despite the relieving action of the moment m_{SP}^m – is not, in order to 'fix' the situation, an additional self-equilibrating loading can always be superposed, in view of the static theorem of the plasticity theory (see Section 6.6.2). This loading consists, on the one hand, of a suitably selected uniform downward load q (kN/m^2), applied only at the two zones considered as being cut out from the slab, and, on the other hand, of the same load q applied upwards on the two zones of the full slab (see Figure 11.15). Thus, the outer region takes in addition the bending moment $m_q^r = q \cdot L^2/8 - 0.25 \cdot q \cdot \lambda \cdot L^2$, while the interior one is relieved by the bending moment $m_q^m = 0.25 \cdot q \cdot \lambda \cdot L^2$. Superposition on the corresponding 'loading' terms leads to the final satisfaction of both the above checks (see Figure 11.15).

Despite the obvious need for more steel reinforcement in the interior region than is required in the case of a uniform distribution of tendons over the whole width of the slab, the layout examined leads generally to a more economical solution. Obviously, the most favourable combination of the parameters λ and ω must be found by means of trials.

11.2.2 *Cantilever slabs*

11.2.2.1 Stress state

The introductory remarks given in Section 11.2 are valid also for cantilever slabs. The uniform load p causes bending proportional to the square of the cantilevered length L, i.e. $m = p \cdot L^2/2$. However, the concentrated load P applied at the free cantilever end causes a total moment $M = P \cdot L$ at the clamped edge, and this is distributed over a length roughly equal to $2 \cdot L$, meaning that the load is transferred to the support at an angle of about 90°. Thus, the bending moment is $m = 0.50 \cdot P$, i.e. it is independent of the length L (Figure 11.16). The moment acting along the longitudinal direction due to the existing curvature in this sense causes, of course, tension on the bottom fibres, and may be considered roughly equal to the bending moment at the clamped side.

It should be noted here that any stiffener or beam located along the free edge somehow increases the distribution width (more favourable bending at the fixed end), while in case of a cantilever of variable thickness the distribution width is limited and the bending moment increases (see Figure 11.16).

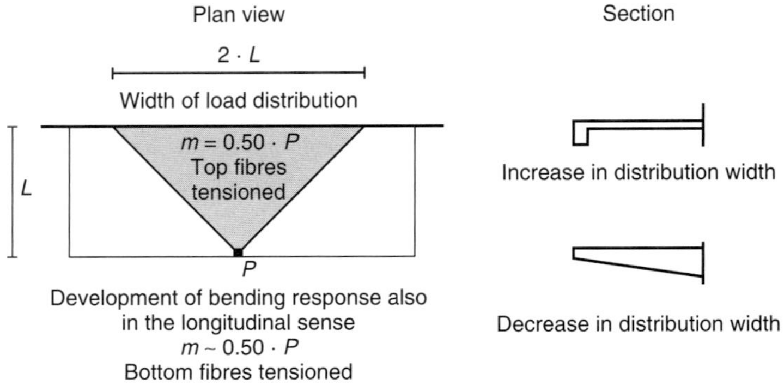

Figure 11.16 Cantilever action under a concentrated load

One should also note the resulting stress state in the region of a cantilever corner, as shown in Figure 11.17. For a uniform load p this stress is $m = 1.50 \cdot p \cdot L^2$, while for a concentrated load P on the external corner it is $m = 1.65 \cdot P$.

11.2.2.2 Design

Cantilever slabs are designed, analogously to the description in Section 11.2.1.2, for bending rather than shearing, with the reinforcement placed at the top side. In a cantilever there is no particular reserve of bearing resistance, because of its 'statically determinate' load-carrying action, and therefore particular attention must be paid to its design, reinforcement, as well as its deformations, as even residential buildings constitute an area where higher live loads may be acting. Moreover, if a cantilever is connected at its clamped end to relatively flexible slabs, its fixed end will be compliant (elastically rotating support), and this will increase the deformation of its free edge, in comparison with a completely fixed support (see Section 2.3.7).

For the calculation of the deflections the same remarks as given for the two-side supported plates in Section 11.2.1.2 are valid, whereby particular attention must be paid to creep. For the deformations due to permanent loads, the principle of virtual

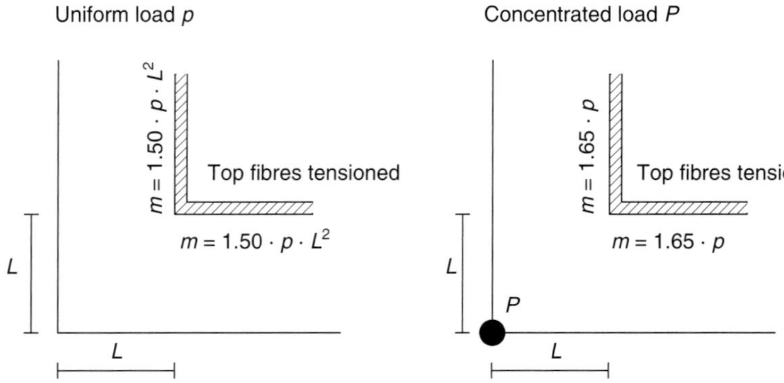

Figure 11.17 Bending state of a cantilever corner under a uniform load and a concentrated load

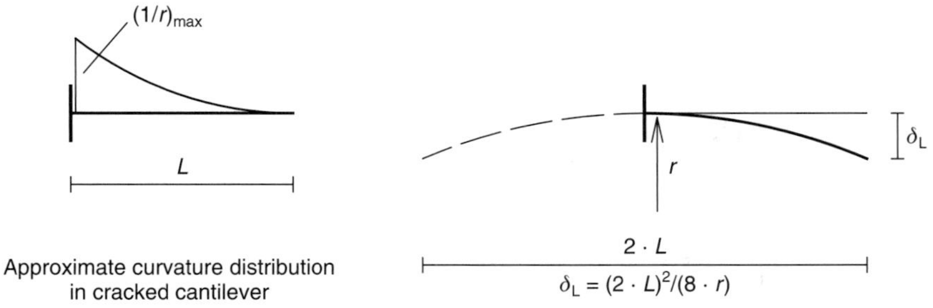

Figure 11.18 Cantilever deformation under a uniform load

work may be applied (see Section 4.2.1.1) by adopting the parabolic distribution of curvatures shown in Figure 11.18. In any case, the expression $\delta_L = (2 \cdot L)^2/(8 \cdot r) = L^2/(2 \cdot r)$ may be used as an acceptable estimate for the deformation δ_L.

The much stricter condition regarding the deformation at the free edge ($< L/250$), which must be respected, can be met with a plate thickness of $d > L/10$. Finally, it is noted that the deflection at the cantilever corner examined previously (see Figure 11.17) for a uniform load is twice that of the adjacent cantilevers.

Where there is a possible need for prestressing, the remarks made in Sections 4.4 and 11.2.1.2 are valid.

11.2.3 Four-side supported plates

Four-side supported plates are the most commonly used type of slabs in building construction. They are supported perimetrically on beams or walls, and are usually arranged orthogonally, being connected to each other in a continuous, monolithic way.

11.2.3.1 Stress state

The four-side supported slabs transfer their loads in both directions, differentiating their load-carrying response depending on whether the load is uniformly distributed or concentrated. In the case of a uniform load, the shorter direction also receives the largest bending response, as has been explored for grids (see Figure 10.2). It should be clarified, however, that, with respect to a uniform load, the concept of 'rigidity' along the x or y sense concerns the distributed load p_x or p_y, respectively, which causes a unit deflection at the midpoint of a characteristic strip of unit width. This rigidity is expressed as $k \cdot (EI)/L^4$, where the factor k is equal to 76.8 or 384, depending on whether the length L refers to a simply supported or to a fixed-end direction, respectively, and $I = d^3/12$. Thus, it may be considered that the load p is 'distributed' to the loads p_x and p_y, i.e. $p = p_x + p_y$, while the requirement for common deflection of the two mutually orthogonal strips requires (according also to Section 10.2) that

$$\frac{p_x}{(\text{stiffness})_x} = \frac{p_y}{(\text{stiffness})_y}$$

Table 11.1 Bending moments of orthogonal plates under a uniform distributed load

L_y/L_x	Perimetrically simply supported plates		Fixed plates			
	$m_x/p \cdot L_x^2$	m_y/m_x	$m_x/p \cdot L_x^2$	m_y/m_x	$m_{ex}/p \cdot L_x^2$	$m_{ey}/p \cdot L_x^2$
1.0	0.037	1.0	0.018	1.0	−0.052	−0.052
1.50	0.073	0.38	0.034	0.30	−0.076	−0.057
∞	0.125	0	0.042	0	−0.083	−0.057

that is

$$\frac{p_x}{p_y} = \frac{(\text{stiffness})_x}{(\text{stiffness})_y}$$

Thus, under conditions of uniform support, in the case where $L_y = 2 \cdot L_x$,

$$p_x = 2^4 \cdot p_y = 16 \cdot p_y$$

As briefly pointed out in Section 11.2, the bending stress state due to a concentrated load is independent of the dimensions of the plate and is proportional to the load.

Tables 11.1 and 11.2, which refer to Figure 11.19, show the bending moments of perimetrically simply supported plates and fixed plates, for both their midpoints and their fixed supports, under a uniform and a concentrated load, for three characteristic aspect ratios. The case of an extremely oblong ground plan has already been examined in the case of two-side supported plates, but it is also included here for purposes of comparison.

It can be shown that the twisting moment m_{xy} (developed only at the free support) is, for the examined side ratios, roughly equal to 4% of the total load of the plate.

With respect to a uniform load, it is observed that the moment m_x at the midpoint of a square plate is barely 30% of the corresponding moment for a two-side supported plate in the case of simple supports, and 43% of the corresponding moment for the case of fixed supports. However, these percentages are increased if the beneficial influence of the twisting moments is ignored, which is equivalent to omitting the corresponding term in the differential equation for the plate (see Section 11.1). In the case of ribbed

Table 11.2 Bending moments of orthogonal plates under a concentrated load

L_y/L_x	Perimetrically simply supported plates		Fixed plates			
	m_x/P	m_y/m_x	m_x/P	m_y/m_x	m_{ex}/P	m_{ey}/P
1.0	0.274	1.0	0.233	1.0	−0.103	−0.103
1.50	0.311	0.83	0.257	0.88	−0.127	−0.034
∞	0.327	0.76	0.26	0.85	−0.170	

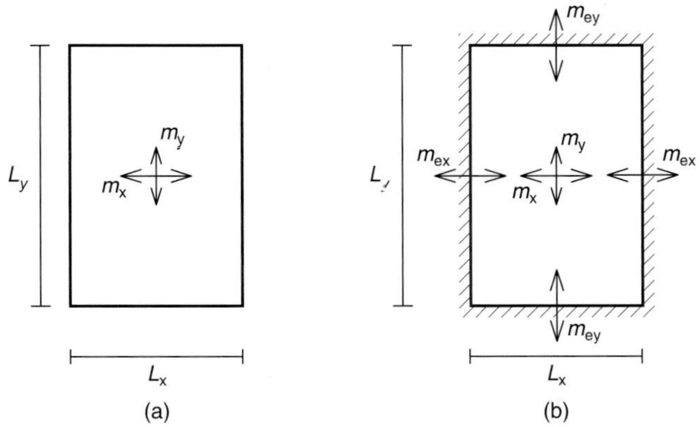

Figure 11.19 Characteristic locations of bending moments in orthogonal plates: (a) simple supports; (b) fixed supports

plates this influence is deliberately not taken into account, as will be examined later. Table 11.3 gives the corresponding data for the case of a uniform load where the twisting moments are neglected. The developing bending moments are, of course, increased, but the ratio of the moments in the two directions remains substantially the same.

An estimate of the reactions on the sides of the perimeter can be obtained by evaluating the load acting on the regions created by the 'bisectors' of the corners with supported sides and the midline of the two opposite longer sides (Figure 11.20). In the case where a fixed side is connected to a simply supported one, the 'bisectors' are set at an angle of 60° to the fixed side. ■

In building construction, four-side supported plates are usually set in a continuous orthogonal arrangement, being supported on beams or, possibly, walls. Thus their support along the length of the beams (or walls) can be considered as a simple one, and only a vertically distributed load is transferred, while the plates themselves are mutually monolithically connected. As the plates receive a permanent load g, as well as a movable live load p, there is a need to determine the most adverse bending moments both at the centre of the plates and over their supports, i.e. in the common contact lines (Figure 11.21).

Table 11.3 Bending moments, with the twisting moments neglected, under a uniform load

L_y/L_x	Perimetrically simply supported plates		Fixed plates			
	$m_x/p \cdot L_x^2$	m_y/m_x	$m_x/p \cdot L_x^2$	m_y/m_x	$m_{ex}/p \cdot L_x^2$	$m_{ey}/p \cdot L_x^2$
1.0	0.077	1.0	0.025	1.0	−0.056	−0.056
1.50	0.128	0.38	0.043	0.31	−0.085	−0.055
2.0	0.142	0.15	0.045	0.06	−0.088	−0.052

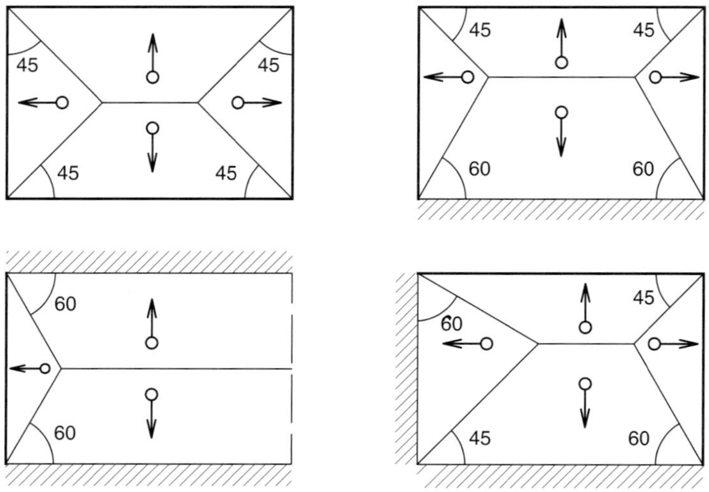

Figure 11.20 Distribution of the reactions at the sides

For this purpose, three distinct loading-layout cases may be considered, namely: (1) loading with $(g + p/2)$, (2) loading with $p/2$, and (3) loading with $(g + p)$.

For each individual plate, the maximum bending moments in the centre can be estimated by superposing the load in case (1), with fixed conditions all over their perimeter, and the load in case (2), assuming simple support conditions over the same support lines. The bending moments at each internal support (tension on the top side) can be estimated from load case (3), with fixed conditions all around, and taking always the mean of the values from each side (see Figure 11.21).

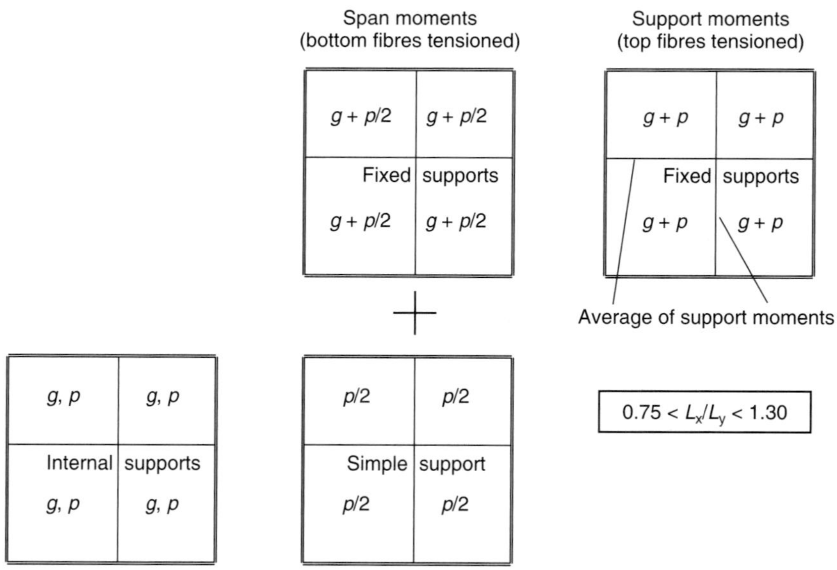

Figure 11.21 Evaluation of the bending moments in continuous plates

11.2.3.2 Design

The reinforcement is determined on the basis of the bending response. The fact that bending occurs in both directions in the centre of a slab implies that the reinforcement bars should be placed in two layers, one just over the other, in the x and y directions respectively. Obviously this results in a differentiated effective depth h (or lever arm z) for the two senses. As is the case along the length of the supports, the bending moments develop only in the transverse direction; they are obviously covered by only one reinforcement layer, placed normally on the top side. To estimate the reinforcement, the equation for two-side supported plates (see Section 11.2.1.2) may be applied for each direction.

With regard now to the deformations, these can be determined through the 'distributed loads' p_x and p_y as has been explained previously:

$$p_x = p \frac{L_x^4}{L_x^4 + L_y^4}$$

$$p_y = p \frac{L_y^4}{L_x^4 + L_y^4}$$

The deformation w at the midpoint of a strip of unit width is

$$w = \frac{1}{k} \cdot p_{x,y} \frac{12 \cdot L_{x,y}^4}{E \cdot d^3}$$

where k is determined according to Section 11.2.3.1. However, the estimated value of w due to permanent loads must be increased by $w \cdot \varphi$ due to creep, and the final result cannot exceed 1/250th of the corresponding span length. This condition is usually met for plate thicknesses $d > L/30$. ■

In the rather rare case where, due to larger dimensions of the plate, the application of prestressing is required, the permanent load g of the plate should be taken up by the deviation forces of the cables. The required deviation forces can be ensured by a uniform parallel arrangement of cables in one direction only, this direction being the shorter one. In this way the cables offer the largest possible deviation forces for the same prestressing force, given that $u = 8 \cdot P \cdot f/L^2$ (kN/m^2) (Figure 11.22). It should also be observed that, because the usually prevailing conditions do not allow a free shift of the 'prestressing front', uniform compressive stresses due to prestressing within the plane of slab must be excluded, as explained previously (see Section 11.2.1.2),

11.2.4 Ribbed plates

If in a slab only the compression zone of the concrete is maintained, with the remainder of the material being supplemented in the form of parallel ribs arranged in both directions, a statically more favourable system is acquired. This arrangement has, of course, a higher structural depth, and hence smaller internal compressive and tensile forces, offering the possibility of using an even smaller thickness for the compressed

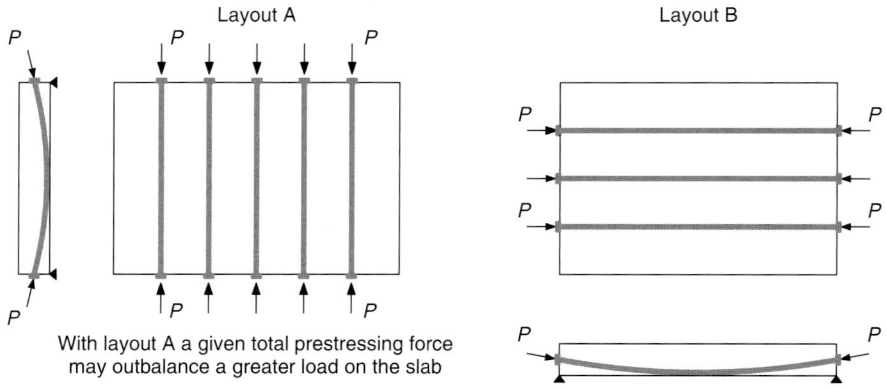

Figure 11.22 Layout for prestressing in a four-sided plate

plate and less tension reinforcement (Figure 11.23). Moreover, the deformations of this structural system will obviously be smaller than those of a solid plate.

If the plate is two-side supported, the parallel ribs are ordered in a dense layout only in the load-bearing direction. However, one up to three transverse ribs still have to be placed in the other direction (depending on the span), for reasons of wider distribution of the eventual live loads, as explained for grids (see Chapter 10).

For an equal axial distance e between the ribs, under the self-weight G of a zone of unit length and width e, and a live load p, the developing bending moment M_r on the rib considered is $M_r = (G + p \cdot e) \cdot L^2/8$, for a two-side supported plate. This moment is, of course, taken by the beam-like rib with depth d. The distance e does not have to exceed 70 cm (see Figure 11.23).

In the four-side supported plate the ribs are arranged, as already mentioned, along the x and y directions. Provided that the same distance e (<70 cm) exists in both senses, the plate may be considered to behave as having a constant thickness. It should be noted that, due to the relatively small thickness of the ribs, which is limited practically to a thickness that allows easy placement of the reinforcement bars in one layer only, their torsional resistance is low, especially after the oncoming cracking. For this reason, it is preferable, at least for preliminary design purposes, to ignore the beneficial influence of the twisting moments and apply what was discussed in Section 11.2.3.1. It is of course clear that the bending moment M_r of the ribs results by multiplying the corresponding tabulated values by the distance e (see Figure 11.23). Moreover, the ribs corresponding to the longer direction will develop the least bending response, as indicated by the ratio m_y/m_x in Table 11.1. ■

In a four-side supported plate, in order to achieve a preferred specific distribution of the bending response, or even for reasons associated with the construction process, it is possible for the rigidities to vary in the two directions, thus creating an *orthotropic plate* (Figure 11.24). This is usually the case if the ribs are ordered along one direction only, and consist of either concrete or of steel profiles, constituting thus a composite structure (see Section 5.6). Of course, the equation for the load distribution considered

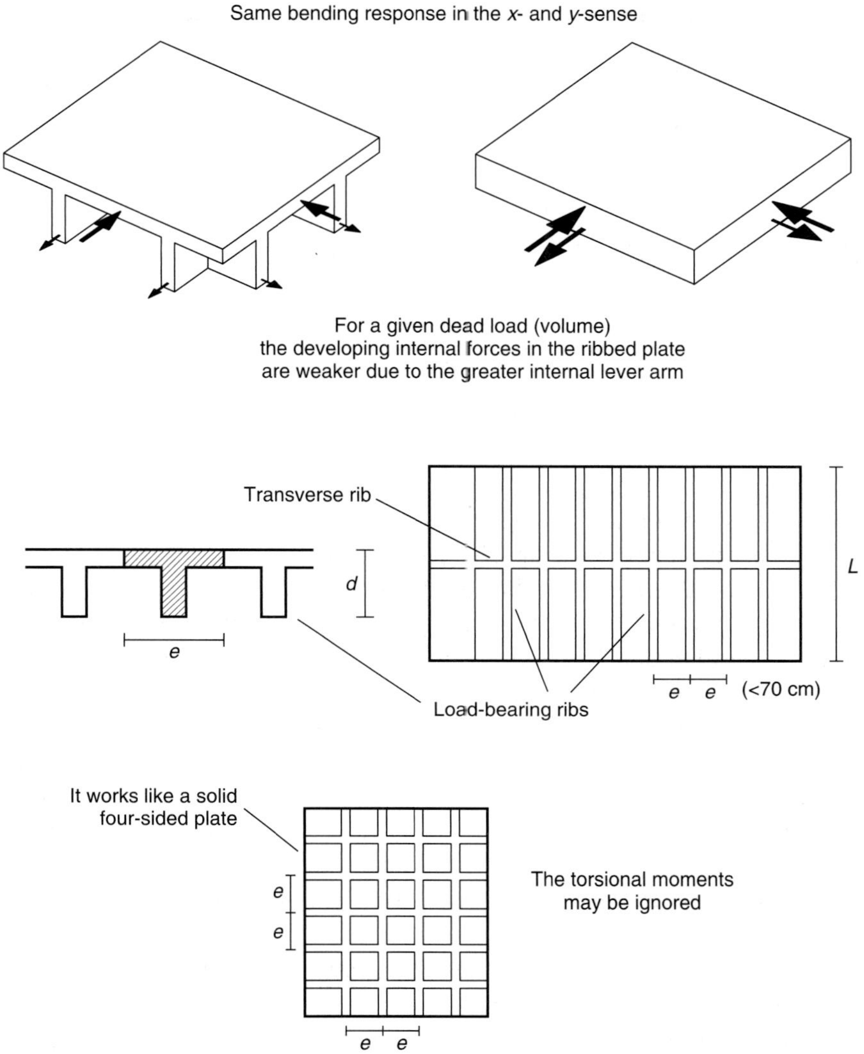

Figure 11.23 The use of a rib arrangement in an orthogonal plate

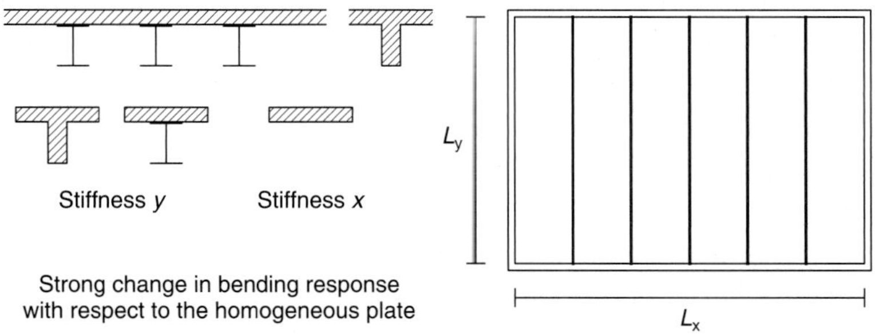

Figure 11.24 Stiffness differentiation in the two directions of a four-side supported plate

Table 11.4 Estimate of the bending moments in the two directions

L_y/L_x	(stiffness)$_y$/(stiffness)$_x$					
	5		10		20	
	$m_x/p \cdot L_x^2$	m_y/m_x	$m_x/p \cdot L_x^2$	m_y/m_x	$m_x/p \cdot L_x^2$	m_y/m_x
1.0	0.0225	5.76	0.0104	13.46	0.0038	37.7
1.50	0.0775	2.22	0.050	4.64	0.028	9.88
2.0	0.119	1.15	0.096	2.43	0.068	5.07

in Section 11.2.3.1 will be applied accordingly. For a total load p this equation is:

$$p_x = p \cdot \frac{(\text{stiffness})_x}{(\text{stiffness})_x + (\text{stiffness})_y}$$

and

$$p_y = p \cdot \frac{(\text{stiffness})_y}{(\text{stiffness})_x + (\text{stiffness})_y}$$

Table 11.4 gives some specific typical results for four-side supported plates, with the beneficial contribution of the twisting moments neglected (Stavridis, 1993). The plates are considered perimetrically simply supported. It can be seen from the values given in the table how radically the distribution of bending moments can be influenced by the stiffness ratio adopted.

The decks of steel bridges may also be included in this category of 'orthotropic plates'. Steel bridges are formed of two parallel main longitudinal beams with the distance between them being equal to the width of the deck. The beams are connected with transverse beams placed at distances of the order of 5 m, thus forming a grid of beams. The deck is formed by a steel plate with longitudinal stiffeners at distances of the order of 30 cm, thus creating a plate, the bending rigidity of which in the longitudinal direction is much higher than the one in the transverse direction (i.e. rigidities ratio of the order of 20). The ribs are usually selected to have closed sections, so that with the existing torsional stiffness the bending response will be limited, as explained previously. As the deck is designed to participate in receiving the intense compressive force of the top flange of the main beams, its contribution to avoiding the lateral buckling of these beams may lead to an increase in its own bending rigidity (see Section 7.5). ■

For oblong areas a skew layout of orthogonal ribs may be used, as discussed in Section 10.2 (see Figure 10.3), in order to achieve an equal distribution of response in all the ribs (Figure 11.25).

Characteristic values of the bending moment M_r of a rib, with respect to the equal distances e, are given in Table 11.5 for a 45° arrangement, for both a simply supported and a fixed plate (Stavridis, 1993). Again, for the reasons already explained, the beneficial influence of the torsional rigidity of the ribs has been ignored.

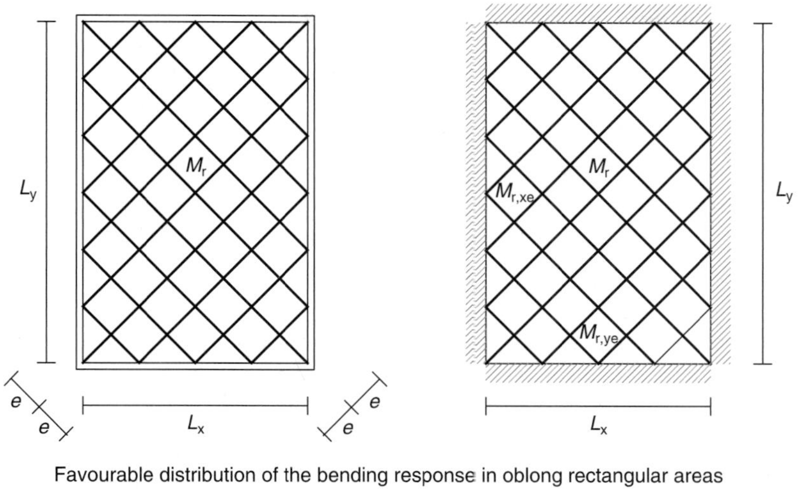

Figure 11.25 The use of a skew layout of ribs in oblong areas

Clearly, the quantity M_r/e corresponds to the moment m of a solid plate with constant thickness. Thus, by comparison with the corresponding values in Table 11.1, it is found that the bending moment of the ribs in the skew layout constitutes for the oblong areas only 40–50% of the corresponding moment developed in the orthogonal arrangement.

11.3 Circular plates

The circular plates used for covering circular areas are either simply supported or fixed at their perimeter. Under uniform load they develop a symmetrical stress state with respect to their centre, which is basically expressed by the principal moments m_r in the radial direction.

For conditions of simple support, the highest span bending moment developed at the centre of the plate can be practically approximated by considering a fictitious square plate with sides 10% longer than the diameter of the circular plate (Figure 11.26). For a plate with diameter D under a uniform load p, $m_r = 0.047 \cdot p \cdot D^2$.

For a fixed plate, the bending response at the centre of the plate is practically identical to that of a fixed square plate that has sides of a length equal to the diameter of the circle. As the cut-out corners of the square plate are substantially non-deformable,

Table 11.5 Estimated bending moments for a skew layout of the ribs

L_y/L_x	Simply supported plate	Fixed plate		
	$M_r/p \cdot e \cdot L_x^2$	$M_r/p \cdot e \cdot L_x^2$	$M_{r,xe}/p \cdot e \cdot L_x^2$	$M_{r,ye}/p \cdot e \cdot L_x^2$
1.0	0.035	0.021	−0.040	−0.040
1.50	0.052	0.029	−0.061	−0.046
2.00	0.069	0.034	−0.072	−0.044

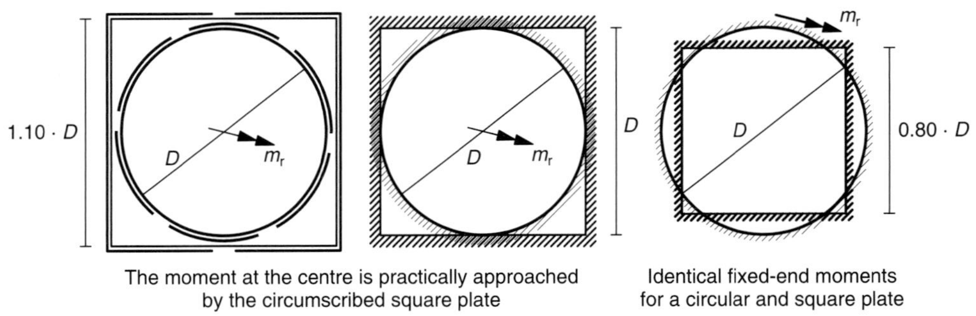

Figure 11.26 Assessment of the bending response in a circular plate

this approximation is roughly valid also in the case of perimetrically simple support conditions (see Figure 11.26). The bending moment at the fixed support is similarly likened to that of a fixed plate with sides equal to 80% of the diameter of the circular plate (see Figure 11.26). Thus, at the centre of the fixed plate $m_r = 0.018 \cdot p \cdot D^2$, while at the clamped support $m_r = -0.031 \cdot p \cdot D^2$.

11.4 Skew plates

Plates with a parallelogram outline and supported on two opposite sides are often used to cover skew transport crossings, and are called *skew plates* (Figure 11.27). The 'skewness' becomes structurally more perceptible as the angle of the straight line that links the corners of the obtuse angles with the supported sides becomes greater (see Figure 11.27). The fact that the direction of this line is followed for the transferring

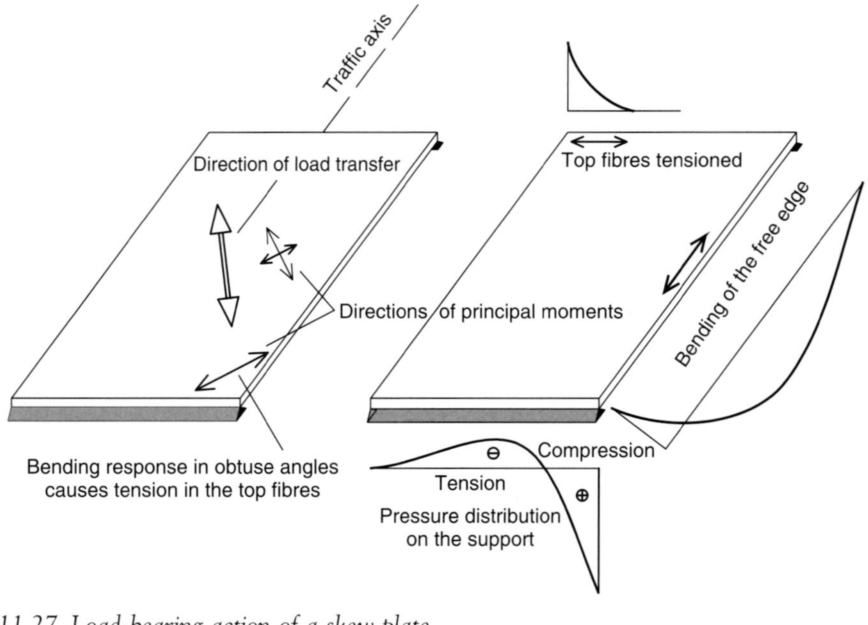

Figure 11.27 Load-bearing action of a skew plate

of loads, it being the shortest (i.e. the stiffest) path to the supports, leads to the conclusion that the directions of the principal moments are also skew with respect to the free edges of the plate, so that its bearing action results differ from those observed in an orthogonal plate. It may practically be considered that the directions of the principal moments that cause tension at the bottom face of the plate are found approximately along the bisector of the obtuse angles. This results in a strongly unequal distribution of the reactions as the 'skewness' increases, with a particular 'accumulation' of forces at the obtuse corners, as there may possibly be a need to anchor the regions at the acute corners, due to the developing downwards reactions.

Moreover, an increasing 'skewness' leads to a longer free edge with larger deformations, while, simultaneously, the conceivable continuance of the free edge zone to the region of the immovable supports at the obtuse corner creates there conditions of fixity. Thus the principal moments that are perpendicular to the 'bisectors' of the obtuse corners cause tension on the top face of the plate. The great variation in bending moments along the free edge, i.e. from the 'negative' moments at the supports to the 'positive' moments at the midlength, signifies the development of large shear forces ($Q = dM/ds$), and this confirms the 'accumulation' of reactions at the obtuse corners of the plate (see Figure 11.27).

A further characteristic of skew plates is that the skew placement of the directions of the principal moments with respect to the plate outline under a uniform load is practically unmodified under a concentrated load. This is of particular importance because it practically allows the direct superposition of the principal moments due to both uniform and concentrated loads. It should be noted, however, that for the final determination of the bending moments in skew plates, the use of suitable computer software is definitely required.

A direct consequence of the skew placement of the principal moments with respect to the free edge is that, apart from the free edge bending according to the developed curvature, it is compelled to receive torsional moments also (Figure 11.28) by following the equilibrium of the cut edge zone. The internal face of this zone in Figure 11.28 has been deliberately shaped as a 'saw' so that only the principal (bending) moments considered are applied to it. As bending in the direction of the diagonal sense is

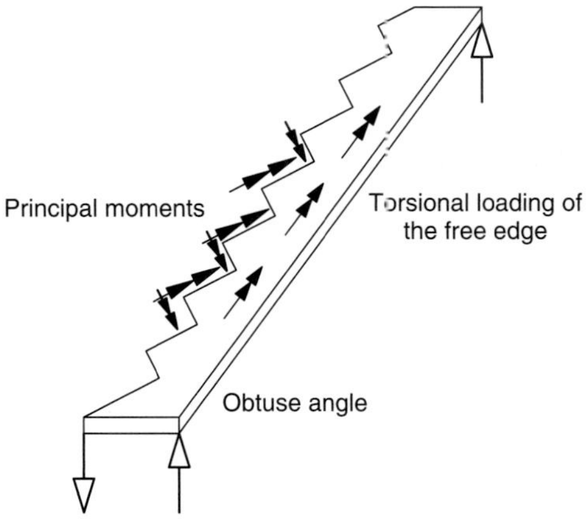

Figure 11.28 Torsional response in the free edge zone

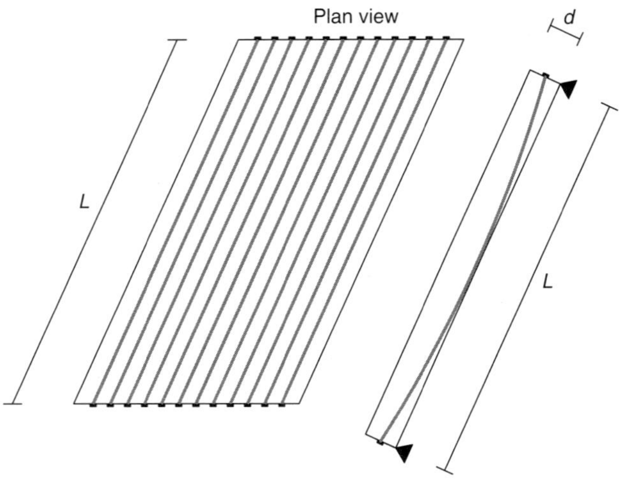

Figure 11.29 Uptake of the loading by equally distributed prestressed cables

stronger than that in the transverse sense, the resultant 'loading' moment vector at each point along the free edge of the plate permanently has the same sign, and consequently creates a torsional loading that must be taken up accordingly. Many skew bridges that do not have a sufficient number of stirrups in their edge zones for the uptake of these torsional moments (see Section 4.2.1.3) have developed cracks due to the above torsional response. ■

For spans longer than 15 m the slab should be prestressed. Moreover, in this case, as in orthogonal slabs, the sensitivity of the longitudinal edge zones with regard to the response and deformations must be taken into account, particularly for the applied vehicle loads. The plate thickness should again not exceed 1.10 m.

The cables are placed parallel to the free sides, usually in a uniform distribution (Figure 11.29). As examined in the case of orthogonal slabs, the full uptake of the permanent load g by the prestressing force P_g per metre, for a skew span equal to L, according to the equation

$$P_g = \frac{g \cdot L^2}{8 \cdot (d/2 - c)}$$

offers, together with the obligatory steel reinforcement, much greater bending resistance than required. For this reason, in order to follow a more economic design, only a percentage of the force P_g is applied, which, when augmented by the 'skewness' of the slab, may even be of the order of 50%.

Note that in a skew plate the full take up of the load g by the cable deviation forces resulting from the prestressing force P_g ensures an equal distribution of the support reactions. The reason for this is that the deviation forces from the cables completely cancel out the load g of the plate, while the remaining equally distributed downward resultants of the prestressing force acting directly on the supports (being totally equal and opposite to the deviation forces) cause uniformly distributed reactions.

As in the case of orthogonal plates, there also exists for skew plates the possibility of arranging the prestressing cables in the edge zones in a direction parallel to the free sides (see Section 11.2.1.2). In this way an even stronger restriction of the required prestressing force is achieved, so that, with an analogously increased percentage of steel reinforcement, the appropriate resistance check can be satisfied (Stavridis, 2001). The procedure for determining the cross-sectional area of the cables (i.e. the prestressing force, by assuming the cable is initially stressed with 70% of its yield stress) follows precisely the same course as that for the case of orthogonal plates (see Section 11.2.1.2). In this way a lower total cost may be achieved for the prestressed and steel reinforcement.

11.5 Flat slabs

11.5.1 State of stress

The load-bearing action of plates in transferring their loads in at least two directions implies that, normally, there should be immovable continuous supports at the ends of these directions to take up the loads transferred. These continuous supports are usually provided by beams of sufficient stiffness, which in turn are supported on vertical elements (columns) and, finally, transfer the loads to the bases of these elements (foundations). It may also (more correctly) be considered that the plates transfer their loads to the horizontal girders of frames, which are formed by the supporting beams and columns (Figure 11.30).

It is intuitively known that a plate can alternatively be supported directly on columns in a raster arrangement, without the use of beams, provided that the plate affords the appropriate rigidity. This arrangement permits the exploitation of the available net space height over the entire area covered. As will be analysed next, such a plate, compared to a plate supported on beams, develops a higher bending response and deformations on one side, as well as a particular stress state on the other, arising from the fact that the plate faces the danger of being punched by the columns. All these constitute, of course, disadvantages which are dealt with simply by increasing the thickness of the plate and by providing additional reinforcement, i.e. by taking measures that lead to a higher cost of construction, despite the simpler formwork

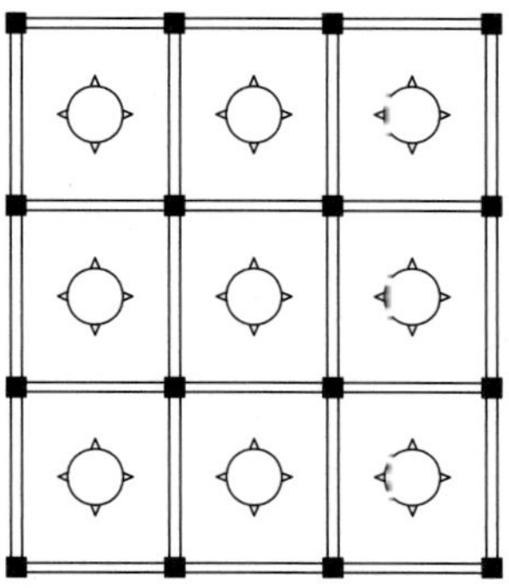

Figure 11.30 The transfer of the slab loads through the beams to the columns

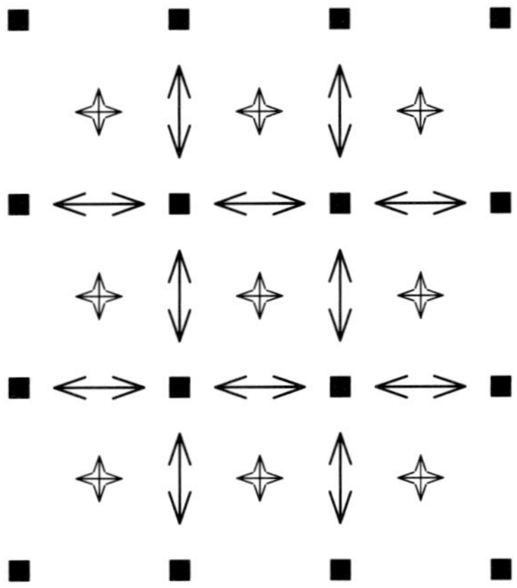

Figure 11.31 The transfer of the slab load to the columns without the presence of beams

required. In any case, the neater aesthetic result offered by this layout has led to a relatively wide application of these plates, particularly in cases with larger spans.

In the orthogonal raster of columns considered (Figure 11.31), the natural tendency of the plate to transfer its loads along the 'most rigid' (i.e. the shortest) path to its supports appears clearly in the trajectories of the existing principal moments shown in Figure 11.32, where the regions between the columns are composed of 'quasi-beams' on which the remaining internal 'orthogonal panels' are supported.

The load-bearing action of such a system may first be considered as shown in Figure 11.33, where each 'horizontal' row of columns constitutes, together with the imaginary strip of width L_y, a one-storey frame with equal spans L_x. In the same way,

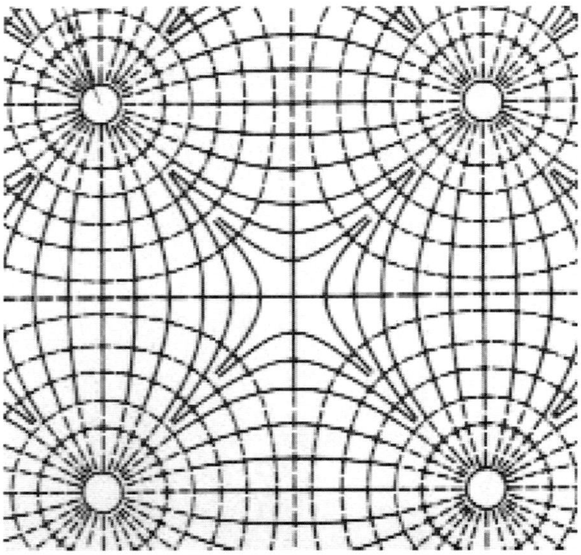

Figure 11.32 Trajectories of the principal moments

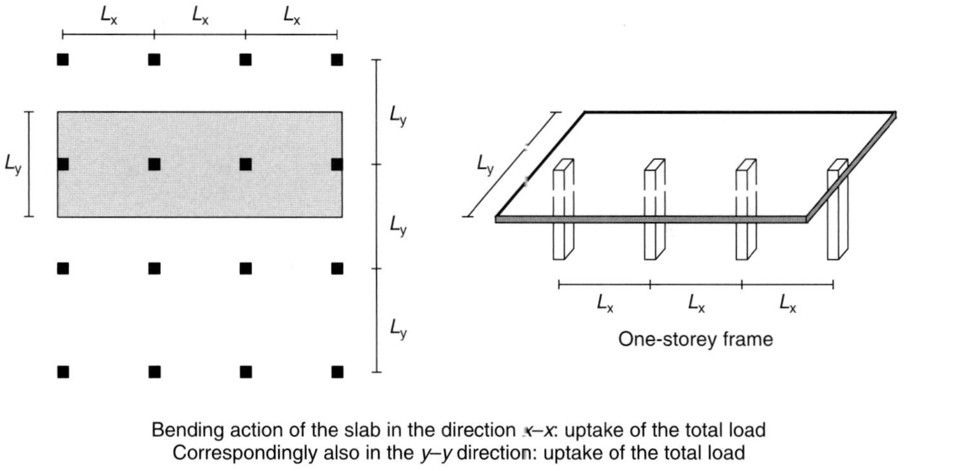

Figure 11.33 Establishment of the action of the frame between the plate and the columns

the action of the plate in the y direction is taken into account. It is understood that the absence of twisting moments along the length of the 'free' edges of this strip, due to symmetry, allows it to be regarded as a free girder. It can be seen, of course, that each span of this frame receives the total load of $q \cdot L_x \cdot L_y$.

The bending response of such a strip is substantially depicted by the form of the moment diagram over the corresponding continuous beam with spans L_x and continuous uniform load $q \cdot L_y$ (Figure 11.34). It is characterised by the corresponding span moments M_F (kN m) and support moments M_S (kN m). These moments, arising from the action of either the frame or the beam, are globally referred to the whole width of the strip and are distributed between the 'quasi-beam' (i.e. an appropriate column strip) and the remainder of the strip. Thus the 'column strip', being more rigid than the remainder strip, will receive the greater portion of the above moments (Franz and Schäfer, 1988).

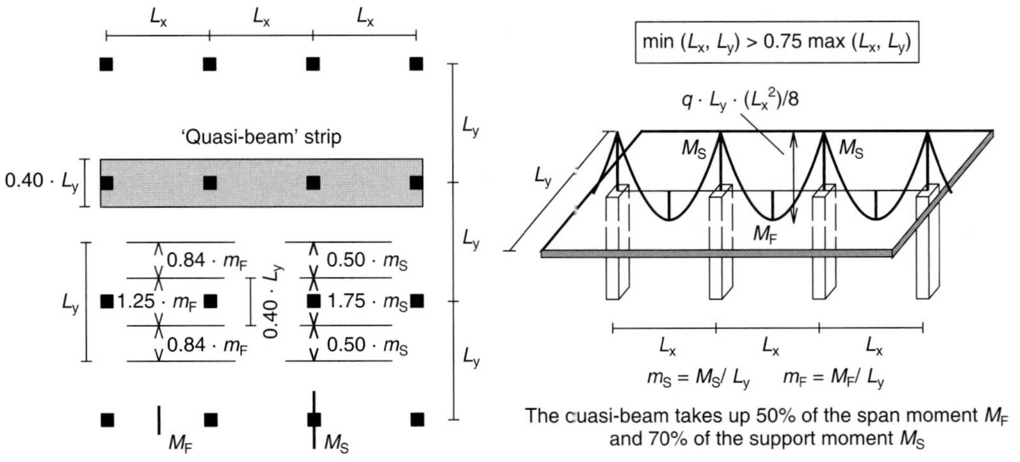

Figure 11.34 Distribution of the bending moments of the strip

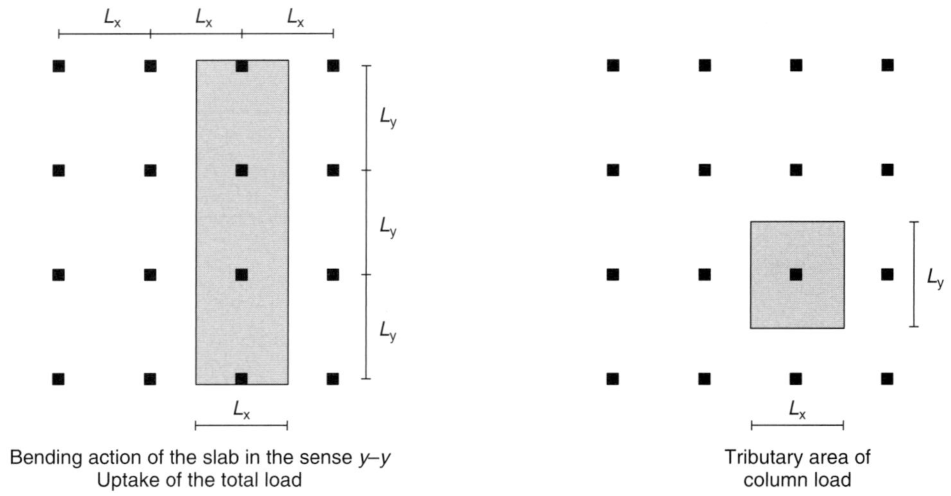

Figure 11.35 Establishment of the action of the frame in the other direction

According to the German Standard DIN 1045, the width of the column strip may be considered equal to 40% of the width L_y, symmetrically arranged with respect to the axes of the column. Thus, with respect to the span moments M_F (kN m) for the column strip region, instead of the 'mean' bending moment M_F/L_y (kN m/m), 125% of this value may be considered, while for the remainder strip width $(0.60 \cdot L_y)$ 84% of the mean value is applied (see Figure 11.34). This means that the moment M_F is equally distributed between the quasi-beam and the remainder strip.

Regarding the support moments M_S (kN m), instead of the mean bending moment M_S/L_y (kN m/m), 175% of this value can be applied on the width of the quasi-beams, while in the remainder strip width $(0.60 \cdot L_y)$ half of the above value is applied. Thus, the column strip region receives globally 70% of the moment M_S, while the remainder strip receives the other 30%.

As the transfer of plate loads occurs also in the transverse *y* direction, the analogous frame with span lengths L_y and strip width L_x should also be considered (Figure 11.35). This frame will again carry over each span the total load of the plate $q \cdot L_x \cdot L_y$, while the magnitudes of the moments M_F and M_S will be distributed in the same manner as in the other direction.

In this way, due to absence of continuous supports, the entire plate load is necessarily taken into account in full, in both directions, in order to ensure equilibrium in each of the two directions. Thus the resulting bending response of the plate is clearly less favourable than the one developed when there is a continuous beam support. Of course, each column practically receives the total load $(q \cdot L_x \cdot L_y)$ of the corresponding contributory area (see Figure 11.35).

It should be clear that the above quantitative estimates concern the preliminary design and are only valid if the existing aspect ratio of each panel is not less than 0.75 (see Figure 11.34). It is obvious that for a complete determination of the state of stress of a flat slab the use of appropriate computer software is required. ■

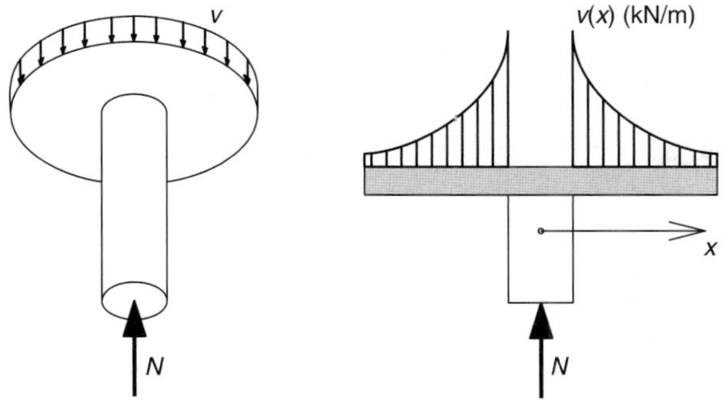

Figure 11.36 Variation in the shear force in the region of a column

As stated at the beginning of this discussion, a basic problem with flat slabs is the danger of punching. The column loads are applied on the slab, practically as upward concentrated loads, and, given their high value, particular attention is required for their safe introduction into the slab. The region around a column presents a more intense increase in the shear forces than is the case in a continuous beam, as shown in Figure 11.36 for the example of a circular column.

Consider a flat slab placed over a column raster with span lengths equal to L. Under a uniform load equal to q the column load is $N = q \cdot L^2$, and the shear force v (kN/m) of the plate at a radial distance x from the column centre will be

$$v = \frac{N - q \cdot \pi \cdot x^2}{2 \cdot \pi \cdot x} = q \cdot \left(\frac{L^2}{2 \cdot \pi \cdot x} - \frac{x}{2} \right)$$

The presence of x in the denominator indicates the particularly intense increase in the shear force, as the circular section shrinks to the outline of the column section.

11.5.2 Design for punching shear

The determination of reinforcement on the basis of the locally considered bending moments does not present any peculiarity in comparison to conventional slabs. However, as previously mentioned, designing for the bending moments only is not enough. The punching shear needs also to be taken into account, as examined below.

As explained previously, the development of large shear forces, and hence shear stresses in the column neighbourhood, induces skew tensile stresses that ultimately cause a perimetric detachment of the slab from the remaining column region over a surface in the form of a cone (Figure 11.37). The failure does not occur at the perimeter of the column but at an angle of roughly 30° to the column face. The reason for this is that in the immediate region where the the load is introduced the plate calculated shear stresses are hazardless, due to the high vertical compressive stresses, while at some distance these relieving stresses practically vanish.

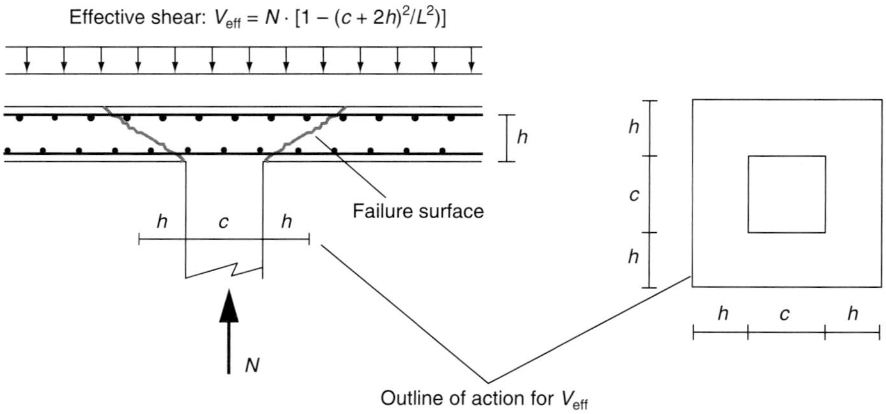

Figure 11.37 Region for checking the effective shear against punching

For additional safety this checking is done at a distance of an angle of 45°. In this position, the transferred 'effective shear' V_{eff} from the slab to the body of the cone through the corresponding failure inter-surface is obviously smaller than the load N of the column due to the presence of the uniform load applied over the column head. It is:

$$V_{eff} = N \cdot k$$

where

$$k = \left[1 - \left(\frac{c + 2 \cdot h}{L}\right)^2\right]$$

In this expression c is the dimension of the column and h is the mean static height of two successive reinforcement layers of the slab in the region over the column head (see Figure 11.37).

Hence, the shear stress τ is directly determined as

$$\tau = \frac{V_{eff}}{4 \cdot h \cdot (c + h)}$$

The ultimate value τ_u of this stress is determined as a function of the ratio μ of the mean value A_s (cm^2/m) of reinforcement in the two top layers to the 'effective' cross-section of the plate $100 \cdot h$ (cm^2), the compression strength of concrete β_c and the yield stress of reinforcement f_{sy}, on the basis of the equation (Herzog, 1996)

$$\tau_u = \beta_c \cdot \frac{1.6 \cdot \mu \cdot (f_{sy}/\beta_c)}{1 + 16 \cdot \mu \cdot (f_{sy}/\beta_c)}$$

The calculated ultimate force N_u of the column according to the above equations is

$$N_u = 4 \cdot \tau_u \cdot h \cdot (c + h)/k$$

If N_u is greater than the design value $\gamma \cdot N$, then sufficient safety against punching is considered to exist. If N_u is smaller than the design value $\gamma \cdot N$, then vertical stirrups

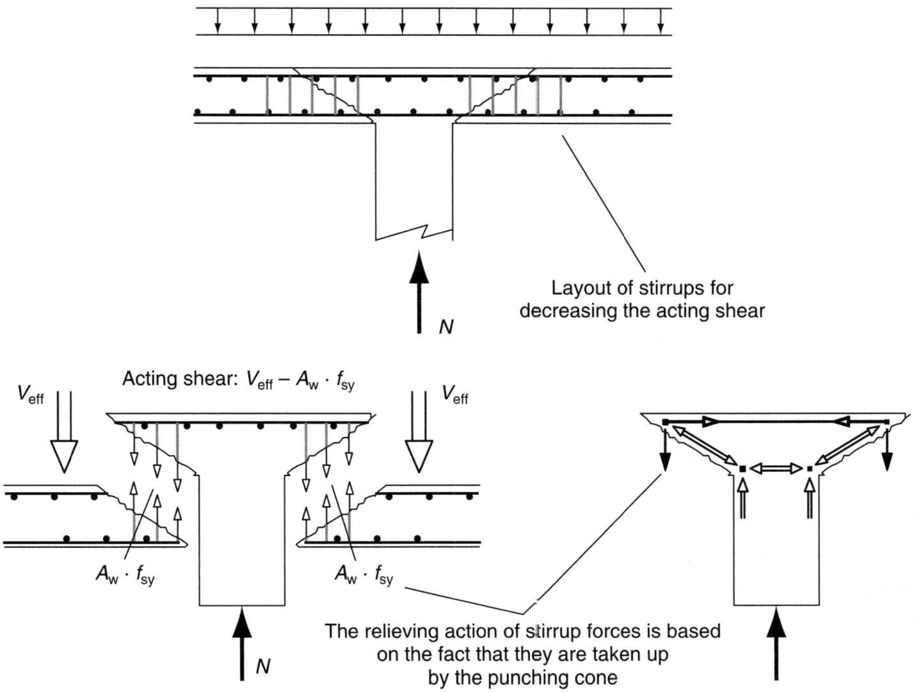

Figure 11.38 Contribution of the stirrups to the uptake of the effective shear force

should be placed in the slab in order to deal with the 'uncovered' column force $(\gamma \cdot N - N_u)$. The action of these stirrups, as shown in Figure 11.38, consists of offering a relieving 'suspension' vertical force through the cracked conoidal surface, at an angle of 30° and towards the loading force $\gamma \cdot V_{eff}$, which is then 'safely' transferred as a compressive force to the remaining column head. Obviously the force $(N_u \cdot k)$, representing the transferable force from the slab through the lateral surface of the cone, should be greater than, or at most equal to, the loading force $\gamma \cdot V_{eff}$, which is reduced by the stirrup forces, with a total cross-sectional area A_w that is cut by the failure surface, being under the yield stress f_{sy}. Thus

$$\gamma \cdot V_{eff} - A_w \cdot f_{sy} = N_u \cdot k$$

and hence

$$A_w = \frac{\gamma \cdot N - N_u}{f_{sy}} \cdot k$$

It should be pointed out that the punching type of failure is a brittle one, i.e. it occurs without previous warning, and its repercussions are particularly unpleasant and dangerous (much worse than, for example, failure of an individual beam or even a part of a plate). For this reason, when designing flat slabs one should avoid exhausting the 'calculation limits'.

11.5.3 Prestressing

Prestressing is particularly suitable for dealing with the bending response and deformations of flat slabs, as well as for increasing their punching resistance.

In order to counteract the bending influence of the load q on a flat slab over a column raster arrangement, the initial idea consists of taking the force up through deviation forces $u_{x,F}$ and $u_{y,F}$ of the prestressed cables in each panel (Figure 11.39). It is reasonable to equally distribute this load to the two directions, i.e. $u_{x,F} = u_{y,F} = 0.5 \cdot q$, and therefore to consider the corresponding sags f of the cables in both senses as practically equal. The corresponding prestressing forces per unit length will then be (see Section 4.3.1.1)

$$P_{x,F} = \frac{0.5 \cdot q \cdot L_{x,F}^2}{8 \cdot f}$$

and

$$P_{y,F} = \frac{0.5 \cdot q \cdot L_{y,F}^2}{8 \cdot f}$$

The profile of each cable, transversally crossing a number of column strips, necessarily follows a path similar to that applied for a continuous beam (see Figure 11.39). This path passes towards the ends of each panel, through an inflection point, to an opposite curvature over the column strips, in both the x and y direction, and continues in this way to the neighbouring panels (see Section 5.4.1).

As explained in Section 5.4.1, the deviation forces of each cable between the column axes constitute a self-equilibrating system, and thus each column strip is acted on by the downward applied deviation loads: $u_{x,S} = u_{x,F} \cdot (L_{x,F}/L_{x,S})$ and $u_{y,S} = u_{y,F} \cdot (L_{y,F}/L_{y,S})$ in

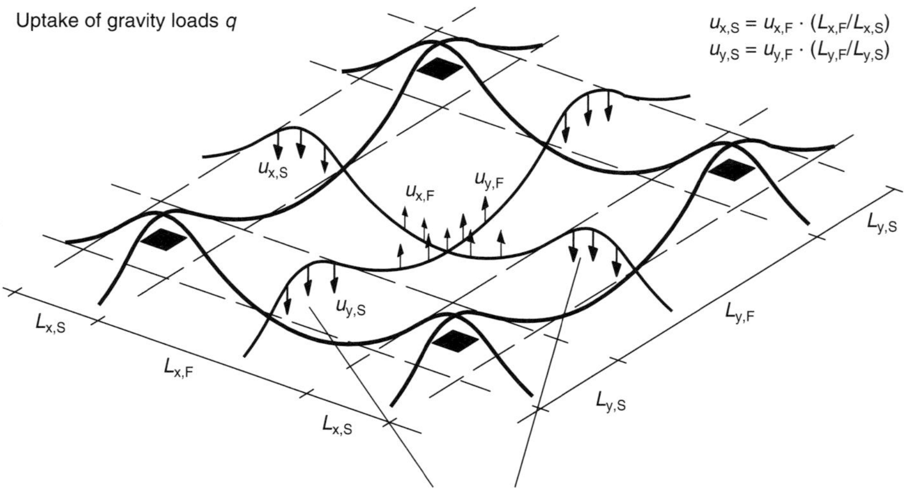

Figure 11.39 Taking up of loads through equally distributed prestressed cables

the x and y directions, respectively. These loads necessarily have to be taken up by prestressing, meaning that additional cables must be placed in the support strip, as shown in Figure 11.39. The distributed prestressing force of these cables per metre width in the corresponding column strip are obtained as

$$P_{x,S} = \frac{u_{y,S} \cdot L_{x,F}^2}{8 \cdot f}$$

and

$$P_{y,S} = \frac{u_{x,S} \cdot L_{y,F}^2}{8 \cdot f}$$

Thus, the total prestressing force, for example in the x direction, is

$$P_{x,F} \cdot L_{y,F} + P_{x,s} \cdot L_{y,S} = \frac{(q \cdot L_{x,F} \cdot L_{y,F}) \cdot L_{x,F}}{8 \cdot f}$$

This clearly corresponds to the uptake of the total load q, not half of it, as was initially intended. Obviously, the same naturally also happens in the y direction and, consequently, the above cable arrangement does not represent an economic solution. ■

A much more reasonable layout of tendons results by placing the cables in the column strips only, making sure that the load q is fully taken up (Figure 11.40). The total load is equally distributed to the deviation forces in the strips along the x and y senses, i.e. $u_x = 0.5 \cdot q \cdot L_y$ (kN/m) and $u_y = 0.5 \cdot q \cdot L_x$ (kN/m), respectively, leading to the total prestressing force

$$P_x = \frac{u_x \cdot L_{x,F}^2}{8 \cdot f}$$

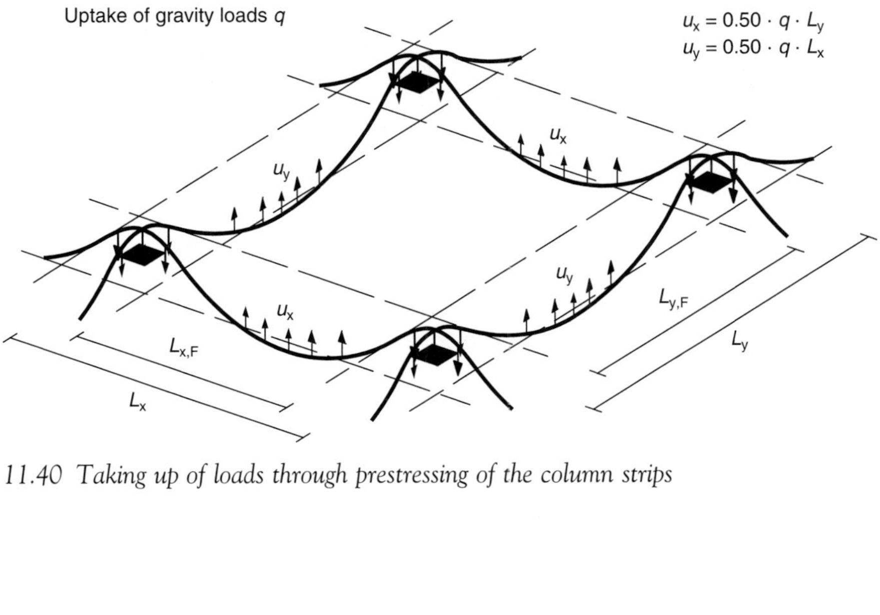

Figure 11.40 Taking up of loads through prestressing of the column strips

and

$$P_y = \frac{u_y \cdot L_{y,F}^2}{8 \cdot f}$$

respectively, for each strip.

In this way, the downwards acting deviation forces, being in equilibrium with those in the panels, are loading the columns directly. Thus, on the one hand, they do not create additional stresses that should be dealt with, as in the previous case; and, on the other hand, they relieve the punching stresses, as will be examined below. It is obvious that with this cable layout the bending moments in the panels are met using steel reinforcement.

With regard to the column strips, it should be noted that the total developing bending moment m_P due to prestressing (i.e. due to the deviation forces) is separated into a statically determinate part m_0 and a statically redundant part m_{SP} (see Sections 5.4.1 and 11.2.1.2). The bending moment m_P should be determined using appropriate computer software. For the statically determinate part, $m_0 = \pm P \cdot e$, where e is the eccentricity of the cable and the sign + or − depends on whether the cable is above or below the neutral axis, respectively. Thus, $m_{SP} = m_P - m_0$. It should be remembered (see Section 5.4.2) that m_{SP} must be added to the 'loading side' of the strength equation, whereas m_0 is automatically taken into account in the determination of the bending resistance of the plate section, whereby the always existing steel reinforcement must also be considered. ■

Regarding now the punching shear, it is clear that, according to the above layout, the deviation forces applied downwards are received directly by the columns, while the deviation forces applied upwards reduce the effective shear (see Section 11.5.2) developed in the region of the punching cone (Figure 11.41). The deviation forces are equivalent to the vertical components of the cable forces, and appear at their intersection points with the punching cone lateral surface, and are equal to $\sum P \cdot \sin\alpha$. Angle α is measured from the point where the cable intersects the failure

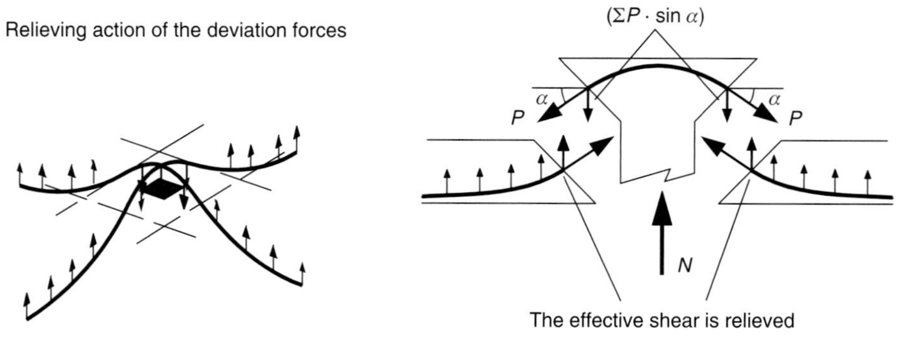

Figure 11.41 Relieving action of prestressing cables against the effective shear

surface and may be (unfavourably) considered equal to 45° (see Figure 11.41). Thus, the punching check formulated in Section 11.5.2 may now be written, without the presence of stirrups, as

$$\gamma \cdot V_{\text{eff}} - \sum P \cdot \sin\alpha < N_u \cdot k$$

and finally

$$\sum P \cdot \sin\alpha > (\gamma \cdot N - N_u) \cdot k$$

a requirement that can easily be satisfied.

11.6 Folded plates

As pointed out at the beginning of this chapter, plates are not only able to carry loads transversely to their plane through bending, but they can also offer much higher stiffness against loads applied within their plane. Exploitation of this property in covering large (mainly orthogonal) areas is possible by connecting oblong plates along their long common sides, thus creating a folded form (Figure 11.42). This structural layout obviously needs a much greater available height than does a single compact plate. Nevertheless, it achieves the covering of comparatively much larger spaces, with a limited plate thickness.

A direct perception of the possibilities of a folded plate can be obtained if a sheet of paper, supported on two opposite sides and deforming excessively because of its small thickness, is folded as shown in Figure 11.43. Its deformation then becomes negligible, even if a small additional change is added.

The load-bearing action of a folded plate is understandable from the typical layout shown in Figure 11.42. It can be seen that this structure is composed of Λ-shaped beams, which, due to their large structural height and their much higher moment of inertia compared to the plane plate, offer a considerably more rigid structure (Figure 11.44).

The above remarks allow the conclusion that the folding edges may offer practically immovable supports to the oblong plates that are formed by each plane surface. It is clear

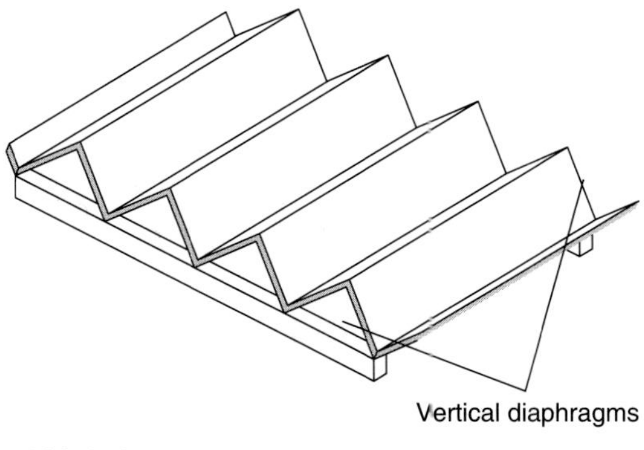

Figure 11.42 Layout of a folded plate

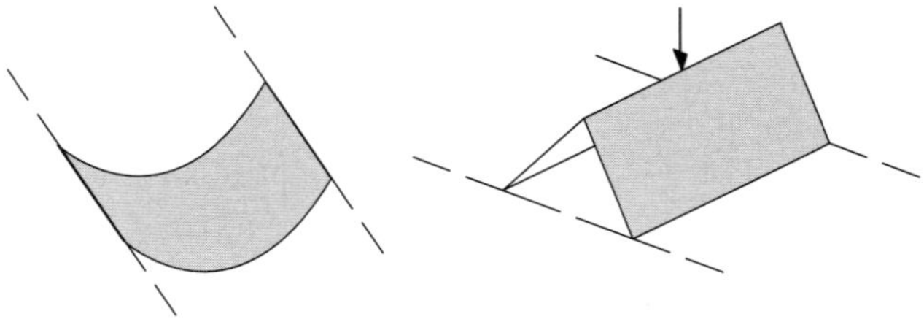

Figure 11.43 Drastic improvement in the bearing capacity of a sheet of paper due to folding

that these plates act in bending as two-side supported slabs, as their length is much greater than their width. It should be noted, however, that the longitudinal edges are 'immovable' only if frontal vertical diaphragms or appropriately formed frames with significant bending rigidity are supplied that are firmly connected to the ends of the plates. The structural action of a folded plate is examined below and is illustrated in Figure 11.45.

First, imaginary supports are considered along each fold, these offering the same vertical upward reactions as those of a corresponding continuous beam representing the individual two-side supported plates. Obviously the bending response of the 'altered' folded system will be identical to that of the continuous one, and thus may be determined directly.

Of course, the fact that the edges of a folded plate are actually free from external forces implies that the above 'reaction' forces on each edge must be applied again, in a second phase, as loads of opposite sense, so that this state of stress may be superposed on the previous one (see Section 3.3.1). In this second phase, each edge load can be analysed along the direction of the adjacent plates, and thus each plate is loaded within its plane by the resultant of the corresponding loads along its two longitudinal sides. These loads are taken up by the plate, which is acting now as a deep beam and transfers the loads to the aforementioned frontal walls through the corresponding shear forces of its end sections. It is clear that the above treatment of the downward edge loads refers to the taking up of these loads by the Λ-shaped folded beams of which the whole folded plate is composed.

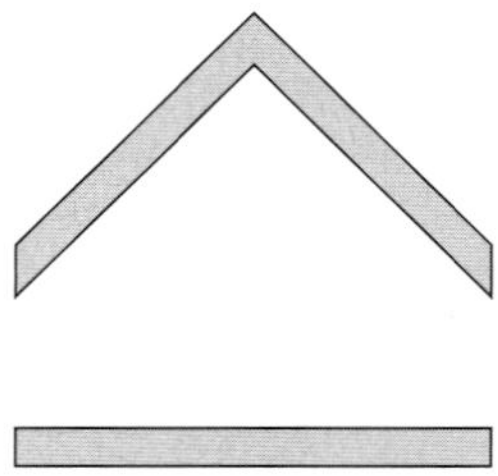

Figure 11.44 Drastic increase in the active moment of inertia due to folding

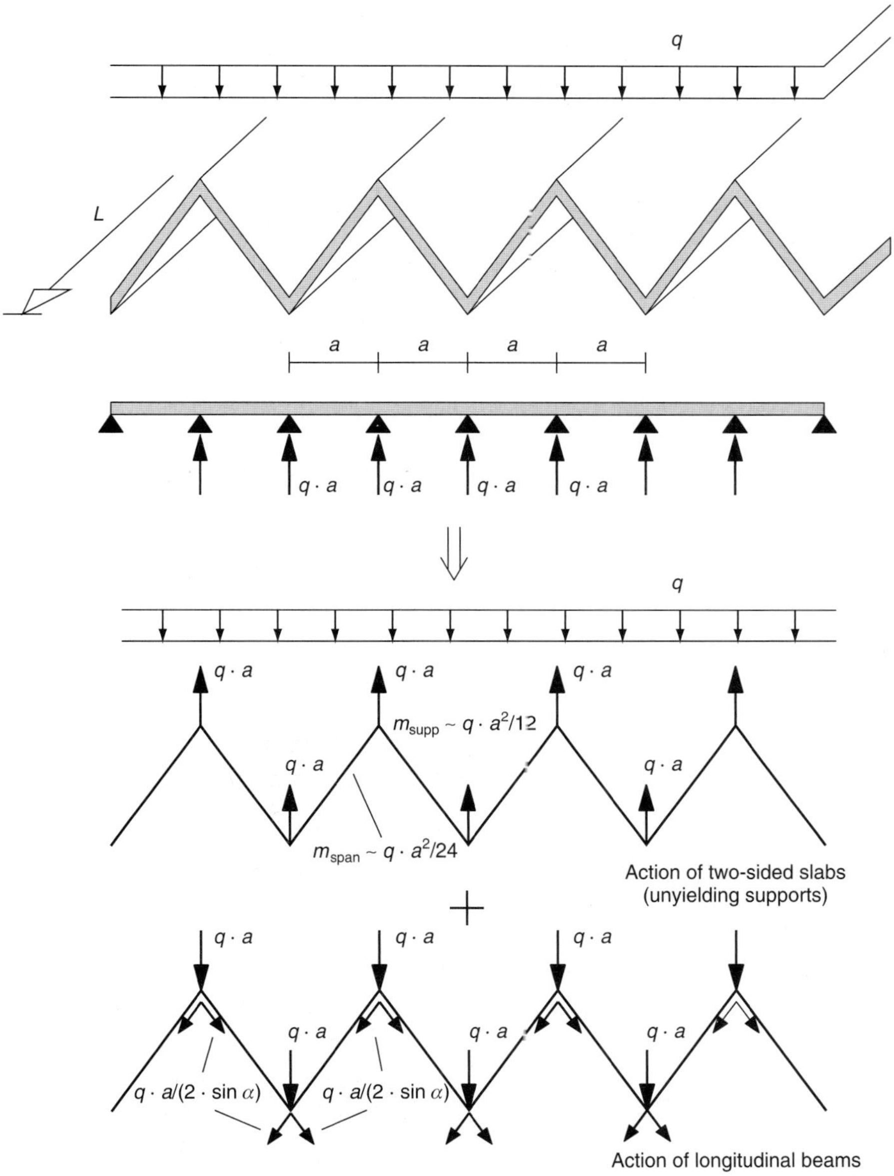

Figure 11.45 Load-bearing action of a folded plate

Thus a folded plate acts in a 'combined' way, i.e. as a combination of a system of rigid deep beams in longitudinal bending together with the formed oblong slabs continuously supported in transverse bending (Figure 11.46).

It should be pointed out that the above approach can be used only for preliminary design purposes. For a more accurate determination of the response the use of a suitable finite-element program is necessary.

As folded plates exploit advantageously the rigidity of their constituent slabs in their own plane, then provided that the necessary space for the required folding height is

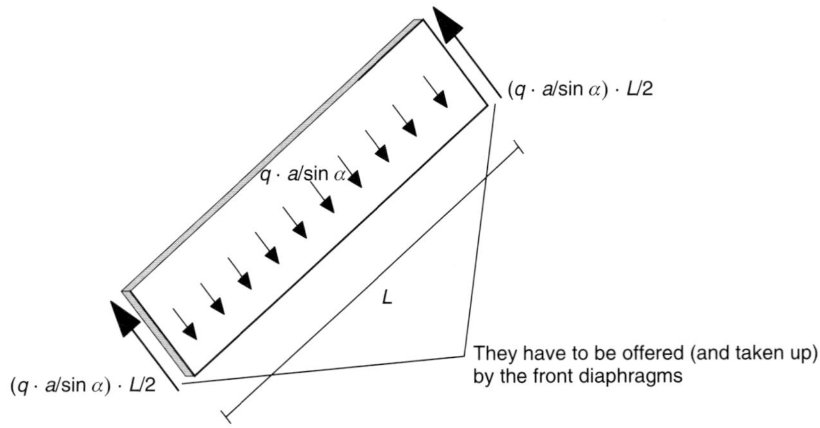

Figure 11.46 Load-bearing action of a plate as a deep beam

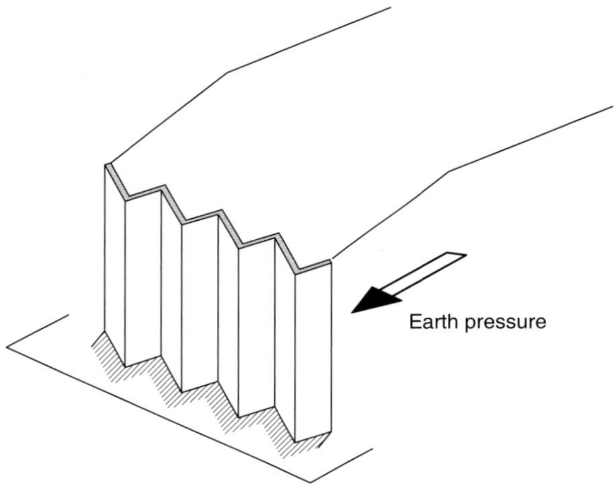

Figure 11.47 The possibility of taking up earth pressures

available, they allow a wide variety of construction layouts, as surface loads can be taken up in vertical as well as in horizontal directions, as, for example, in retaining walls (Figure 11.47).

References

Franz G., Schäfer K. (1988) *Konstruktionslehre des Stahlbetons*, Band II, B. Berlin: Springer-Verlag.

Herzog M. (1996) *Kurze baupraktische Festigkeitslehre*. Düsseldorf: Werner Verlag.

Menn C. (1990) *Prestressed Concrete Bridges*. Basel: Birkhäuser Verlag.

Nervi P.L. (1956) *Structures*. New York: McGraw-Hill.

Stavridis L.T. (1993) Static and dynamic analysis of orthotropic rectangular plates [in German]. *Stahlbau* **62**, 73–80.

Stavridis L.T. (2001) Alternative layout for the prestressing of slab bridges. *Journal of Bridge Engineering* ASCE **6**(5), 324–332.

12

Shells

12.1 Introduction

Covering spaces of large dimensions which are about 15 m long in both directions with a concrete plate is inefficient in practice, and becomes impossible for larger dimensions. In the case of a beam, a single span (without involving prestress) has length limitations that are overcome by adopting a structure of the so-called funicular form (i.e. an arch). In exactly the same way, the flexural plate also has limited dimensions, which can be dealt with by adopting a monolithic type of thin structure, formed by an appropriate curved surface of the same basic outline (Figure 12.1). This type of structure is called a 'shell'.

The contour of the shell boundary does not necessarily lie on a plane. It may be represented by any properly shaped spatial curve, whose projection on the horizontal plane corresponds to the boundaries of the floor space which needs to be covered. Of course, the shape of the shell is affected by its contour, and this shape, in turn, has a major influence on its static behaviour. A typical shell may have a ratio of base length to thickness of the order of 500.

The morphological options for the problem of covering large spaces using shells are essentially unlimited, and the aesthetic result is always of importance.

12.2 The membrane action as a basic design concept

A shell is always referred to by its middle surface, and for this reason it is necessary to define some fundamental geometric attributes that play an important role in its load-carrying behaviour.

It is known that there is only one line passing from a point O of a surface that is perpendicular to it. Two planes are considered now to pass from that line, having arbitrary orientations and being perpendicular to each other. These two planes intersect the tangent to the surface plane at point O along the straight axes x and y, and the surface of the shell along the curves $z = f_x(x)$ and $z = f_y(y)$, where z is measured in the normal direction, as shown in Figure 12.2.

Each of these two curves has at the common point O a specific radius of curvature and a curvature $1/R_x$ and $1/R_y$, respectively. If the radii of curvature R_x and R_y are found on the same side with respect to the surface, the surface is considered to be of positive (Gaussian) curvature, otherwise it is of negative (Gaussian) curvature.

Copyright Thomas Telford Limited © 2010
All rights reserved

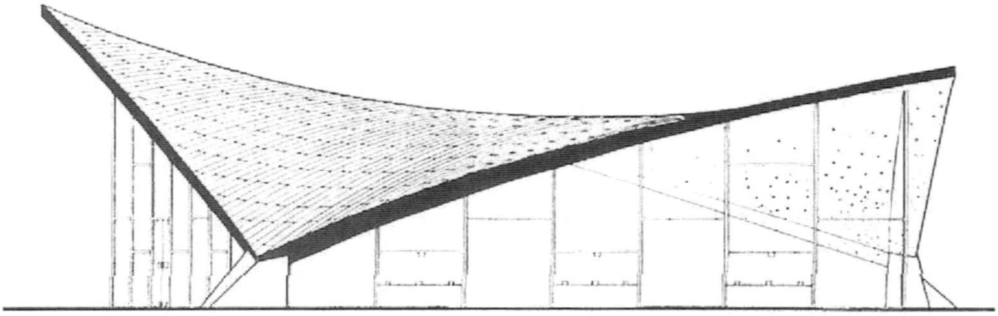

Figure 12.1 Hyperbolic paraboloid shell for covering a rectangular space

Whether the surface has positive or negative curvature is very important for the static behaviour of the shell.

At any point O of a surface, there is always a pair of mutually perpendicular planes which produce curvatures corresponding, simultaneously, to the maximum and minimum values that they can achieve as the pair of planes rotates about the normal line at point O. These curvatures are called *principal curvatures*, as they are in the same directions as the principal plane.

Let a point O_1 be considered on the curve f_y at a small distance Δy from O, and proceed with the intersection of the shell surface and a plane passing from point O_1 that is parallel to the plane defined by the curve f_x (see Figure 12.2). This intersection is another curve, and if its tangent at point O_1 forms an angle α with the direction O_1x, then the quantity $\alpha/\Delta y$, i.e. the rate of change of the angle α, is called the 'twist' t_{xy} of

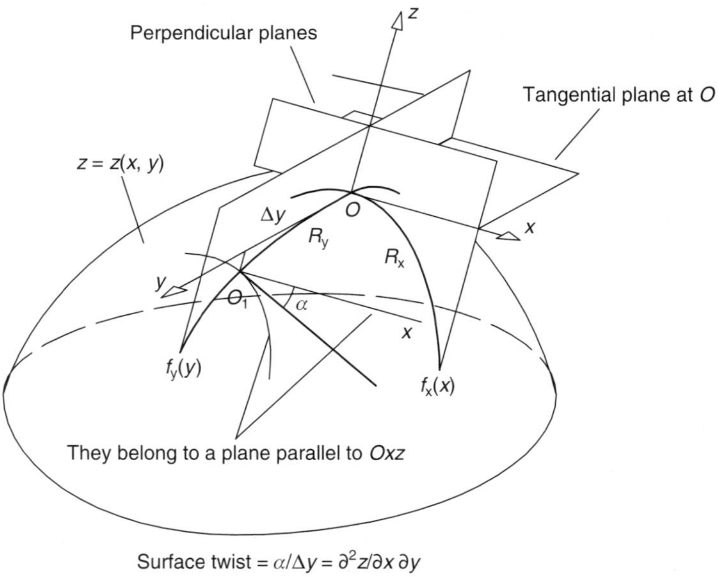

Figure 12.2 Surface geometry

Figure 12.3 Creation of a funicular membrane

the surface at point O with respect to the coordinate system Oxy, and

$$t_{xy} = \frac{\partial^2 z}{\partial x \cdot \partial y}$$

where $z = z(x, y)$ is the function describing the surface of the shell.

It becomes clear that the twist of the surface with respect to the directions of the principal curvatures is always zero, since in that case the angle α vanishes. ■

It should be recalled here that in the case of an arch which carries the loads only through compressive internal forces, its shape is determined by considering the inverted shape of a fictitious cable which is subjected to exactly the same loads. Exactly the same procedure may be applied in the case of a shell.

More specifically, observation reveals that, if a membrane takes only tension and no bending at all and is firmly supported along the contour of the shell, application to the membrane of the vertical loads which will be eventually received by the shell will generate a specific shape for the membrane (Figure 12.3). If this load is inverted and applied to a shell having the same specific shape as the deformed membrane, the shell will be stressed exactly like the membrane, but with opposite sign, meaning solely compression. Considering its mirror image with respect to the horizontal plane leads to the solution of the problem (Figure 12.4). This state of the shell is called the *membrane state of stress*, and plays a dominant role in design, since the described approach achieves uniform utilisation of the fibres of the section (see Section 2.2.7).

Adoption of the specific directions x and y is necessary in order to consider the membrane-carrying mechanism for a distributed load q acting along the z direction – the normal at point O of the surface – through the cut-off quadrilateral element of the shell, as shown in Figure 12.5.

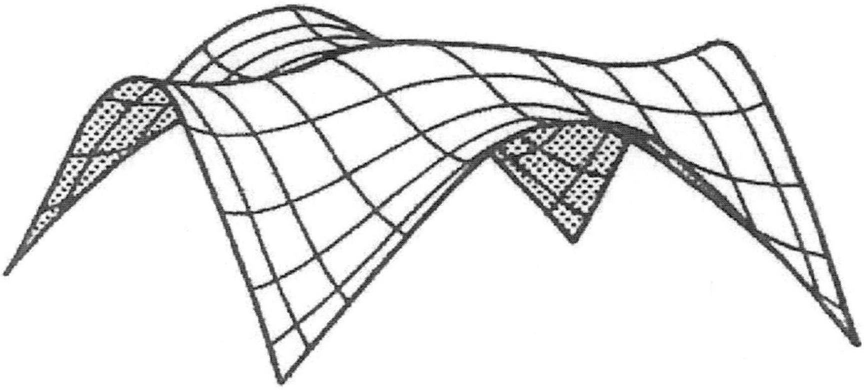

Figure 12.4 Shell formation by inverting the funicular membrane

The state of stress of the membrane consists of an axial membrane force N, which is normal to the element side and tangential to the shell surface, and of a tangential shear force S acting along the side. Both forces refer to a unit side length.

Equilibrium of the quadrilateral element in the z direction reveals that the load p is taken over by the membrane forces N_x, N_y and S, according to the following relation (see Figure 12.5):

$$p = \frac{N_x}{R_x} + \frac{N_y}{R_y} + 2 \cdot S \cdot t_{xy}$$

The first two terms on the right-hand side of this equation are in direct correspondence with the equilibrium equation of cables, as studied in Section 2.2.7 (see Figure 2.42). The third term refers to the vector resultant of the shear forces S along the z axis.

It is clear that the shear force S does not contribute in carrying the load when the twist of the surface at the point in question is zero, and, consequently, there is no reason for its

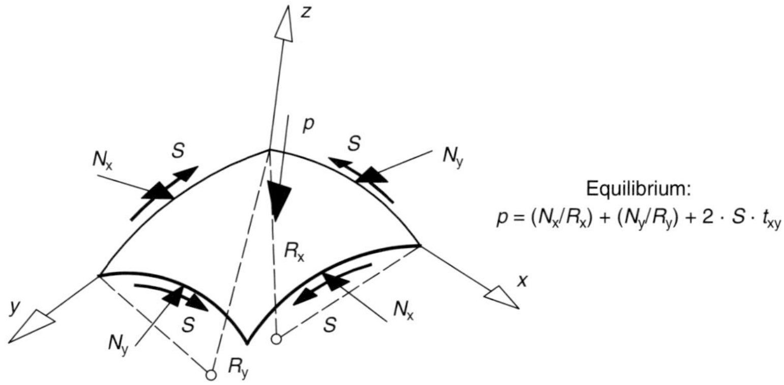

Figure 12.5 Membrane equilibrium

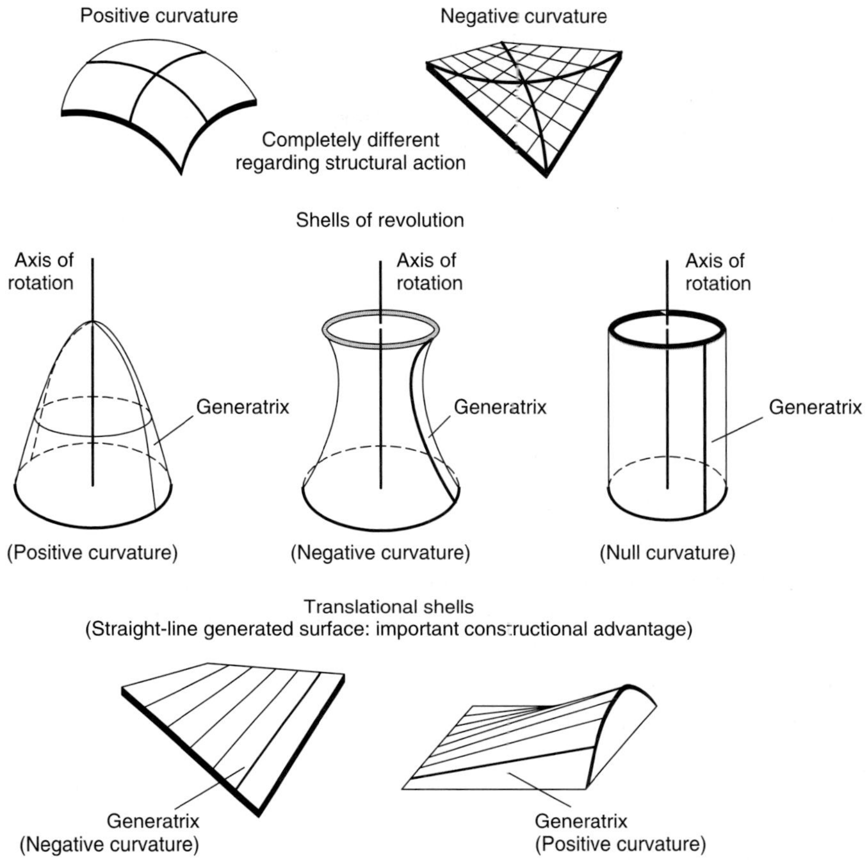

Figure 12.6 Different ways of generating shells

development ($S = 0$). Obviously, this happens when the directions x and y correspond to those of the principal curvatures. ■

Although shells can exhibit unlimited morphologies, they can be placed into two categories with respect to their basic geometric configuration: shells generated by revolution and shells generated by translation (Figure 12.6). The first category refers to shells whose midsurface is generated by revolving a planar line, either straight or curved, which is called the 'generator', about a constant axis. The shells of the second category are produced by sliding a planar line (straight or curved generator) on another constant line, referred to as the 'directrix', maintaining its plane always parallel to itself. This distinction must be combined with that of positive or negative curvature.

For any distributed load p there is one and only one membrane state of stress (N_x, N_y and S) which is in equilibrium with the load, and its determination can always be considered as a statically determinate problem. It should be pointed out that the uniqueness of the membrane solution is attributed to the existence of curvature. If, for example, a plate is subjected to in-plane loads, the developed membrane state

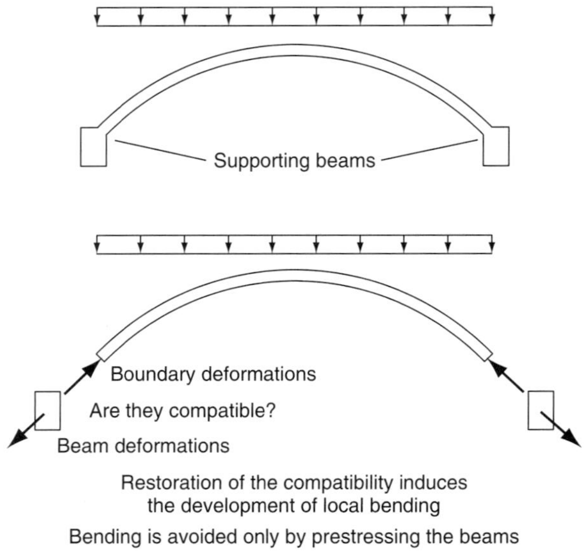

Figure 12.7 The need for displacement compatibility at the boundary of the shell

of stress cannot be determined simply through equilibrium considerations, because there are three unknown quantities, and there are only two available equilibrium equations for the plane case. This plane stress problem is addressed using the theory of elasticity. ■

The construction of a shell usually requires an arrangement of beams along its boundaries, which should be designed to carry the opposite of the forces applied to the shell at the boundary (Figure 12.7).

The shell, by developing the membrane action under the applied loads and the forces on its boundary required for equilibrium, exhibits some specific deformations. However, the same boundary forces of the shell are acting in the opposite direction on the beams or, more generally, on the supporting structural elements, causing other deformations there. The question that arises is whether these two different types of deformation are compatible. If this is not the case, then the membrane state of stress for the shell is not possible, and re-establishment of the displacement compatibility requires one of the following two procedures.

According to the first procedure, additional appropriate forces are applied to the shell boundary and exactly opposite forces to the supporting structure, in order to make the final deformations equal. Application of such forces to the shell will cause bending, although bending is not acceptable for a shell whose natural state of stress is that of a membrane. Since the displacement compatibility is restored by introducing these boundary forces, this means that the problem at hand turns into a 'statically indeterminate' one, and the bending response will be dependent on the shell thickness. Of course, the thickness does not affect the membrane action of the shell. In many cases, this bending of the shell occurs only in a small region within the neighbourhood of the boundary.

In the second procedure, prestressing is introduced to the supporting structure, so that its displacements are the same as those developed on the shell. This solution, whenever applicable, is statically preferable. It is also reasonable to apply prestress to the shell instead of the supporting structure in order to achieve compatibility for the displacement between the shell boundary and the supports. ■

The design of a shell, i.e. the selection of its shape and dimensions, can and must be based on the membrane action, in order to utilise to the maximum its material. Nevertheless, in cases of marked incompatibility between the deformations of the shell boundary and the supporting system, the state of stress due to bending should also be considered in the selection of dimensions, as was previously explained, although for shells with positive curvature this is confined within a small area close to the boundary. However, a detailed computation of the stresses requires appropriate computer software.

Obviously, given that shells covering large areas are subjected to high compressive forces, particular effort should be made to avoid failure due to buckling, which may possibly become the dominant factor in the determination of the thickness. Since the critical compressive stress and the critical pressure applied to the shell are, in all cases, proportional to the modulus of elasticity E of the material, i.e. of the concrete, and given that there is always uncertainty about the implementation of the analytical results due to constructional defects, creep of concrete, etc., it is logical to adopt a reduced value for E, in order to increase the desired safety factor.

The basic characteristics of the load-carrying behaviour of the most common types of shells are presented in the following sections.

12.3 Cylindrical shells

Cylindrical shells of revolution are used either for storing liquids or granular materials (reservoirs or silos). Their longitudinal axis may either coincide with the vertical direction, in which case they rest on a circular concrete base, or may be horizontal in order to cover orthogonal areas, using only a part of their ring section.

12.3.1 Cylindrical shell under constant internal pressure

The perception of the load-carrying behaviour of a cylindrical shell which is supported at its lower circular end and is loaded internally by constant pressure p_r along its height is of particular importance, as will also be seen later for other shells of revolution (Figure 12.8).

The internal pressure p_r, considered initially as a self-equilibrating system acting on the free shell, produces the pure membrane tension N_θ in the ring, while the straight-line generators, due to their infinite radius of curvature, do not develop any axial forces. All the foregoing may be directly concluded from the membrane equilibrium equation given in Section 12.2, which yields: $N_\theta = p_r \cdot R$, where R represents the radius of the cylinder.

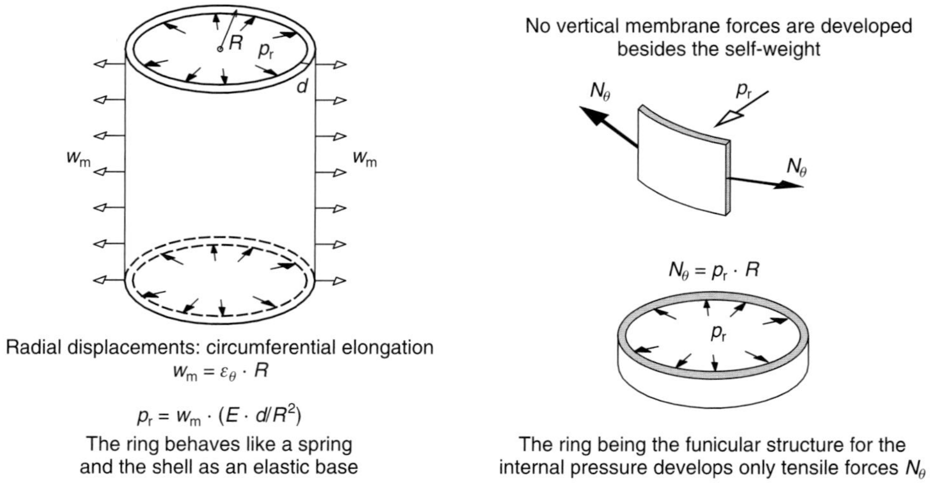

Figure 12.8 Load-carrying behaviour of a free cylindrical shell under internal pressure

This tensile state of stress results in an angular elongation strain ε_θ, which depends on its thickness since $\varepsilon_\theta = N_\theta/E \cdot d$, and, consequently, its radial outward displacement w_m is equal to $w_m = \varepsilon_\theta \cdot R$. These last relations yield

$$p_r = w_m \cdot \frac{E \cdot d}{R^2}$$

This expression reveals that the pressure p_r is taken by the shell through membrane action, just as in the case of an elastic spring base having stiffness $k = (E \cdot d/R^2)$ (see Section 17.3.2.2). This stiffness is offered by the fictitious consecutive shell rings through the whole height of the shell, and is proportional to their thickness d and inversely proportional to the square of the shell radius R. Obviously, w_m represents the deformation of the fictitious spring base (see Figure 12.8). ■

It is now well understood that the fixed support around the circular bottom end of the shell is not compatible with the development of w_m. As a consequence, bending moments and radial shear forces should act along the boundary, in order to make its radial displacement vanish and to guarantee the zero slope of the generators at the clamped support (Figure 12.9).

Otherwise, the shell may be considered to consist of vertical straight beams (having, of course, a bending stiffness) and to be supported elastically by the horizontal shell rings along the shell height (Figure 12.9). These vertical beams may be considered as fixed or hinged at their base. In case they are not supported there, the constant pressure p_r is carried through the membrane action of the rings, much like springs with the aforementioned stiffness k. The fact that all along the vertical beams there is an imposed constant displacement

$$w_m = (p_r \cdot R^2)/(E \cdot d)$$

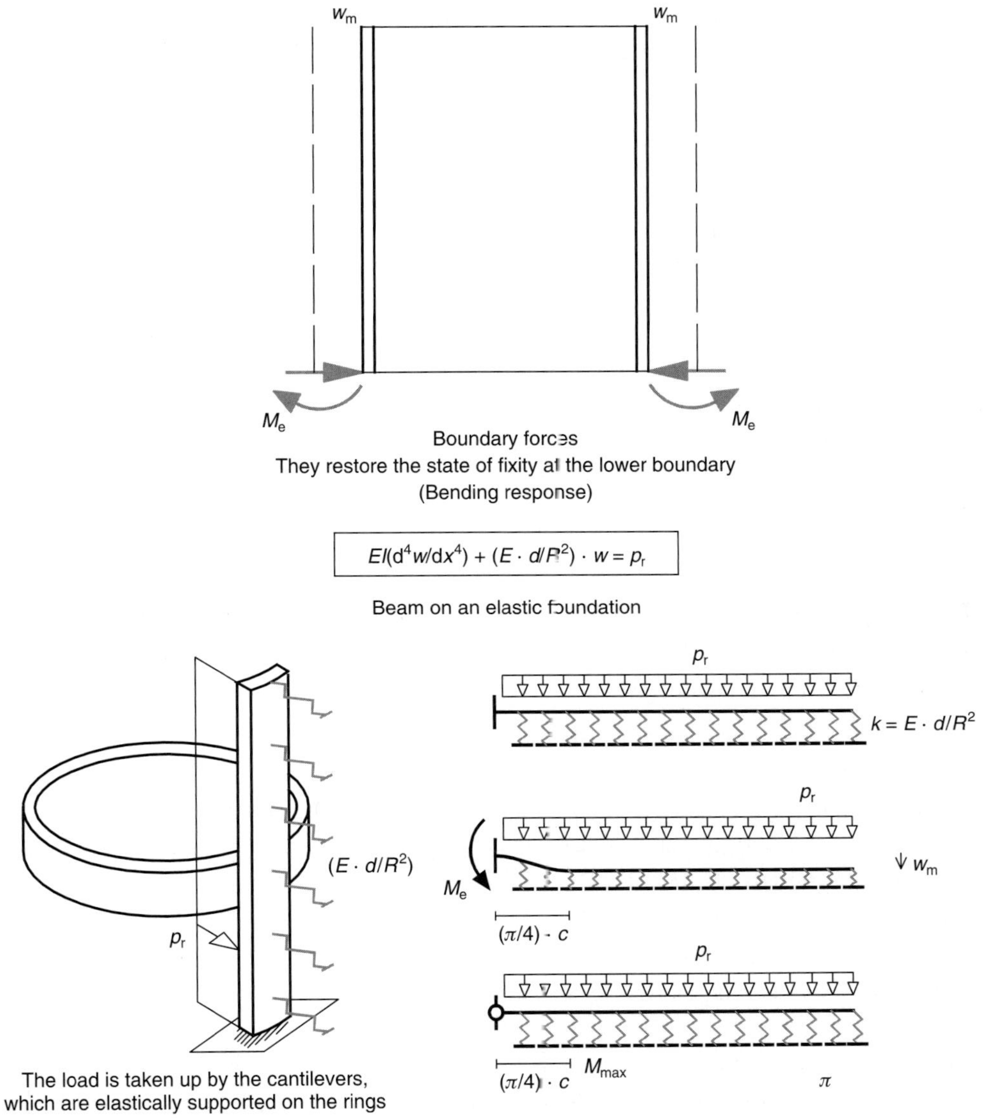

Figure 12.9 Load-carrying action of a cylindrical shell fixed at its base

leads to the conclusion that these beams remain stress-free. If, however, their lower end is supported, the developed transverse displacements will obviously change, resulting in the activation of their bending stiffness. Thus, the load p_r is considered to be carried by these beams as if these were resting on an elastic foundation (see Section 17.3.3.1) with the modulus of the subgrade reaction $k = (E \cdot d/R^2)$ (see Figure 12.9).

The above consideration means that the load p_r is carried by the membrane action of the shell ring on the one hand and the bending resistance of the vertical beams on the other, both being expressed through the developed displacement w. So,

$$EI \cdot \frac{d^4w}{dx^4} + \frac{E \cdot d}{R^2} \cdot w = p_r$$

In this way, the well-known equation of a beam of unit width resting on an elastic foundation with the subgrade modulus $k = (E \cdot d/R^2)$ can be directly recognised (Billington, 1965; see Section 17.3.3.1).

If the lower shell boundary is fixed, as previously stated, the required uniformly distributed radial forces and moments have to cancel not only the radial displacement w_m due to membrane action but also the rotation of the generatrix at its base (see Figure 12.9). It is clear that the bending moment resulting at the lower boundary of the shell will be identical to the bending moment M_e at the fixed end of the corresponding beam on an elastic foundation, i.e. (Timoshenko, 1956)

$$M_e = \frac{2}{c^2} \cdot EI \cdot \frac{p_r}{k}$$

where

$$c = \sqrt[4]{\frac{4 \cdot EI}{k \cdot 1}}$$

Physically, the quantity c, which is called the 'characteristic length' (see Section 17.3.3.1), relates to the distance $(c \cdot \pi/4)$ from the fixed end of the beam (equivalently from the end of the shell) over which the bending disturbance is extending $(c = 0.76 \cdot \sqrt{R \cdot d})$, while the rest of the beam remains essentially free of stress, due to the imposed constant displacement w_m (see Figure 12.9).

It is interesting to show that the above expression for M_e, after the necessary substitutions, yields the bending moment of a cantilever having a length c and carrying the load p_r, i.e.

$$M_e = p_r \cdot c^2/2 = 0.29 \cdot p_r \cdot R \cdot d$$

which, obviously, applies tension to the inner fibres of the cylinder (see Figure 12.9).

It is useful to point out that the term p_r/k in the initial expression for M_e is equal to w_m, and thus may also be written as

$$M_e = 0.29 \cdot E \cdot d^2 \cdot \frac{w_m}{R}$$

In the case where the lower boundary of the shell is hinged, the maximum bending moment M_{max}, which applies tension to the external fibres, occurs, by analogy to the beam on an elastic foundation with one hinged end, at a distance $c \cdot \pi/4$ from the hinged end (Figure 12.9). This moment is given by (see Timoshenko, 1956)

$$M_{max} = \frac{0.64}{c^2} \cdot EI \cdot \frac{p_r}{k}$$

or

$$M_{max} = 0.092 \cdot E \cdot d^2 \cdot \frac{w_m}{R} = 0.092 \cdot p_r \cdot R \cdot d$$

12.3.2 Cylindrical tanks

These are acted on by a permanent internal hydrostatic pressure p_r perpendicular to their surface and growing linearly with depth H, i.e. $p_r = H \cdot \gamma$, where γ is the specific weight of the liquid stored (Figure 12.10).

The bending moment M_e at the fixed boundary of the shell can be approximately assessed according to the previous section, as the fixed-end moment of a cantilever having a length c and subjected to the uniform load $(H - c) \cdot \gamma$, i.e.

$$M_e = (H - c) \cdot \gamma \cdot c^2/2$$

The maximum tensile force N_θ occurs at a distance $x = c \cdot \pi/2$ from the bottom, thus resulting in $N_\theta = \gamma \cdot (H - x) \cdot R$ (see Figure 12.10). ■

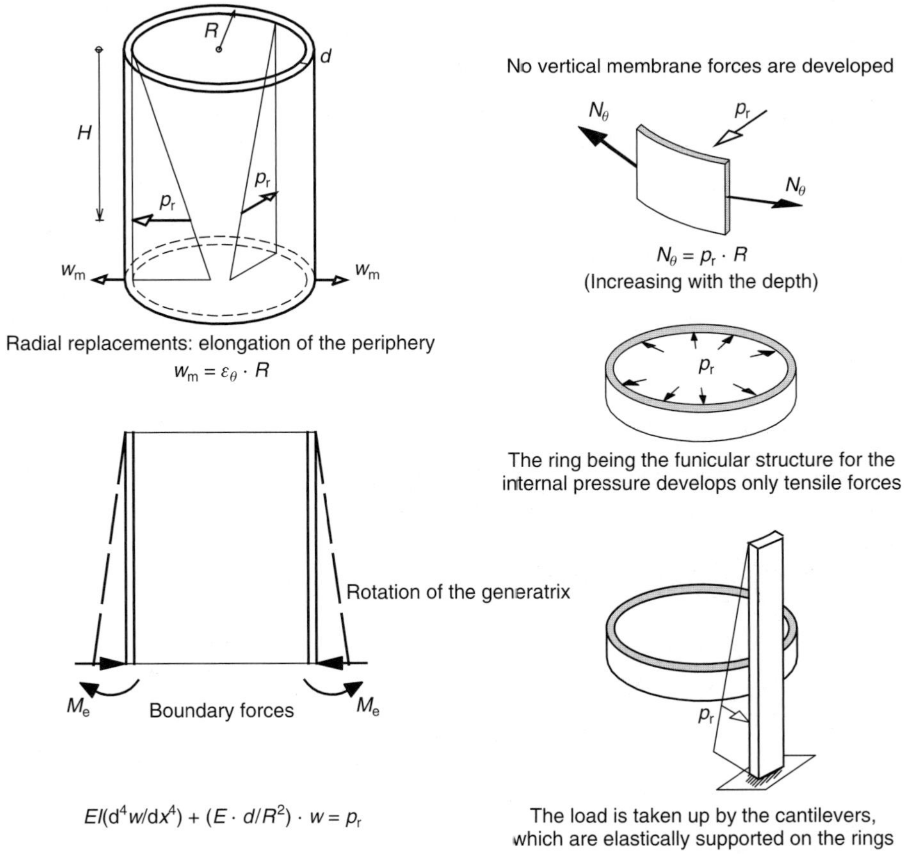

Figure 12.10 Loading-carrying behaviour of a cylindrical liquid tank

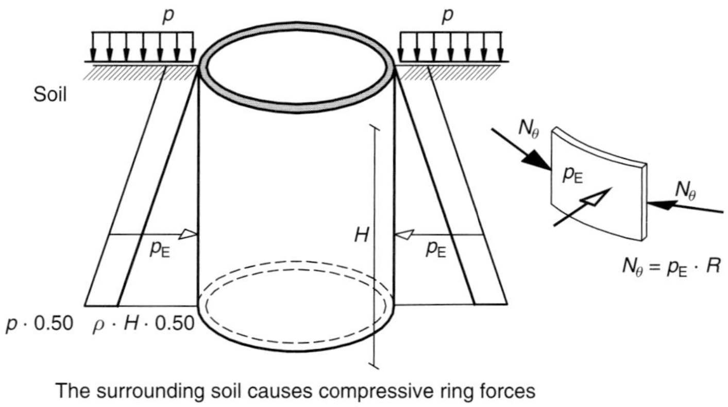

Figure 12.11 Cylindrical shell under lateral soil pressure

Such a shell may also be loaded by the inward pressure of the possibly surrounding soil, while being empty inside (Figure 12.11). This pressure p_E will correspond to the lateral earth pressure at rest, which also depends on the possibly acting live load p on the soil surface (see Section 17.2.1). Thus,

$$p_E = \rho \cdot H \cdot 0.50 + p \cdot 0.50$$

where ρ is the specific weight of the soil and H is the depth, measured from the soil surface.

It is obvious that the lateral pressure p_E will produce a state of stress corresponding to the previously considered hydrostatic pressure and constant internal pressure, acting in the opposite sense. It is clear that in the last expression, the value $0.50 \cdot \rho$ has been used instead of γ.

However, the fact that the shell will be subjected to a ring compression raises the problem of the risk of buckling. ■

The critical external pressure p_D of a closed cylindrical shell may be assessed from the relation (Pflüger, 1966)

$$p_D = 0.62 \cdot E \cdot \frac{R}{H} \cdot \left(\frac{d}{R}\right)^{2.50}$$

Any additional load acting along the longitudinal axis of the shell and producing a compressive stress σ_x must be taken separately into account.

The critical longitudinal compressive stress σ_{xD} may be assessed as (Pflüger, 1963)

$$\sigma_{xD} = \frac{0.48 \cdot E \cdot d/R}{\sqrt{1 + \dfrac{R}{100 \cdot d}}}$$

The coexistence of these two compressive states of stress can be regarded as secure, if the following condition is satisfied (Flügge, 1960):

$$\frac{\sigma_x}{\sigma_{xD}} + \frac{p_E}{p_D} \leq 1 \quad ■$$

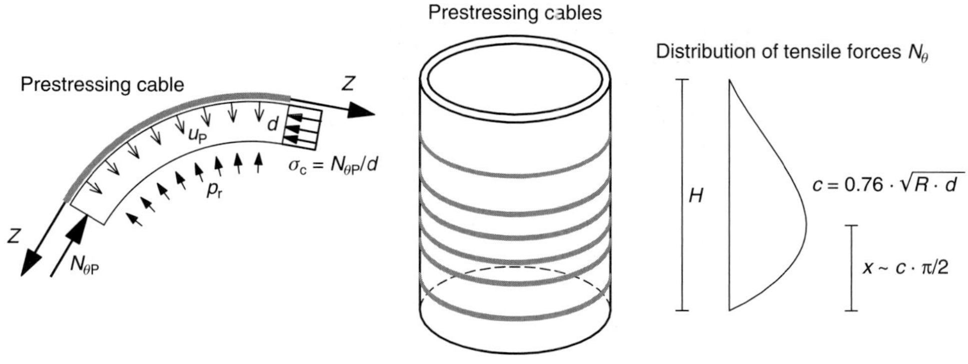

Figure 12.12 Peripheral prestressing of a cylindrical tank

The development of tensile ring stresses (N_θ/d) in a cylindrical tank, may possibly be problematic due to special requirements regarding leakage risk. Specifically, these stresses should clearly be lower than the concrete tensile strength σ_{cz}. Thus, a peripheral prestressing is often applied consisting either of circular tendons or of high-strength wires tightened along the perimeter over the whole height of the shell (Figure 12.12).

If Z is the prestressing force of a cable circling the shell, referring to the unit length of shell height and assuming its radius to be essentially equal to R, then the inward-acting deviation forces $u_P = Z/R$ cause the compressive ring force $N_{\theta P} = u_P \cdot R$, and, consequently, the developed compressive stress σ_c in the shell equals (see Figure 12.12)

$$\sigma_c = N_{\theta P}/d = Z/d$$

The prestressing force, which entirely takes over the internal pressure p_r (i.e. $u_P = p_r$), equals $Z_p = p_r \cdot R$. Prestressing with this force will lead to approximately zero stresses in the shell wall. However, in order to take additional tensile stresses into consideration, due to the influence of concrete shrinkage as well as of other secondary factors, a precompression σ_c of the order of 500–1000 kN/m^2 needs to be provided through an additional prestressing. The required additional prestressing force Z_d for this purpose is, according to the above relation, equal to $Z_d = \sigma_c \cdot d$. Thus, the applied prestressing force Z in the circular cables – after losses – should be

$$Z = Z_p + Z_d = p_r \cdot R + \sigma_c \cdot d$$

The circular cables should be placed in the external region of the wall thickness, which should in principle not be thinner than 25 cm. Moreover, the distribution of prestressing force Z over the height should in principle follow the corresponding distribution of the tensile forces N_θ (see Figure 12.12).

12.3.3 Silos

These also have a circular section, and are used for storing granular material. Their ratio of height to diameter is considerably greater than that for liquid tanks, because the internal pressure of the content in the case of tanks increases linearly with depth,

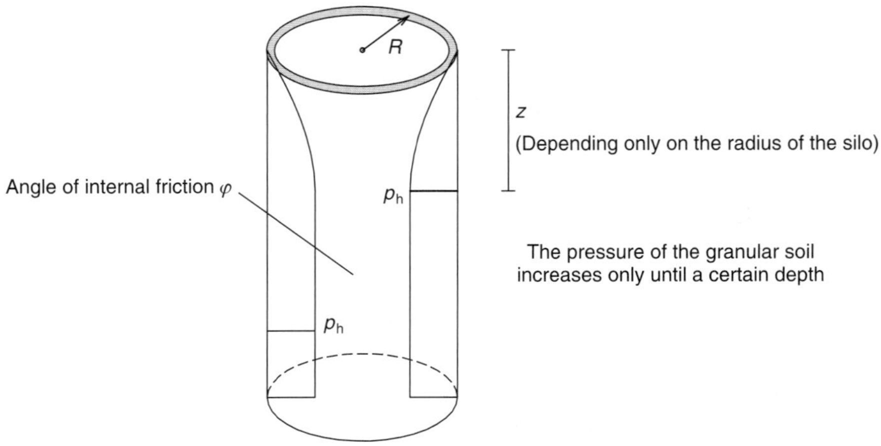

Figure 12.13 Distribution of internal pressure in a silo containing granular material

whereas in silos containing granular material with an angle of internal friction φ the internal pressure increases linearly only up to the depth (Ciesielski *et al.*, 1970)

$$z = \frac{2 \cdot R}{4 \cdot \tan(0.6 \cdot \varphi)}$$

and then remains constant to the bottom (Figure 12.13).

The maximum internal pressure which can be developed is

$$p_h = \gamma \cdot z = \frac{\gamma \cdot R}{2 \cdot \tan(0.6 \cdot \varphi)}$$

12.3.4 Barrel shells

A long, thin cylindrical shell with a closed circular cross-section placed horizontally and supported only at its two extreme fronts in such a way that its profile remains undeformed (e.g. through an appropriate stiffener or a suitable circular wall) presents notable stiffness towards transverse loads, which allows it to bridge spans multiple times longer than its diameter by developing exclusively membrane forces (Figure 12.14).

Assuming that the shell is acted on only by its self-weight g and considering the basic equation given in Section 12.2, which expresses how a load perpendicular to a shell surface can be taken over by its membrane forces, it can be deduced that, at the highest and lowest points of the cross-section, the developed ring forces N_θ – tensile or compressive – are $N_\theta = g \cdot R$. These ring forces vanish at the ends of the horizontal diameter, given that the perpendicular component of g to the surface is zero there. Furthermore, it can be seen that the bending moment M_0 over the span L, which is equal to

$$M_0 = (g \cdot 2R\pi) \cdot L^2/8$$

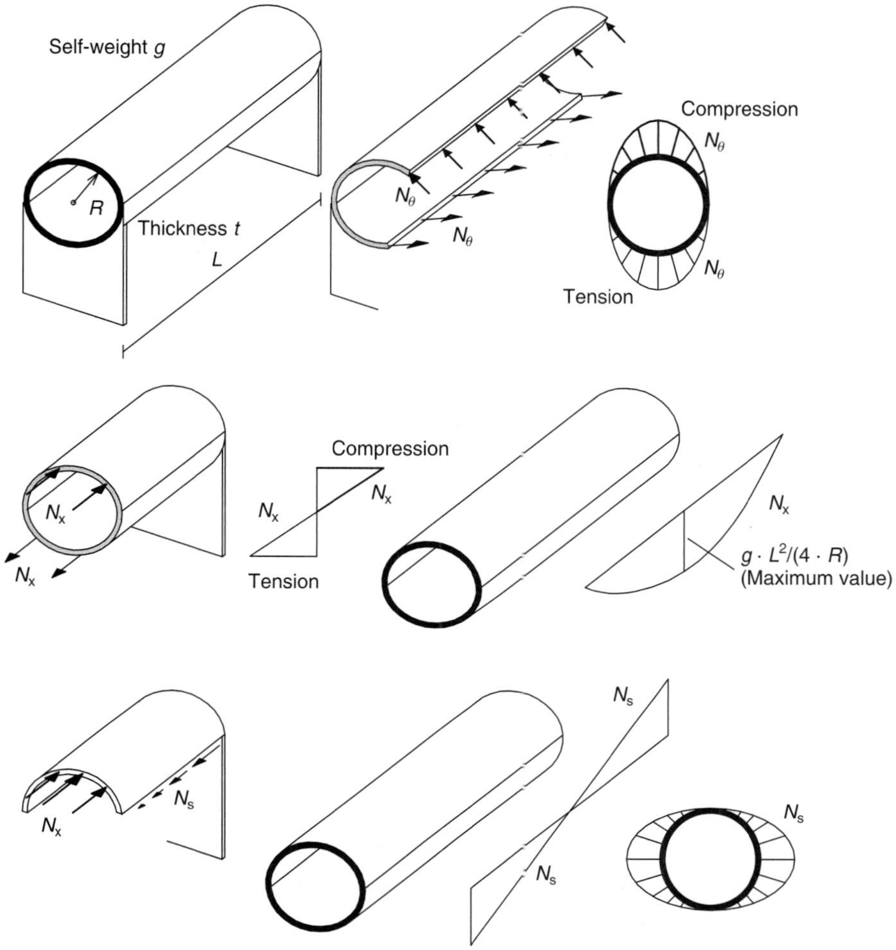

Figure 12.14 Membrane load-carrying action of a closed beam-like cylindrical shell

can be provided through longitudinal membrane forces N_x, whose maximum value may result on the basis of an approximate value of moment of inertia, namely

$$I = \pi \cdot d \cdot R^3$$

Thus,

$$N_x = (M_0 \cdot R/I) \cdot d = g \cdot L^2/(4 \cdot R)$$

This maximum value appears as compression at the upper edge of the profile and as tension at the lower edge of the profile, while at the ends of the horizontal diameter N_x becomes null (see Figure 12.14).

It is now clear that longitudinal membrane shear forces are also developed along the generatrices, arising from the longitudinal variation of N_x. Thus, given that N_x vanish at the horizontal edges, by considering the horizontal equilibrium of the upper section cut out from, say, the right half of the closed shell, membrane shear forces N_s must be

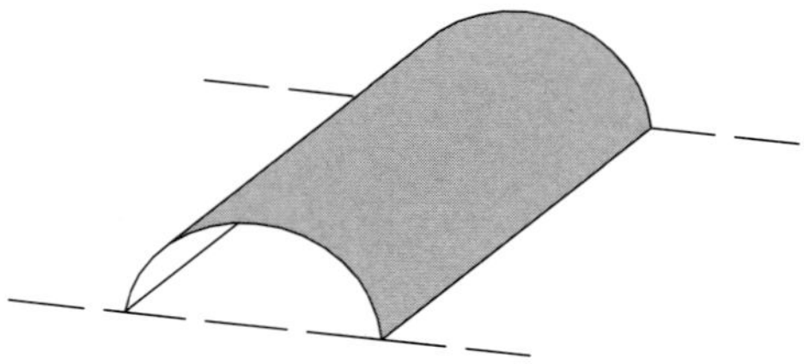

Figure 12.15 Appearance of stiffness with the formation of a cylindrical cross-section

developed in order to balance the compressive forces N_x acting over the cross-section at the middle (see Figure 12.14). These shear forces obtain their maximum value at the edges of the shell, given that, acting along the periphery of the circular edges, they represent the only possibility of equilibrating the total vertical load on the shell. According to Cauchy's theorem, the longitudinal shearing forces at the edges have the same value. This value is equal to $N_s = 2 \cdot g \cdot (L/2)$. The above state of stress of the cylindrical shell is obviously a pure membrane state (see Figure 12.14). ■

The form of a closed cylinder may not be suitable for covering an orthogonal area, unless cut by a horizontal plane, and its resulting upper part is taken, again under the condition that its edges retain their cross-section, to be undeformable. A sheet of paper fashioned appropriately illustrates this directly (Figure 12.15)

In order to now examine whether it is possible for a pure membrane state of stress to be developed in such a structure, under the action of gravity loads, those membrane forces should at first be applied along its free edges which would be correspondingly developed in a closed cylinder. If it is considered that the examined barrel constitutes a semicircle, given that the ring forces are at this location null, as previously stated, then the only forces which can be offered externally are the longitudinal shearing ones N_s, which may be offered by a horizontal beam (Figure 12.16). However, such a beam, acted on by the opposite longitudinal shearing forces N_s, is tensioned, and its elongation is not compatible with the absence of strain at the longitudinal fibres of the shell, given that $N_x = 0$, as previously explained. It can be seen that for the restoration of compatibility, additional longitudinal shearing forces should be introduced at the free boundary of the shell – together, of course, with the corresponding opposite forces on the beam – which will obviously cause tension in the lower regions of the shell and which, together with the unavoidable action of self-weight of the beam along the free edge of the shell, will lead to a deviation from the pure membrane state, i.e. they will cause (transverse) bending.

If the intersection of the full cylinder with the horizontal level is made at a higher level in order to maintain the membrane state, apart from the longitudinal shearing forces N_s of the free boundary, the ring forces N_θ should also be provided (see Figure 12.16). However, the gap in the longitudinal sense will now be greater, due to the presence of the compressive force N_x at the boundary of the shell, and therefore the beam should be able to provide, through its stiffness, the required inclined forces N_θ. Thus,

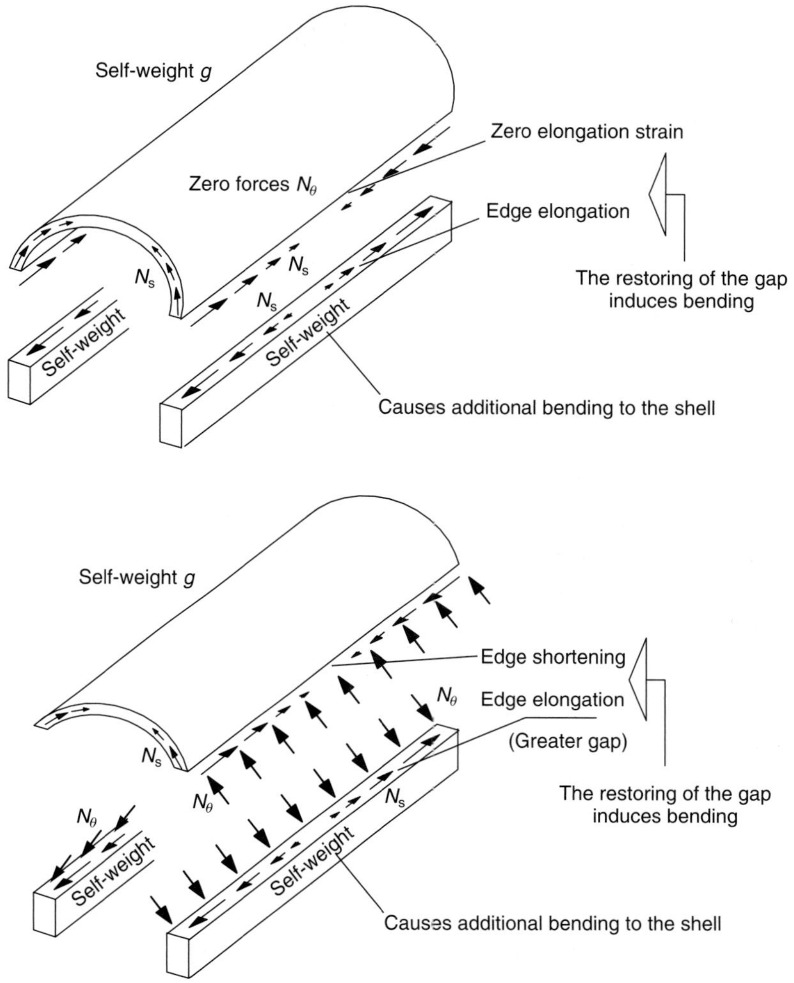

Figure 12.16 Consequences of membrane action in a barrel shell

with the additional exertion on the free edge of the shell of the self-weight of the beam, the deviation from the membrane condition will be even more intense.

Certainly, the shell and beam system may be designed by taking into account the introduced bending state of stress for the restoration of the gap, as may be found using appropriate computer software. However, it may often be desirable to maintain, as far as possible, the membrane state of stress, and this can be done by prestressing the edge beam.

As shown in Figure 12.17, the compressive force of prestressing is able to eliminate the gap between the two longitudinal edges (between the shell and the beam), while, on the other hand, both the ring forces N_θ and the self-weight of the beam may be taken over by the cable deviation forces. ■

The barrel shells can be divided into long and short ones, depending on the ratio of the length L to the width b of their ground plan.

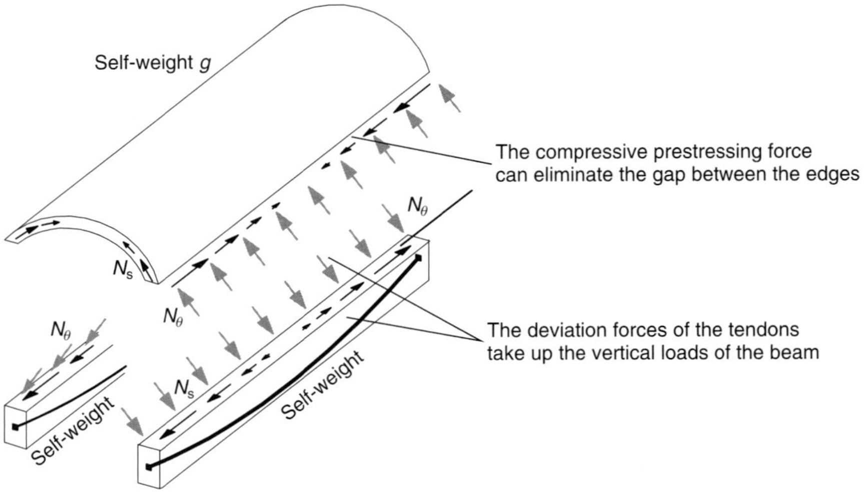

Figure 12.17 Prestressing the edge beam of a barrel shell

The long barrel shells have a ratio L/b greater than 2, and they may be designed at a preliminary stage as thin-walled beams, according to the 'classical' theory of beams. On the basis of the maximum bending moment M_0, the compressive force N_x, occurring at the top of the middle section, can be determined from the relation $N_x = M_0 \cdot y_o \cdot d/I$ (kN/m), whereas the tensile force Z, which will be taken over by a corresponding reinforcement, may be considered equal to $Z = M_0/z$, where z is the lever arm of the internal longitudinal forces, estimated approximately as $z = 0.90 \cdot h$ (Figure 12.18).

The shearing forces N_s, which are applied to the end sections of the span and ensure the equilibrium with the vertical loads, result from the vertical shear stresses (see Section 2.2.1) $f_s = (V \cdot S)/(I \cdot b)$ as $N_s = (f_s/\cos\alpha) \cdot d$. These forces show their maximum value at the centroidal axis of the section.

Of course, apart from these forces, ring forces N_θ are also developed, which have their maximum value at the top of the arch. This value, for the preliminary design needs, may be considered to be, according to the membrane state, $N_\theta = g \cdot R$ (see Figure 12.18).

A possible additional support of the shell at an intermediate position implies a similar treatment to the continuous beam.

At this point, regarding the longitudinal bending response of the long barrel shell, an alternative approach can be considered, based on the results of Section 12.3.1. Specifically, by considering the vertical load on the barrel shell as playing, approximately, the role of constant internal pressure, but in the opposite sense to a closed cylindrical shell, direct use may be made of the results.

By setting, approximately, $g = p_r$, the following expressions for the bending moments M_e at a clamped-end support and the 'span' bending moment M_{max} in the case of a simply supported shell (both per unit length) may be considered to be, respectively,

$$M_e = 0.29 \cdot g \cdot R \cdot d$$

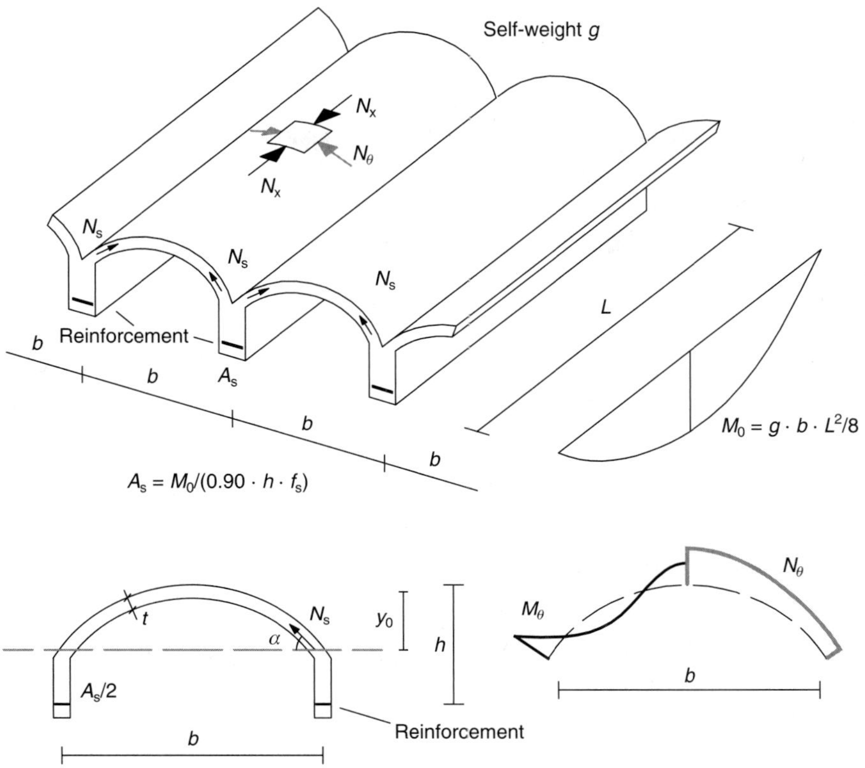

Figure 12.18 Long barrel shells as beams with an additional transversal response

and

$$M_{max} = 0.092 \cdot g \cdot R \cdot d$$

It is clear that these values may be regarded only as approximate (see Section 12.3.1).

As explained above, the deviation from the membrane state of stress also means, for the shell, the development of transversal bending moments M_θ causing tension eventually at both the top and bottom fibres (see Figure 12.18). An approximate assessment of their values is given by the expression $g \cdot (b/4)^2/8$.

However, these transversal moments may be very limited if care is taken to preserve the curved profile as undeformable, through the insertion of a number of transverse ribs, i.e. arches of significantly greater stiffness than that offered by the thickness of the cross-section of the shell only.

North light shells should also be included in the long cylindrical shell category, as shown in Figure 12.19, which, regarding lighting, show definite functional advantages compared with the full barrel shell, when they are used for covering industrial areas.

These shells may be designed at a preliminary stage in the same way, namely as thin-walled beams according to theory, taking into account the fact that their principal axes are skew. These beams develop a certain torsional response, as will be explained in Chapter 13, but this is not of significance in their preliminary design. However, for the validity of the above approach, it is important to ensure the undeformability of

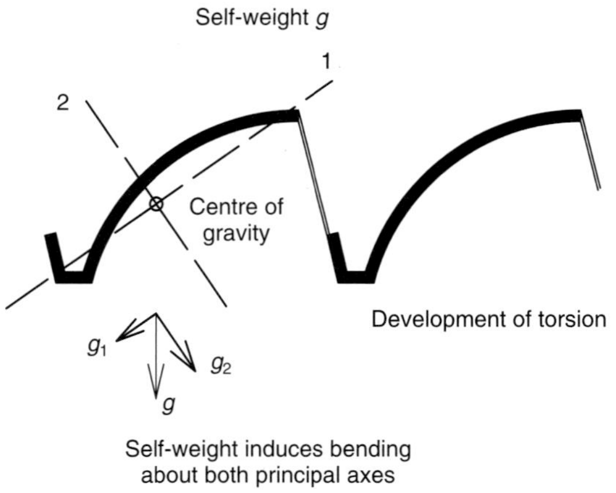

Figure 12.19 Formation of north light shell beams for covering an orthogonal layout

the cross-section, e.g. with transversal ribs. Moreover, the ring forces N_θ are necessary in order to maintain the equilibrium, and may be assessed as before.

In short barrel shells having a ratio L/b less than 2, the distribution of longitudinal membrane forces N_x may not be based on the theory of bending. Of course, the free edges will develop tension, but tension may also occur in the top region. The ring forces N_θ can be considered, as in the case of long shells, and the transversal bending moments may also be assessed here according to the last expression. However, for the final design of these shells, appropriate computer software should be used. ■

The risk against buckling should in no case be neglected, given that this is the factor which, in preliminary design, determines the thickness of the shell. Here, particular attention must be paid to both longitudinal and ring compressive stresses.

For the longitudinal (bending) compressive stresses, the following stress limit may be considered as critical (Seide, 1981):

$$\sigma_{xD} = \frac{0.58 \cdot E \cdot d/R}{\sqrt{1 + \dfrac{R}{100 \cdot d}}}$$

while for the transverse compression, the pressure p_D can be considered to be the critical loading, as established for the closed cylindrical shell in Section 12.3.2. Both criteria should be examined. ■

Cylindrical shells are particularly suitable for the aesthetically satisfactory covering of square as well as triangular ground plans, by using a layout of intersected vaults.

Figure 12.20 shows the covering of a square ground plan with four intersecting barrel vaults, one based on each side. The arches created along the two diagonals receive the practically constant ring forces $N_\theta = g \cdot R$ from both their sides. The resulting 'V' cross-section of arches gives them a clearly greater bending stiffness than that which the shell thickness alone can afford.

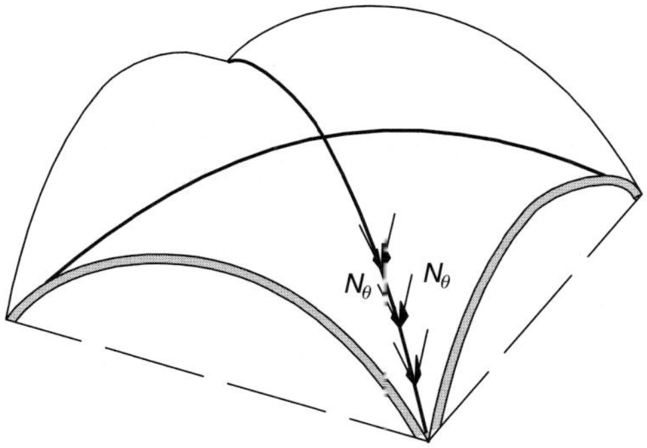

The arches take up the vertical components of N_θ

Figure 12.20 Intersecting barrel vaults for covering a square area (cross-vault)

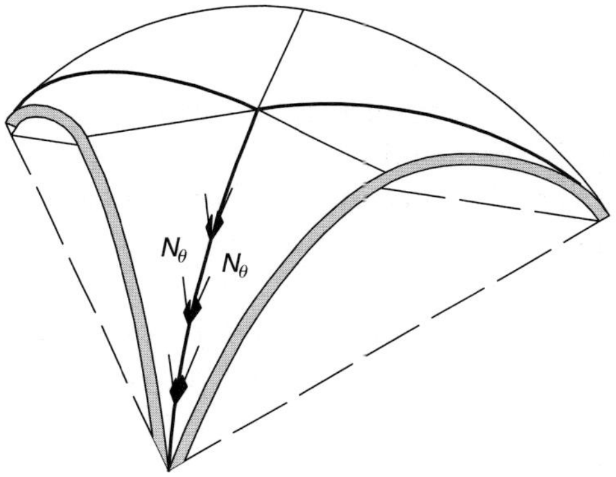

Figure 12.21 Intersecting barrel vaults for covering a triangular area

The same logic is followed for covering a triangular ground plan, as depicted in Figure 12.21. The previous remarks are again valid.

The preliminary design of these shells can be made on the basis of the above-mentioned buckling criteria, as well as of the ring forces N_θ according to the previously exposed membrane state. However, it bears repeating that, for ground plans of large dimensions, the buckling criterion is predominant.

12.4 Dome shells

12.4.1 Non-shallow shells

With their middle surface being part of a spherical surface, dome shells are suitable for covering not only circular areas but also various polygonal plans. The corresponding form is derived by cutting off the spherical surface with planes passing through every linear segment of the periphery of the base (Figure 12.22). Of course, in the case of a circular base, the shell boundary will be a circle, but in all other cases it will consist of consecutive arch-like segments, representing the trace of the sphere on the previously

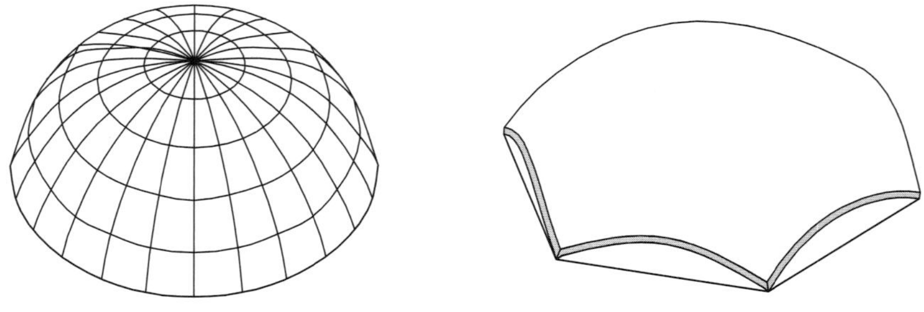

Figure 12.22 Design alternatives for covering areas with spherical shells

mentioned planes. It is clear that these planes are not necessarily vertical, and a small outward slope leads to a more aesthetically appealing result.

Spherical shells are particularly appropriate for carrying vertical distributed loads by membrane action, something which is always aimed for via their appropriate design. However, the degree to which this may be accomplished depends on the support conditions at the boundary of the shell, as will be examined later.

The rotational symmetry of a shell suggests the consideration of, for equilibrium purposes, a surface element formed by two adjacent meridian and parallel curves (Figure 12.23). The membrane forces N_φ and the membrane ring forces N_θ act along the parallel and meridian curves and in a direction perpendicular to them. Moreover, if the vertical loading is also axisymmetric about the same axis as the shell, then it is easily concluded that the membrane shearing forces along the edges of the element, either meridian or parallel, vanish.

The membrane forces N_φ acting along a parallel curve can always be determined from the equilibrium condition in the vertical direction of the typical cut-off element (Figure 12.24). If a represents the radius of the parallel circle, φ is the angle between a shell radius to the periphery and the vertical axis of symmetry, and V is equal to

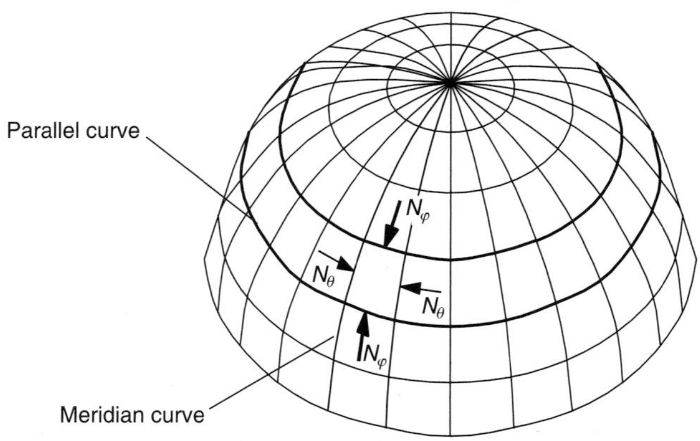

Figure 12.23 Membrane forces in a spherical shell

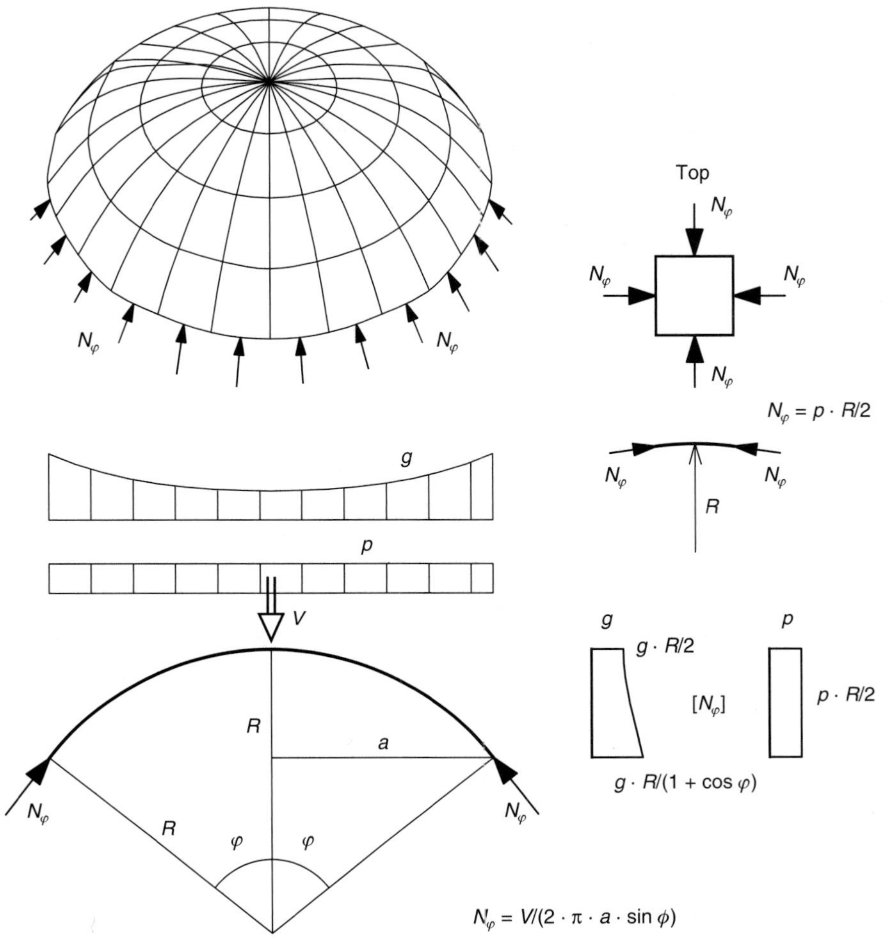

Figure 12.24 Determination of the meridional membrane forces

the total load applied to the upper shell portion being examined, then

$$N_\varphi = V/(2\pi \cdot a \cdot \sin\varphi)$$

It is now assumed that the shell is subjected to a uniform load which is axisymmetric about the vertical axis. Clearly, the value of N_φ for an element at the top of the shell which is under uniform loading p – which may also represent the weight of the shell element – cannot be determined from the last equation, but instead from the membrane equation of equilibrium in Section 12.2. So, application of this equation to the four-sided shell element with its centre being located at the top of the shell yields the relation $2 \cdot (N_\varphi/R) = p$, since in the absence of twist it is $S = 0$. Equivalently,

$$N_\varphi = p \cdot (R/2)$$

From the above two equations, it can easily be concluded that the last value of N_φ is constant over the entire shell, if a uniform constant load p is applied over the horizontal projection of the shell (see Figure 12.24).

It should be noted that, although the self-weight g of the shell is not uniformly distributed over the horizontal projection of the shell, its value increasing towards the edges, it can nonetheless be considered uniform in the neighbourhood of the top of the shell, exactly as the load p. Therefore, the meridional forces N_φ due to the self-weight g of the shell are expressed, according to the previous relation, as

$$N_\varphi = R \cdot g \cdot \frac{1}{1 + \cos\varphi}$$

The ring force N_θ forming a right angle with N_φ is derived from the general membrane relation given in Section 12.2 by taking into account the corresponding principal curvatures at the point under consideration. It should be noted that, whereas the meridional curve of radius $R_\varphi = R$ refers to a principal curvature at that point, the corresponding parallel circle of radius a, whose tangent at each point provides the line of action of the ring force N_θ, does not represent the other principal curvature at the point under consideration (see Figure 12.24). Instead, it corresponds to a circle of radius $R_\theta = a/\sin\varphi = R$, which is a tangent to the parallel circle at that point, and N_θ belongs to both these planes as a common tangent to both circles (see Figure 12.23). Given that $S = 0$, the following relation can be written:

$$\frac{N_\varphi}{R_\varphi} + \frac{N_\theta}{R_\theta} = p_{\mathrm{r}}$$

where p_{r} is the component of the distributed load normal to the surface of the shell. So, the ring force N_θ may be expressed as

$$N_\theta = R \cdot \left(p_{\mathrm{r}} - \frac{N_\varphi}{R}\right)$$

The developed ring force N_θ, due only to the self-weight g of the shell, is evaluated from the expression (Billington, 1965)

$$N_\theta = R \cdot g \cdot \left(\frac{1}{1 + \cos\varphi} - \cos\varphi\right) \quad \text{(positive values refer to tension)}$$

In addition, for a uniformly distributed vertical load p,

$$N_\theta = p \cdot R \cdot \cos(2 \cdot \varphi)/2$$

The development of ring forces N_θ can easily be understood by considering the equilibrium of a narrow ring cut off from the shell and subjected to the applied meridional forces N_φ along its upper and lower edges (Figure 12.25).

It is well known that as long as the ring belongs to the upper shell region, the inward horizontal component of N_φ at the lower edge prevails over the outward one of the upper edge, resulting in a compression of the ring (see Section 2.2.7 and Figure 2.44). However, in the lower shell region, the outward component of N_φ at the upper edge dominates, therefore leading to tensile ring forces N_θ. The exact value of the angle φ for which the transition from positive to negative values of N_θ takes place is derived

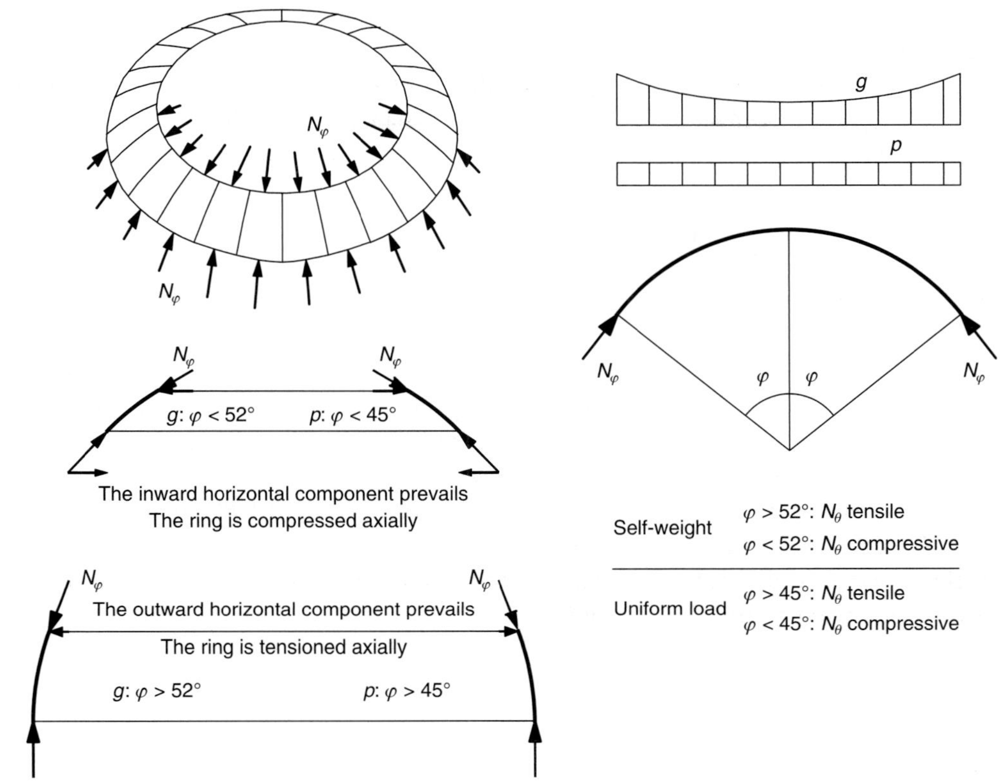

Figure 12.25 Behaviour of the ring membrane actions

from the above expressions for N_θ, as $\varphi = 51.50°$ for the self-weight and $\varphi = 45°$ for the uniform load p. ■

In the case where the shell base is a horizontal circular plan, the meridional membrane forces N_φ at its boundary also have to act on the element that supports the shell, but in the opposite direction, i.e. outwards (Figure 12.26). This element may normally be a circular beam (ring) which, under the action of the horizontal components of N_φ, will

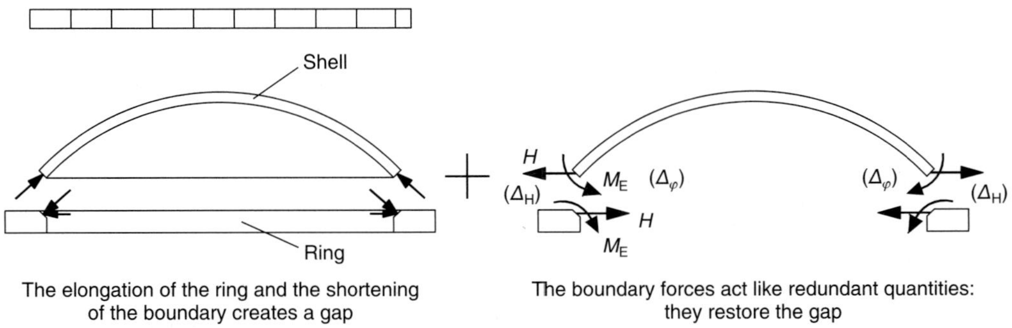

Figure 12.26 Interaction between the shell and the ring beam

develop only tension without bending, since it represents the funicular form for such a radial uniform loading (see Section 2.2.7 and Figure 2.42). Consequently, its fibres tend to elongate, and do not conform to the deformation of the shell boundary as is implied by its ring forces N_θ. If the ring beam lies over the limit of the approximately 52° angle, the shell boundary will shorten ($N_\theta < 0$), whereas under this limit it will elongate ($N_\theta > 0$). Hence, in both these cases there will be a gap in the deformation of the two systems, although in the latter the gap will be less. Even in the case of a hemispherical shell, where the horizontal component of N_φ vanishes and the ring beam is not loaded within its plane, the existence of the tensile ring shell forces N_θ leads again to a gap between the ring beam and the shell.

Re-establishment of the deformation compatibility requires application of appropriate additional forces on the shell boundary, which also act on the ring beam but in the opposite sense. These forces are suggested by those which the ring beam can sustain according to its stiffness, and these, obviously, are horizontally distributed radial forces H as well as radially distributed torsional loads M_E (see Figure 12.26). This means that the shell will be subjected to equal and opposite boundary loads, with the torsional loads M_E of the beam clearly representing bending moments for the shell. An analytical examination reveals that, due to the double curvature of the shell, this disturbance of the membrane state does not extend significantly into the body of the shell. Rather, it is restricted to its boundary region, and thus does not have an essential influence on the preliminary design of the shell. Despite this, a thorough examination of the stress state and deformation of the boundary region is necessary in order to have a better understanding of its response (Billington, 1965).

12.4.1.1 Shell boundary

Generally, the shell boundary exhibits, under its loads, a radial horizontal displacement Δ_H and a rotation of angle Δ_φ (see Figure 12.26).

Due to its own weight g and under a membrane state of stress it is

$$\Delta_H = \frac{R^2 \cdot g}{E \cdot d} \cdot \left(\frac{1}{1 + \cos\varphi} - \cos\varphi \right) \cdot \sin\varphi$$

$$\Delta_\varphi = \frac{2 \cdot R \cdot g}{E \cdot d} \sin\varphi$$

It is well known that Δ_H is produced by the strain $\varepsilon_\theta = N_\theta / E \cdot d$ of the horizontal boundary ring, which has the radius $R \cdot \sin\varphi$ and is subjected to the action of N_θ. Consequently,

$$\Delta_H = \varepsilon_\theta \cdot (R \cdot \sin\varphi)$$

It should be noted that Δ_H is inwards when $\varphi < 51.50°$ and outwards when $\varphi > 51.50°$. The development of Δ_φ results in an increase of the end slope at the boundary of the shell.

According to the previous analysis, for a constant vertical load p uniformly distributed all over the shell,

$$\Delta_H = \varepsilon_\theta \cdot R \cdot \sin\varphi = \frac{N_\theta}{E \cdot d} \cdot R \cdot \sin\gamma = \frac{R^2 \cdot p}{2 \cdot E \cdot d} \cdot \cos(2 \cdot \varphi) \cdot \sin\varphi$$

Due to a uniform horizontal force H (kN/m) in the outward radial direction,

$$\Delta_H = \frac{2 \cdot R^2 \cdot \sin^2\varphi}{E \cdot d \cdot c} \cdot H$$

$$\Delta_\varphi = \frac{2 \cdot R^2 \cdot \sin\varphi}{E \cdot d \cdot c^2} \cdot H$$

where $c = 0.76 \cdot \sqrt{R \cdot d}$, according to Section 12.3.1. Here, Δ_H is directed outwards, whereas the tangent slope at the edge is decreased due to Δ_φ.

Due to a uniformly distributed bending moment M_E applying tension to the inner boundary fibres of the shell,

$$\Delta_H = \frac{2 \cdot R^2 \cdot \sin\varphi}{E \cdot d \cdot c^2} \cdot M_E$$

$$\Delta_\varphi = \frac{4 \cdot R^2}{E \cdot d \cdot c^3} \cdot M_E$$

where Δ_H is directed outwards, whereas the tangent slope at the edge is decreased due to Δ_φ.

In the above expressions, it should be noted that the values of Δ_H and Δ_φ are reciprocally equal, according to the Betti–Maxwell theorem (see Section 2.3.5).

12.4.1.2 Ring beam

It is already known that a ring beam (section data: A_R, I_R) subjected to a distributed radial force H develops only an axial force $N_R = H \cdot R$ and no bending at all, due to its funicular behaviour under that loading (see Section 2.2.7 and Figure 12.27).

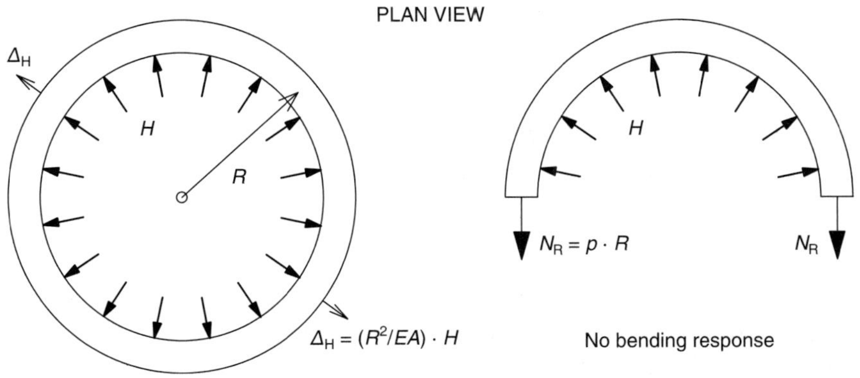

Figure 12.27 Behaviour of a ring beam under internal pressure

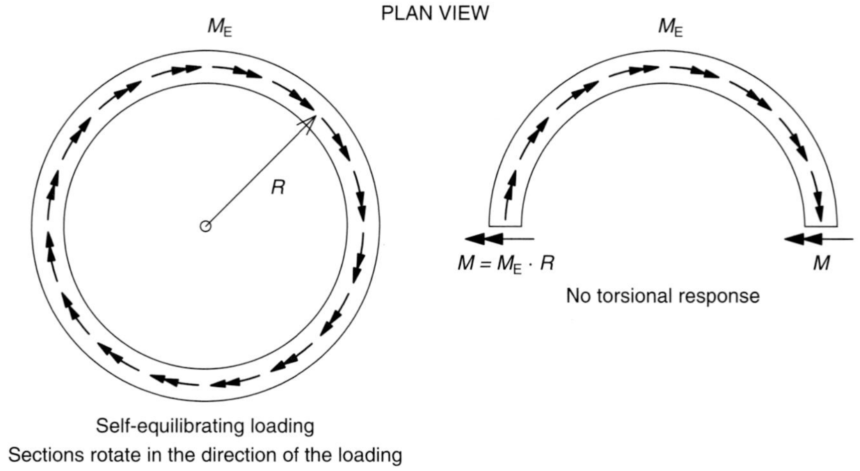

Figure 12.28 Behaviour of a ring beam under a distributed torsional load

The strain in the ring beam is $\varepsilon_R = N_R/(E \cdot A_R)$, the elongation of its circumference is $\varepsilon_R \cdot 2\pi R$, and its radial displacement is

$$\Delta_H = \varepsilon_R \cdot R = \frac{R^2}{E \cdot A_R} \cdot H$$

On the other hand, the application of a torsional load M_E uniformly distributed along the entire length of the periphery simply causes an equally directed rotation Δ_φ of its sections,

$$\Delta_\varphi = \frac{R^2}{E \cdot I_R} \cdot M_E$$

and a constant bending moment $M = M_E \cdot R$, without creating any torsional moments, as shown in Figure 12.28. Under these conditions, the centroid of the beam section does not undergo a radial displacement Δ_H.

In view of the above results, the evaluation of the magnitudes H and M_E becomes possible by imposing compatibility of the displacements at the shell boundary and the ring beam, following a procedure identical to the force method, which was presented in Chapter 3.

However, it should be noted here that the displacements Δ_H have a much stronger influence on the magnitudes H and M_E than do the rotations Δ_φ. ■

At this point it should be emphasised that the bending response of the shell edge is essentially identical to that of a cylindrical shell of the same thickness, which circumscribes the spherical shell and is tangential to it at their boundaries (Figure 12.29). The cylindrical shell in question obviously has the same radius as the spherical one, i.e. equal to R. This result, which can be confirmed through a more detailed analysis, has clearly to do with the fact that the characteristic length c appears in the above expressions for the deformations of the spherical shell boundary.

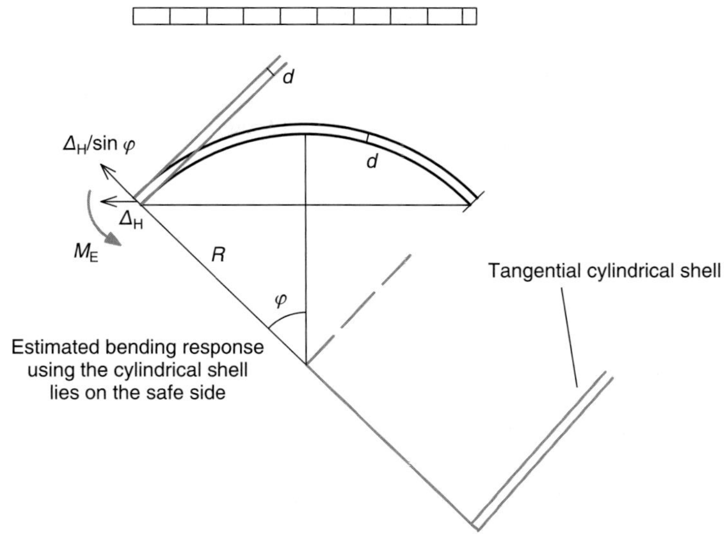

Figure 12.29 Identical bending behaviour at the boundary of a spherical and a cylindrical shell

This length was first introduced in the examination of the behaviour of cylindrical shells (see Section 12.3.1).

In this way, and for $\varphi > 20°$, the bending moment at the spherical shell boundary can be approximately determined according to the results of Section 12.3.1, through the projection ($\Delta_H/\sin\varphi$) of the horizontal displacement Δ_H (due to its membrane state) on its radial direction. The value of the projection coincides with the radial displacement of the cylindrical shell that is tangentially equivalent.

For a spherical shell with a fixed boundary, the moment is expressed as

$$M_E = 0.29 \cdot E \cdot d^2 \cdot \frac{\Delta_H}{R \cdot \sin\varphi}$$

while for a hinged boundary it is

$$M_{max} = 0.092 \cdot E \cdot d^2 \cdot \frac{\Delta_H}{R \cdot \sin\varphi}$$

Δ_H in the above equations takes the value which corresponds to the prevailing loading, which is either the self-weight or a uniform load.

The resulting distributed thrust H (kN/m) acting on the shell boundary and directed outwards can be determined directly from the corresponding equations of the previous discussion. ■

As has already been pointed out, the resulting bending disturbance of the pure membrane state in a spherical shell is, in any case, restricted to a relatively narrow region by its boundary, having a width of the order of magnitude of the characteristic length c.

Prestressing of the ring beam can restore the incompatibility of the deformations. This, additionally, solves the problem of taking on the tensile stresses in the case of a concrete beam, and also restricts the dimensions of its section (Figure 12.30).

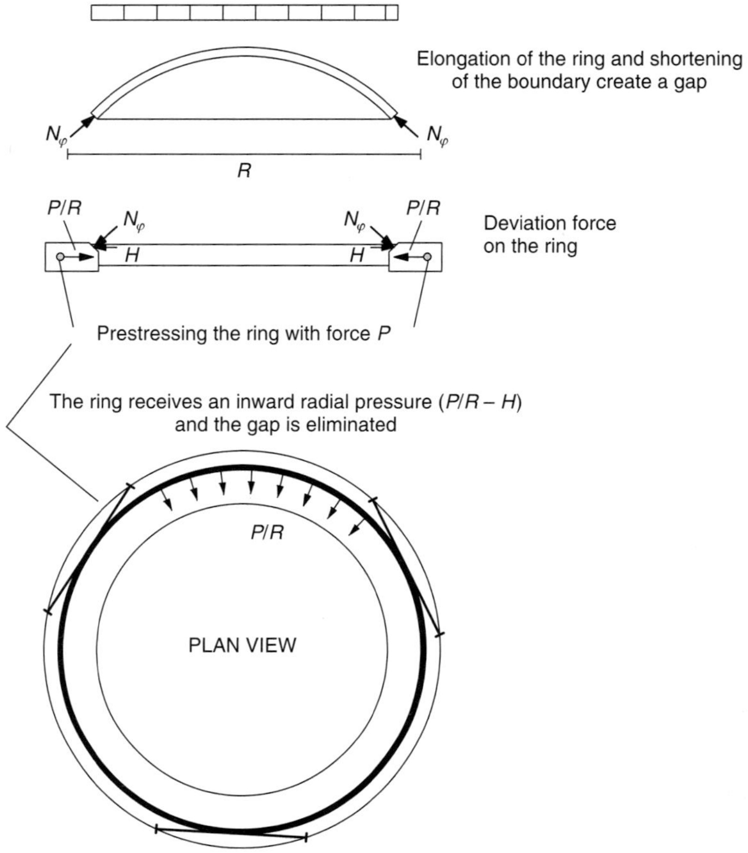

Figure 12.30 Prestressing of the ring beam

By placing a circular tendon prestressed with a force P in a concrete ring beam, the ring will be automatically subjected to an inward radial uniform pressure P/R. If this pressure is greater than the already existing pressure H, the ring will develop an axial compressive force $N_R = (P/R) \cdot R - H \cdot R = P - H \cdot R$, and consequently it will exhibit an inward radial displacement (see above):

$$\Delta_H = \frac{N_R}{E \cdot A_R} \cdot R = \frac{R^2}{E \cdot A_R} \cdot \left(\frac{P}{R} - H\right)$$

The prestressing force P can be selected in conjunction with the other structural parameters, so that, according to the previous analysis, the resulting gap between the ring beam and the shell boundary vanishes (see Figure 12.30).

12.4.2 Shallow shells

In order to confine the covered volume, as well as for functional or aesthetic reasons, a restriction of the dome height is often necessary, and thus the shell is characterised as

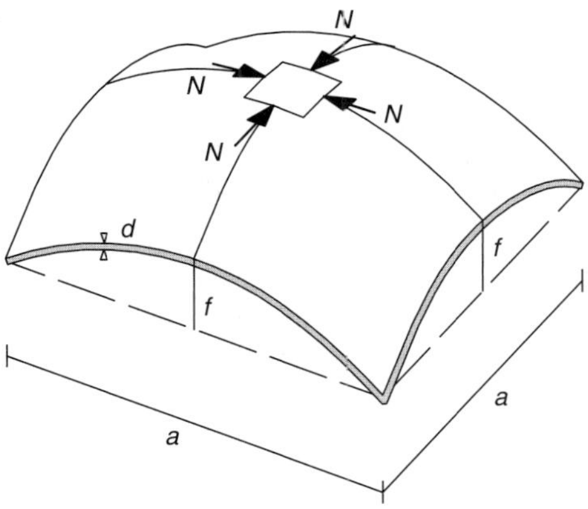

Figure 12.31 Structural layout of a shallow shell

shallow. This implies a decrease in curvatures and, consequently, an increase of the membrane compression forces. In such a shell, where the height f is less than about one-fifth of the covered span a, the radius of curvature can be approximately expressed as

$$R = \frac{f^2 + (a/2)^2}{2f}$$

or simply $R = (a/2)^2/(2 \cdot f)$, if $f < a/10$ (Figure 12.31).

Such a shell can cover an orthogonal, usually square, area, where the resulting boundary curves are sections of the spherical surface cut by vertical planes corresponding to the four sides of the shell base.

Due to the self-weight g, the resulting meridional membrane forces N_φ along any 'parallel' are, according to the relations given in Section 12.4.1,

$$N_\varphi = \frac{\pi \cdot a^2 \cdot g}{2 \cdot \pi \cdot a \cdot \sin\varphi} = g \cdot \frac{R}{2}$$

So, it can be concluded that this force remains constant essentially over the entire shell surface. On the basis of this result and according to the analysis in Section 12.4.1, the ring force N_θ, which is obviously compressive, is equal to the meridional one: $N_\theta = g \cdot (R/2)$.

In the case of an orthogonal base, the above result may also be confirmed by considering that the shallow shell consists of shallow arches which transfer their load in both directions through compressive forces. These compressive forces N (per unit of width) are equal to

$$N = \frac{g}{2} \cdot \frac{a^2}{8 \cdot f} = \frac{g \cdot R}{2}$$

according to the previous approximate expression for the radius of curvature. ■

Table 12.1 Estimation of the bending moment ($M_e/p \cdot d^2$) *at the fixed boundary of a shallow shell*

	a/d			
f/d	100	150	200	250
0.2	4.20	5.20	5.90	6.50
0.1	12.80	16.70	19.90	22.80
0.05	33.50	43.80	53.20	62.00

In such a shell, having essentially the form of a curved orthogonal plate, the bending response should be taken into consideration even during the preliminary design phase, especially when its boundary is restrained from rotation. An estimation of the bending moments developed at the boundary can be made on the basis of an analogy to the 'tangential' cylindrical shell, as previously explained (see Section 12.2.1 and Figure 12.29). This bending moment is found to be

$$M_e = 0.29 \cdot p \cdot R \cdot d$$

and causes tension of the upper fibres. Note that, because of the shell shallowness, the self-weight can be considered essentially as a uniformly distributed load.

This result lies, generally, on the safe side, i.e. the bending moment is overestimated. However, in order to estimate directly the influence on the bending response of the ratios *f/a* and *a/d* using more accurate analytical methods (Stavridis and Armenakas, 1988), it is useful to provide Table 12.1, which gives shallow shells with square bases and fixed boundaries. The table values refer to the dimensionless magnitude $M_e/(p \cdot d^2)$. In contrast to plates, the dependence of the bending response on the shell thickness is clear here. Poisson's ratio has been assumed to be null.

It should be noted that, in the case where the lateral transverse displacement of the boundaries is not constrained – which means that no forces N_φ can develop at the boundary – the values in Table 12.1 must be doubled. ■

In the compressed dome, and particularly in the shallow dome, the danger of buckling deserves special attention. Of course, the existence of double curvature is a relieving factor which increases the critical load. This can be estimated using the relation (Schmidt, 1961)

$$p_D = 0.15 \cdot E \cdot (d/R)^2$$

From this relation it is seen that an increase in the radius of curvature decreases the load-carrying capacity of the shell, whereas an increase in the shell thickness increases the critical load. One way to increase the critical load is to place transverse ribs along the mutually orthogonal directions, which increases the average shell thickness, as is often seen in older structures. ■

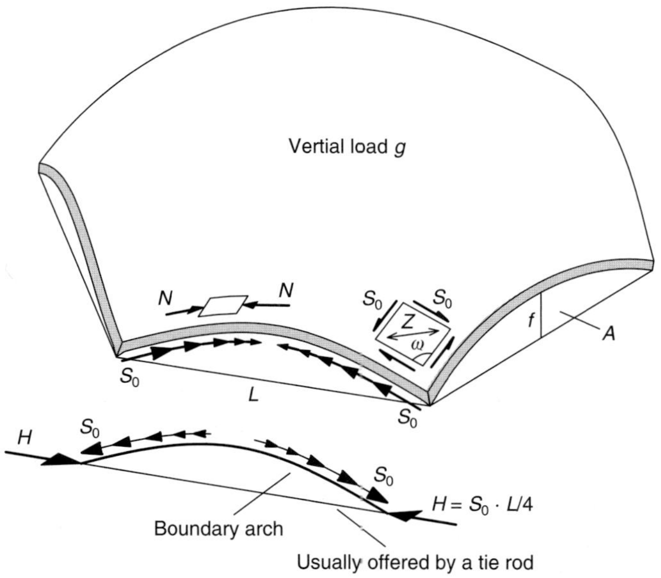

Figure 12.32 Load-carrying behaviour at the boundary of a shell with a polygonal base

As previously mentioned, a spherical dome can cover a triangular, square or polygonal area. The sections of the shell surface on the corresponding planes passing through the sides of the polygonal base create arch-like openings with favourable lighting conditions (Figure 12.32).

For the proper structural design of the shell boundary, it should be taken into account that forces cannot be applied in a direction which is normal to the boundary and tangential to the shell, while shear forces should be applied along the boundary, taking up the whole vertical load on the shell, if equilibrium has to be maintained through only a membrane state of stress.

The fact that membrane forces N_φ acting transversely to the arch-like boundary cannot exist means that the vertical load in the boundary regions will be taken only by the membrane forces N, which is parallel to the arch-like edges (see Figure 12.32). If R_b represents the radius of curvature of the corresponding arch, then according to the basic membrane relation given in Section 12.2

$$N = g \cdot R_b$$

Shear forces have to vanish along each axis of symmetry of the shell. So, in the arch-like boundary segment with span length L and height f, they vary from zero at the crown to a maximum of S_0 at its ends. Clearly, the integral sum of their vertical components over all the boundary arches must equilibrate the total load $g \cdot A$ of the shell, where A denotes the area of the polygonal base. Assuming a linear distribution of the shear forces along each boundary arch, the approximate maximum value is directly obtained as

$$S_0 = \frac{1}{k} \cdot \frac{3 \cdot g \cdot A}{4 \cdot f}$$

where k is the total number of boundary arches (see Figure 12.32) (Salvadori and Levy, 1967).

These shearing forces can be directly applied to the shell by an arch of suitable section, which may be adjusted to the shell boundary. Such an arch cannot transmit forces other than those lying in its own plane; therefore, it has to take on the shearing forces of the shell boundary along its axis and transfer them to the ends of the arch, which are also the support points of the shell. These forces are obviously directed towards the ends, and consequently produce an outward horizontal thrust H. This thrust can be estimated on the basis of the adopted linear distribution of shearing forces as $H = S_0 L/4$, and has to be borne either by a fixed support or through a tie connecting the ends of each boundary arch (see Figure 12.32).

At each shell corner, the developed membrane shear forces S_0 produce a tensile force Z acting in a direction perpendicular to the angle and equal to

$$Z = S_0/\tan(\omega/2)$$

This tensile force has to be supported by appropriate reinforcement.

12.5 Hyperbolic paraboloid shells

12.5.1 Overview

These shells belong to the translational shells category. Their midsurface is generated by a downwards curved parabola (generatrix), which 'slides' on another fixed parabola curved upwards (guide), keeping its plane always parallel to itself. The created surface has the form of a 'saddle' and, obviously, has a negative curvature, in contrast to the shells examined previously (Figure 12.33).

Such shells, apart from their impressive aesthetics resulting from the alternating curvature, show a characteristic of particular constructional importance because of their negative Gaussian curvature, i.e. they may also be generated using straight lines.

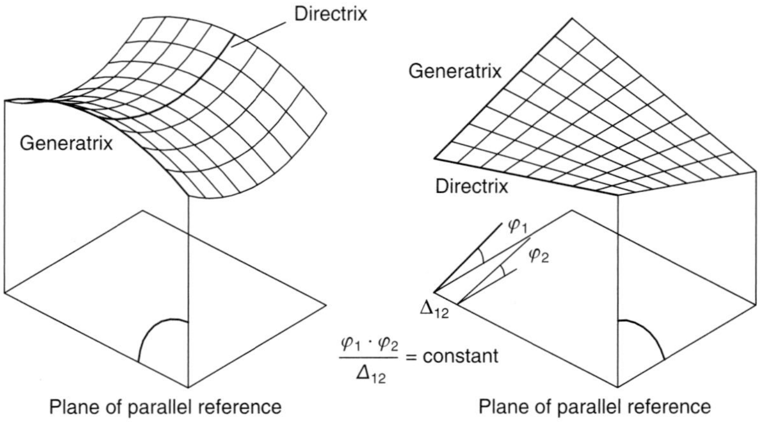

Figure 12.33 Creation of hyperbolic paraboloid shells

Thus, if a straight line, regarded as a generatrix, moves in space with one of its edges, sliding over another fixed straight line, keeping itself parallel to a fixed vertical plane on the one hand and with a constant rate of variation of its slope over the travelled distance by its sliding end on the other, then this line will produce the same surface as that previously described (see Figure 12.33). This will be explained in detail immediately below. What will, however, be pointed out first is that an understanding of this 'dual' nature of the hyperbolic paraboloid shell is of paramount importance for the comprehension of its load-carrying potential, as well as for the design of these shells in general, as they are able to cover orthogonal (and other) layouts with an unlimited variety of morphologies. In all cases, the possibility of creating the surface by using exclusively straight-line segments is a crucial constructional advantage which gives these shells a clear economic precedence.

As an understanding of the geometry of this shell is necessary for understanding its structural behaviour, it is appropriate first to examine it through the following analysis.

12.5.2 *Perception of the geometry*

The analytic description of the creation of the surface by the two parabolas is based on Figure 12.34.

In the considered orthogonal system *Oxyz* the 'generatrix' parabola with a span and rise equal to $2a$ and f_1, respectively, has its plane parallel to the plane *Oxz*, while the 'guide' parabola curved in the opposite direction lies in the plane *Oyz*, having a span and rise (sag) equal to $2b$ and f_2, respectively. The coordinate z of the created surface is determined from the relation

$$z = \frac{x^2}{2R_1} - \frac{y^2}{2R_2}$$

where R_1 and R_2 are the radii of curvature of the two parabolas, assumed to be constants, and equal to $R_1 = a^2/2 \cdot f_1$ and $R_2 = b^2/2 \cdot f_2$, respectively. It is assumed that both parabolas are 'shallow' in the sense described in Section 12.4.2.

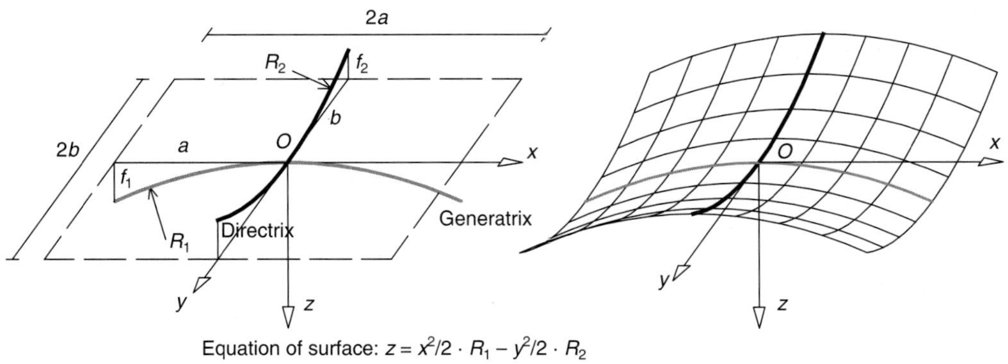

Figure 12.34 Creation of a shell through a parabolic generatrix

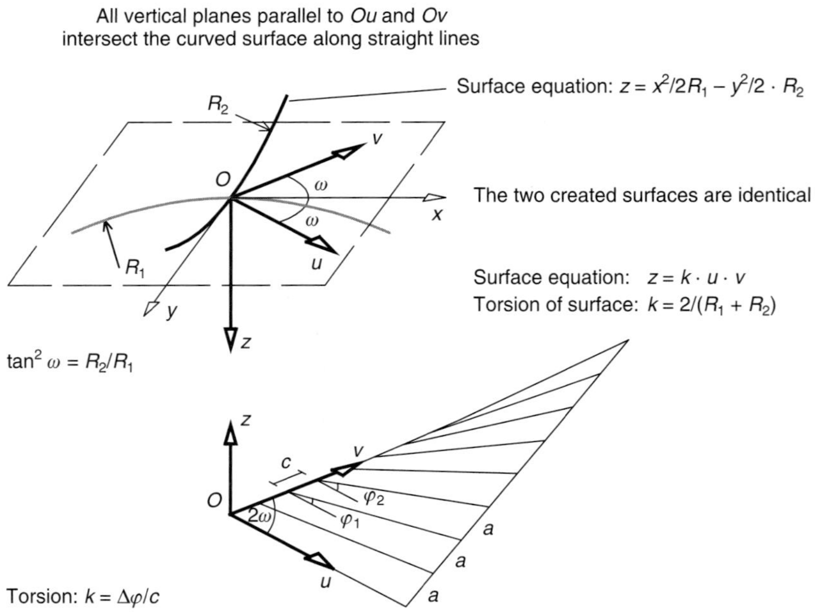

Figure 12.35 Creation of a shell through a straight generatrix and the appearance of twist

It can be seen that the so-created surface covers a rectangular ground plan of dimensions $(2 \cdot a) \cdot (2 \cdot b)$. Every section of this surface with a vertical plane is a parabola, while a section with a horizontal plane produces a hyperbola.

If a vertical plane is now considered whose trace on the horizontal plane Oxy forms an angle ω with the x axis such that $\tan^2 \omega = R_2/R_1$, then this plane intersects the shell along a straight line. In addition, if the symmetric of this plane with respect to the x axis is considered, its intersection with the shell is also a straight line. In other words, the x axis bisects the angle $2 \cdot \omega$ formed by the traces of the above two planes over the horizontal plane Oxy (Figure 12.35).

If these two traces are considered as skew coordinate axes Ou and Ov, then it can be proved that the coordinates z of the surface may also be obtained from the expression

$$z = k \cdot u \cdot v$$

where $k = 2/(R_1 + R_2)$.

It is clear that the intersection of the surface with a plane parallel to Ouz or parallel to Ovz is a straight line. The same shell can be assumed to be created if a straight line Oa, initially coinciding with axis Ou, is transposed in space while remaining parallel to the plane Ouz, with its edge always lying on the axis Ov. During this line movement, the ratio of $\Delta\varphi$, the change in the angle of its slope φ relative to the plane Ouv, to the corresponding distance travelled by its edge O on the axis Ov should be kept constant and equal to k (see Figure 12.35). It is clear that k represents the constant twist of the surface relative to the coordinate system Ouv considered:

$$k = \partial^2 z/\partial u \cdot \partial v \quad \blacksquare$$

The transition from one geometry to the other may also take place.

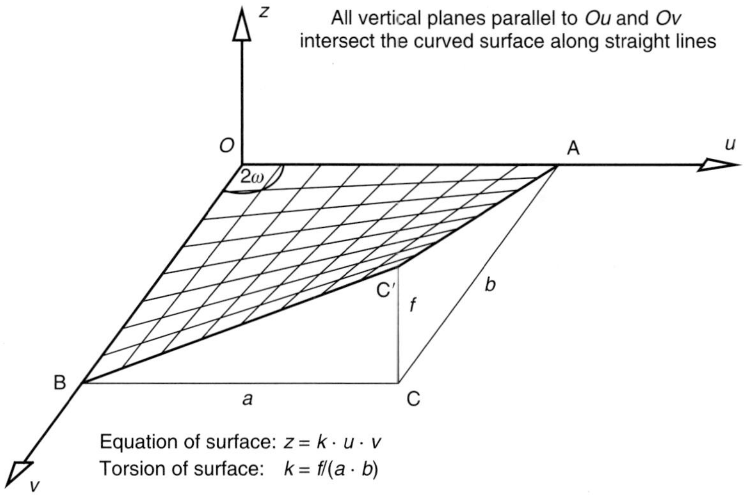

Figure 12.36 Creation of a hyperbolic paraboloid surface over a skew ground plan

Assume that the shell is intended to cover a parallelogram ground plan *OACB* with sides *a*, *b* enclosing an angle $2 \cdot \omega$, while the skew axes *Ou* and *Ov* coincide with the two sides of the parallelogram (Figure12.36).

Corresponding to the edge C with skew coordinates $u = a$, $v = b$, a point C′ is defined with coordinate $z = f$ and connected through a line segment to the two other edges A and *B*. The hyperbolic paraboloid shell can be created if the segment *OA* moves parallel to the plane *Ouz*, inclined towards the lines *OB* and *AC*′. The coordinates *z* of the so-formed surface are $z = k \cdot u \cdot v$, where $k = f/a \cdot b$.

It is clear that any vertical plane parallel either to the axis *Ou* or to *Ov* intersects this surface along a straight line, thus making it possible for the surface to cover any parallelogram ground plan.

This same surface may now be depicted through a system of mutually orthogonal parabolas with opposite curvatures in the following way (Figure 12.37).

If the axis *Ox* is considered as the bisecting line of the angle $2 \cdot \omega$, and the axis *Oy* is drawn perpendicular to it, then each of the planes *Oxz* and *Oyz* intersects the surface along parabolas. The intersection with the plane *Oxz* shows an upward curvature with a radius

$$R_1 = 2 \cdot \cos^2 \omega / k$$

while the intersection with the plane *Oyz* shows a downward curvature with a radius

$$R_2 = 2 \cdot \sin^2 \omega / k$$

Thus, the coordinates *z* of this same surface referred to the orthogonal system *Oxy* can be expressed through the radii of curvature R_1 and R_2, as previously shown.

From the above analysis, it is clear that a hyperbolic paraboloid shell with equal radii of curvature $R_1 = R_2 = R$, being referred to the orthogonal axes *Oxy*, can also be considered as created from the straight generatrices representing its intersections with

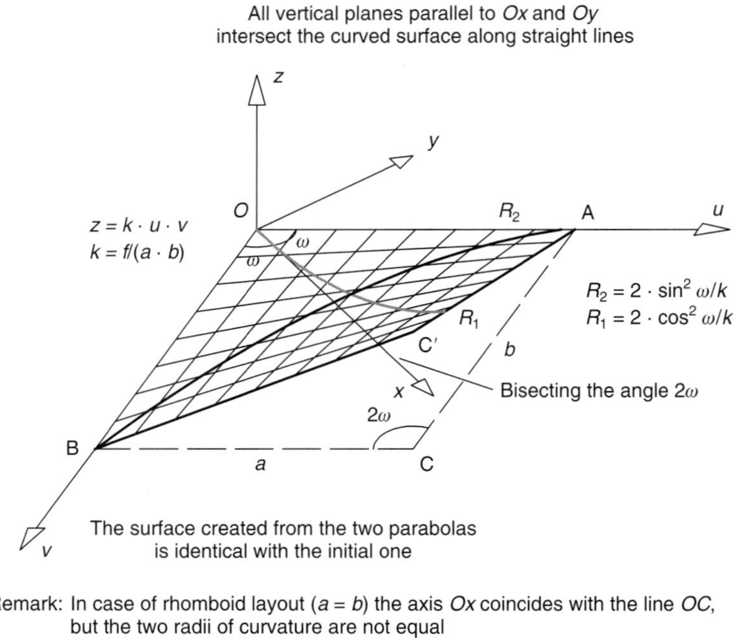

Figure 12.37 Restoration of parabolic generatrices in a shell over a rhomboid ground plan

those vertical and mutually perpendicular planes Ouz and Ovz which bisect the right angles of the axes Ox and Oy. The twist k of the surface corresponding to the equation $z = k \cdot u \cdot v$ is, according to the above, equal to $k = 1/R$.

In order to acquire a more direct understanding of the load-carrying behaviour, shells with equal radii of curvature in both the x and y directions will now be considered.

12.5.3 Considerations of equilibrium

In the orthogonal layout shown in Figure 12.38, $a^2/2 \cdot f_1 = b^2/2 \cdot f_2 = R$ (see Section 12.5.2). The shell, as can be seen from the figure, extends between the four parabolas corresponding to the sides of the orthogonal base.

The twist t_{xy} of the surface with reference to the system Oxy is zero, as can be deduced from the expression $\partial^2 z/\partial x \cdot \partial y$, which implies that the tangential shear forces S in the directions x and y are also zero. The equilibrium of a curved quadrilateral element (see Figure 12.38) shows that the load p may be carried by the compressive forces, as well as by the tensile forces N_x and N_y. This is clear from the fact that the load p causes compression of downward-curved parabolas, making them act as 'arches', while simultaneously it applies tension to the upward-curved parabolas, making them behave as 'cables'. From the equilibrium equation in Section 12.2, it can be deduced that the membrane forces N_x and N_y have a common value equal to $p \cdot R/2$.

However, in order for this membrane state in the shell to exist, the above axial membrane forces should be appropriately introduced at the boundaries of the shell. If

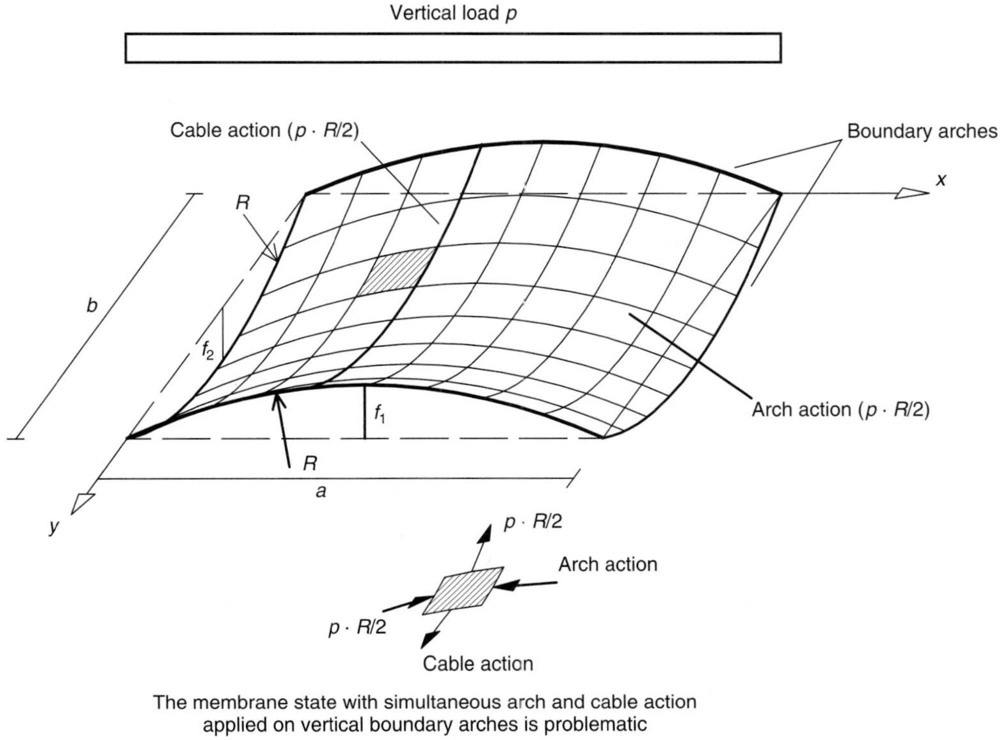

Figure 12.38 Load-carrying action of a hyperbolic paraboloid shell with parabolic edges

the shell is unyieldingly supported along its parabolic boundaries, then this is feasible. However, if the shell is supported on, for example, vertical parabolic arches, these are not positioned to carry forces perpendicular to their plane, and thus the purely membrane state of stress is impossible. The very fact, of course, that these arches are able to carry vertical loads means that vertical shear forces will necessarily be introduced in the shell, thus leading unavoidably to bending.

Another support possibility is the existence of unyielding supports along the ends of downward-curved (i.e. compressed) parabolic 'arches', while the edges of the upward-curved ones are free (Figure 12.39). Then, the load p has to be supported entirely by the compressed parabolas, which will develop a compressive force N_x equal to $p \cdot R$, double the previous value (see Section 12.2).

Of course, unyielding supports may also be considered only at the edges of the tensioned parabolas, the sides of the other direction being left free. Then, the load p going through the tensioned parabolic trajectories with the axial force $N_y = p \cdot R$ should be carried by prestressed tendons embedded in the concrete (see Section 9.1). ■

It is now of interest to examine whether the above-discussed case of a shell carrying its load p through compression paths (arches) can be realised without transferring transversal loads to the edge-supporting arches, which has been previously shown as

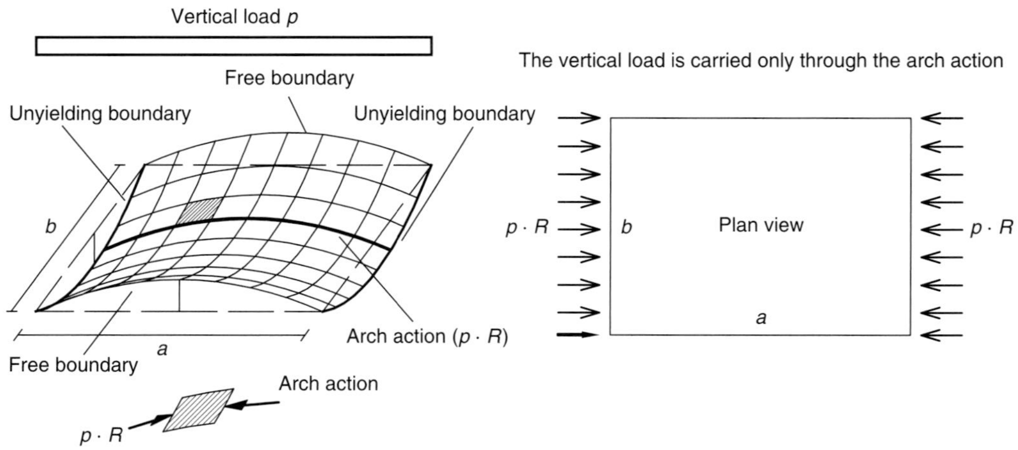

Figure 12.39 Feasible membrane equilibrium using arch action and free opposite boundaries

problematic, while ensuring that a pure membrane state develops (see Franz and Schäfer, 1988). This is achieved on the basis of the remark that the cuts of the shell with vertical planes mutually perpendicular and with an angle of 45° to the axes x and y are straight lines, as mentioned previously (see Section 12.5.2).

First, it is again assumed that the load p is carried by the axial compressive force $N_x = p \cdot R$ in one direction only, as in Figure 12.39, keeping the two other opposite edges free from transverse loads. This state of stress will be superposed on another one, as explained in the following two cases.

In the first case, where the ground plan is assumed to be square, a self-equilibrating loading is further considered, as in Figure 12.40. This loading consists of a distributed load $p \cdot R$ acting in the same sense perpendicularly to the opposite edges of the compressed parabolas, being equilibrated by tangential shearing forces applied at the two other opposite sides.

As is evident from Figure 12.40, the normal forces applied to the two boundaries may be analysed in two perpendicular directions at 45° as tensile or compressive forces, which are transferred directly to the neighbouring boundary where they are balanced by the shearing forces acting there. Thus, this additional loading is self-equilibrated within the shell itself, and therefore able to be superposed on the former state of stress. It is now obvious that one boundary of the compressed parabolic paths will be free of forces, whereas on the opposite side an external compressive force $2 \cdot p \cdot R$ is required through an appropriate support (see Figure 12.40). It is also clear that the existing arches, at the other two opposite sides, are able to receive – and to transmit – the longitudinal shearing forces without problem. It can also be seen that the required compressive force $N_x = p \cdot R$ is applied to the free edge, due to the simultaneous tensile forces inclined by 45° which meet on the boundary (see Figure 12.40), so that the overall compressive action of the parabolic shell arches is not disturbed.

The second case, where one side in the ground plan is twice as long as the other, also allows the development of a purely membrane state by considering the application of

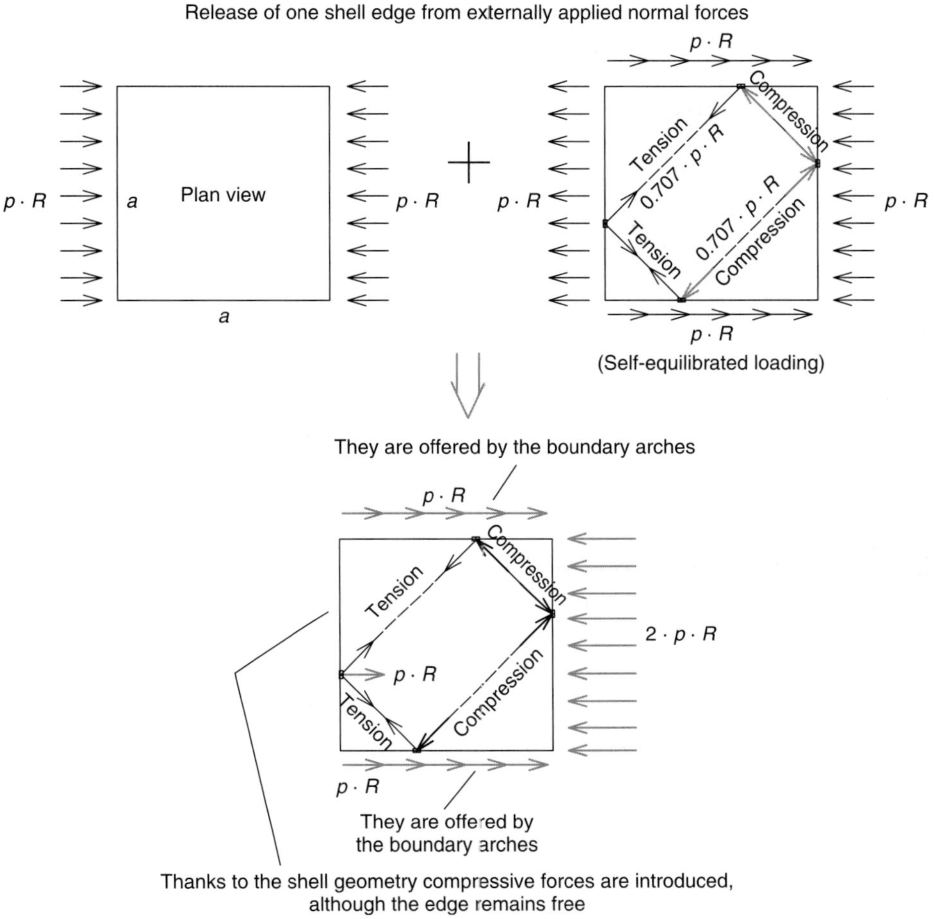

Figure 12.40 Feasible membrane equilibrium in a square layout with a free edge

additional self-equilibrating loading tensile forces $N_x = p \cdot R$ to the shorter sides of the shell (Figure 12.41).

This load, which eliminates, of course, the initially obtained corresponding compressive forces, may be considered as coexisting with the self-equilibrated shearing forces on each oblong side of the shell. The analysis of longitudinal tensile forces at both short sides in components at an angle of 45° makes the establishment of the internal equilibrium possible. This is achieved due to the selected directions being implemented in straight lines on the shell, which allows the direct transfer of forces between the oblong boundaries, where they are balanced by the shearing forces acting there. These shearing forces can be imposed by the corresponding boundary arches while the short sides remain free, as the tensile forces balancing there provide the compressive forces $N_x = p \cdot R$, leading to the compressive carrying action of the arches in the oblong sense (see Figure 12.41).

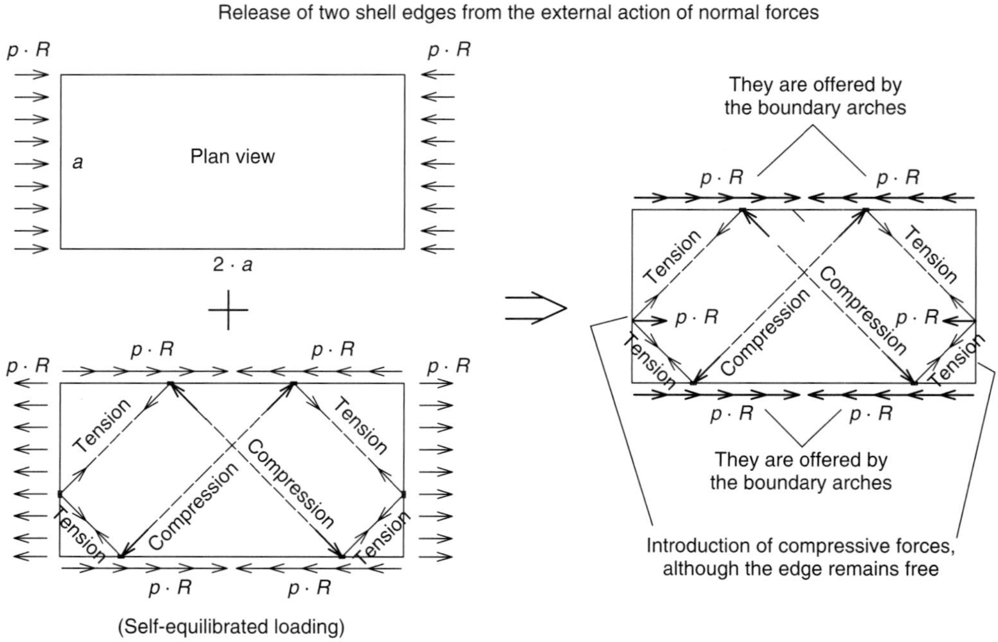

Figure 12.41 Feasible membrane equilibrium in an oblong layout with two free edges

12.5.4 Hypar shells with straight edges

To cover orthogonal areas, hypar shells are usually formed with straight boundaries. The shell is necessarily referred to a system of axes *Oxy* comprising two adjacent sides of the orthogonal base *OABC* (Figure 12.42). With the origin *O* kept fixed, the three other edges can be moved vertically at any distance, as shown in Figure 12.42. The shell is created through the constant displacement of side *OA′* in parallel to the plane *Oxz*, inclined towards the sides *OB′* and *A′C′*, so as to result in position *B′C′*.

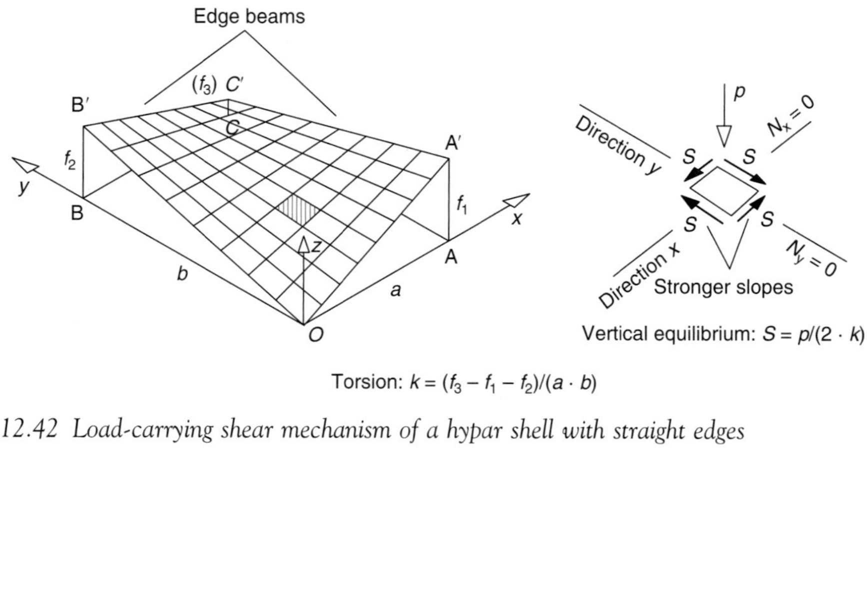

Figure 12.42 Load-carrying shear mechanism of a hypar shell with straight edges

The twist k of the surface thus created is

$$k = \frac{f_3 - f_1 - f_2}{a \cdot b}$$

The segments 'f' are considered signed quantities with respect to the axis z. According to Section 12.5.2 and given that $2 \cdot \omega = 90°$, both surface curvatures (i.e. '$1/R$') are equal to k. These, understandably, appear along the skew directions at an angle $2 \cdot \omega/2 = 45°$ relative to the axis Ox.

It should be noted that, in this case, the equation of the surface is not expressed by $z = k \cdot x \cdot y$ but by

$$z = k \cdot x \cdot y + (f_1/a) \cdot x + (f_2/b) \cdot y$$

which does not interfere at all with its geometry, as the twist of the surface according to the expression $\partial^2 z/\partial x \cdot \partial y$ remains equal to k.

Considering now the equilibrium of an orthogonal shell element in the system Oxy under the vertical load p, it can be seen that the only membrane mechanism through which this load may be carried is the tangential shearing forces S acting along the sides of the element (see Figure 12.42). This happens because the shell element presents null curvatures about the two orthogonal axes (i.e. x and y), and, consequently, any developing internal forces N_x and N_y cannot offer a component opposing the load, because they are collinear for any corresponding pair of opposite element sides. Thus, it follows that the above forces cannot be developed, so that the basic equation for membrane equilibrium in Section 12.2 yields the following constant value for S:

$$S = \frac{p}{2k}$$

Indeed, the difference in tilt at opposite sides of the element allows the forces S to offer an upward component, which equilibrates the vertical load p (see Figure 12.42).

These forces S must necessarily be applied at the straight edges of the shell too, which is possible only if edge beams are arranged along each boundary.

It should be noted here that the previous conclusion about zero membrane forces N_x and N_y is strictly valid only if the load p acts perpendicularly to the shell surface. Since the acting load is vertical, it is clear that for small slopes of the surface the above conclusion is essentially valid. 'Small slopes' is synonymous with the existence of 'small curvatures', which in turn means that the shell should be 'shallow', and this is the case if the edge beams have a slope not exceeding 18°. For steeper slopes, the vertical load p has a non-negligible component in the direction of the straight generatrices, leading to the development of axial forces in their direction and, consequently, the membrane shearing forces are no longer constant. However, for most cases of roofing, the hypar shells used can be considered 'shallow'.

Of course, irrespective of the above impact of the equilibrium consideration for the axes x and y, it should not be overlooked that, due to existing curvatures of 45° in both oblique senses, axial compressive and tensile forces N that are equal in magnitude

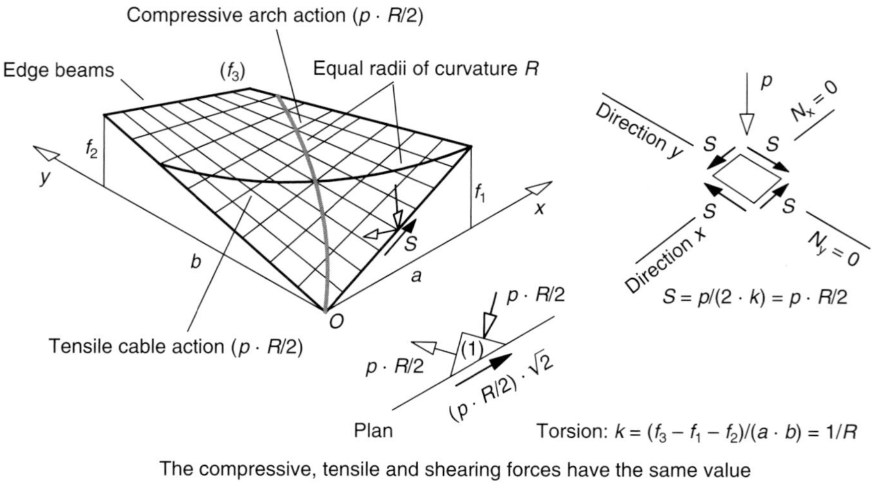

Figure 12.43 Load-carrying mechanism through the arch and suspension action of a shell

are developed, which, according to Sections 12.5.2 and 12.5.3, are

$$N = \frac{p \cdot R}{2} = \frac{p}{2 \cdot k}$$

i.e. equal to the membrane shear forces (Figure 12.43).

Thus, the shell always works as a group of compressed 'arches' and suspended 'cables' both developing the same axial force. These forces reaching the boundary may be considered as being applied to the vertical sides of unit length of the orthogonal triangular element, as shown in Figure 12.43, the hypotenuse of which is $\sqrt{2}$. The equilibrium of this element, given that the component of forces N perpendicular to the boundary is zero, confirms the necessity for the membrane shearing force to act along the boundary and to be equal to N. This is exactly the shearing force found above, running through the whole shell, which should, therefore, also be applied at the boundary.

The edge beams which transmit these shear forces to the shell are obviously subjected to the opposite forces, and contribute their sum throughout the boundary length to each support of the shell.

As shown in Figure 12.44, the forces on the edge beams accumulate to their lowest points, where they are taken up by appropriately designed columns that transfer the total load of the shell to them. The edge beams develop a progressively increasing compression towards their low supports, and in all cases care must be taken to carry their self-weight, usually by providing a continuous support. ■

The question arises, however, of whether the edge beams that will carry the above shearing forces can be designed free of intermediate support. In other words, is a layout of boundary beams which are supported (fixed) only at their corresponding lower ends and remain free all along their length, and, through the use of large cantilevers, provide an imposing aesthetic result, possible?

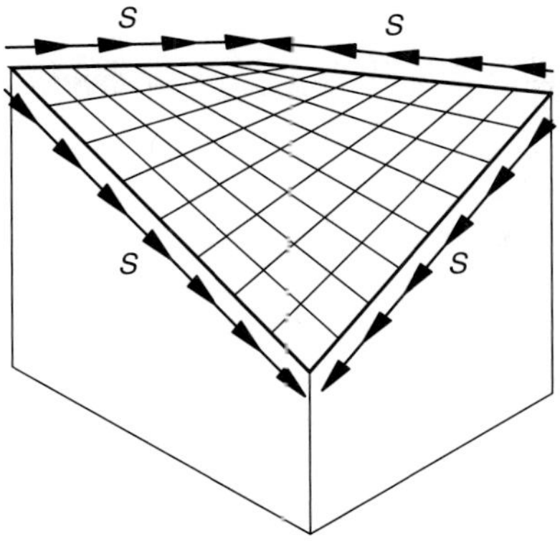

Figure 12.44 Forces acting on the edge beams of the straight boundaries

The fact is that, although according to the membrane state these cantilevers do not carry any part of the shell vertical loads but only longitudinal loads, they nevertheless have a significant self-weight, which they are not able to carry as long they are acting as cantilevers. The question, therefore, which essentially arises is whether the shell can participate in carrying the self-weight of the edge beams and, if so, with what consequences for its state of stress.

The answer to this question is in principle positive, and is based on the fact that there are straight generatrices between the edge beams which can act as cables. These 'cables' can act against the tendency of the cantilevers to deform, by opposing tensile forces along the boundary, which relieve the cantilever from the response due to its self-weight. These forces are carried essentially unchanged from a given boundary edge to the opposite one (Schlaich, 1970).

As shown by the model in Figure 12.45 – which is provided with an optional support at the high-point C to take into account a possible asymmetric loading – the system of the two adjacent 'cantilevers' *OAB*, which is subjected to the vertical loads of their self-weight, is also acted on by the forces of the tensioned cables at every discrete level. These cable forces Z counteract the tendency of the 'cantilevers' to tilt about the axis *AB*, as they cause an opposite bending moment to M_0 due to their self-weight, thus considerably relieving them. For simplicity, the structure is assumed to be symmetrical about the vertical plane passing through the axis *AB*.

An approximate estimation of the tensile cable forces can be made by assuming that the cantilevers are elastically undeformable and behave like rigid bodies rotating freely about the hinge axis of their support *AB* (Figure 12.46). It can be seen that the system of cables actually consists of prestressed tendons anchored at the opposite edge beams.

Thus, for each level *i*, the forces Z_i are determined on the basis of the angle ω between the cables considered in plan view, the difference Δz_i of their ends, their projection length Δx on the vertical plane *OC*, the corresponding cable length L_i, the distance e_i of the corresponding resultant R_i of forces Z_i at each node from the support axis *AB*

Carrying the weight of the boundary beam through the shell

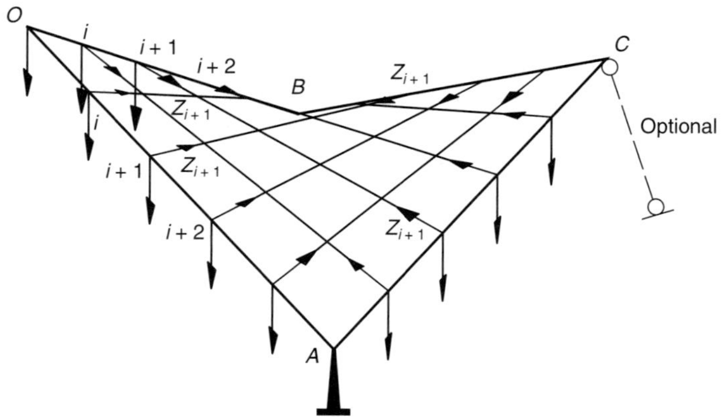

The cable forces relieve the cantilevers, and in this way strongly decrease the bending action of the loads

Symmetry about the vertical plane *AB*

Figure 12.45 Self-weight of the edge beams is taken up by 'cable action'

and, finally, the cross-sectional area F_i of the 'cables', which may at first represent the tributary shell strips between them.

Thus (Schlaich, 1970),

$$Z_i = M_0 \frac{K_i}{2 \cdot \cos(\omega/2)} \cdot \sum_i \frac{1}{K_i \cdot e_i}$$

where

$$K_i = \frac{F_i \cdot l_i}{L_i^3} (\Delta x \cdot \sin \gamma + \Delta z_i \cdot \cos \gamma)$$

and

$$\Delta x = \frac{(AB)}{2 \cdot \tan(\omega/2)}$$

As mentioned previously, these tensile forces Z_i should be taken up by prestressing (Figure 12.47). By introducing the appropriate prestressing forces along the straight generatrices, a purely membrane state in the shell may be feasible in practice, given that each compressive force applied at the shell boundary is not spread out but remains in the 'strip' of the corresponding generatrix. In addition, the absence of curvature in the tendons essentially eliminates the friction losses. ■

The synthesis of individual shells with a square ground plan in wider forms makes the covering of larger areas with a small number of supports possible. With the appropriate juxtaposition of shell parts, edge beams are needed only at the free boundaries in practice, because at the 'interior' straight boundaries of each shell, the required shearing forces S for the equilibrium are provided mutually by the adjacent shells.

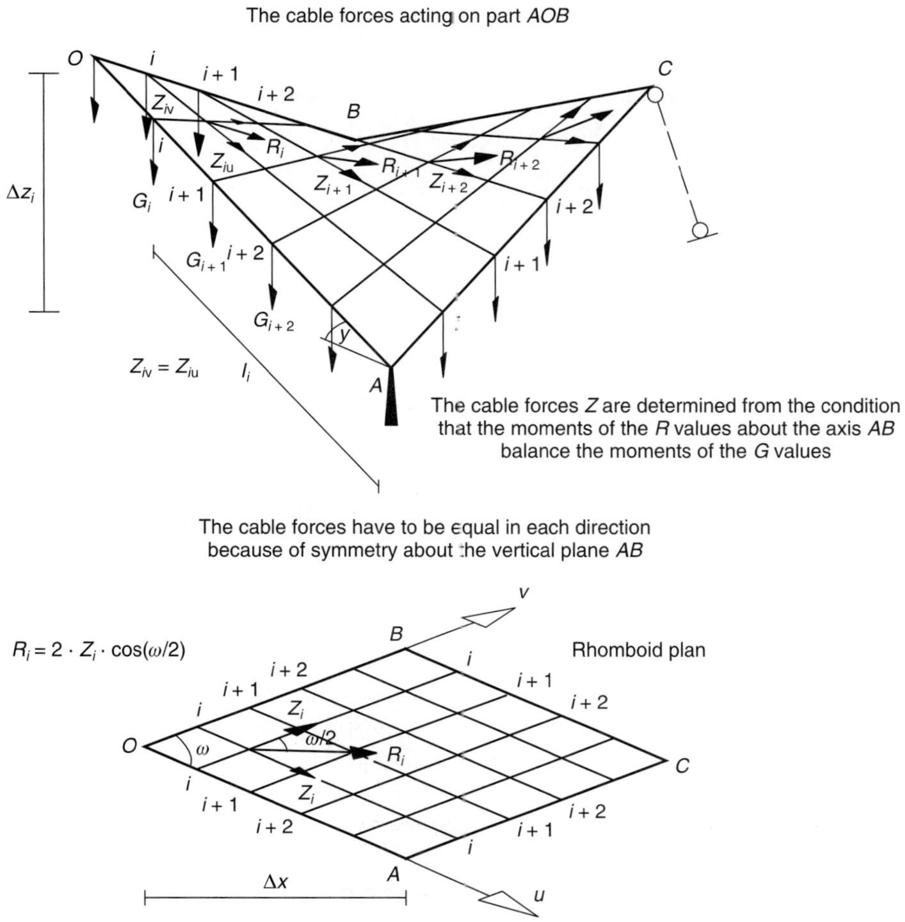

Figure 12.46 Assessment of the required tensile (prestressing) cable forces

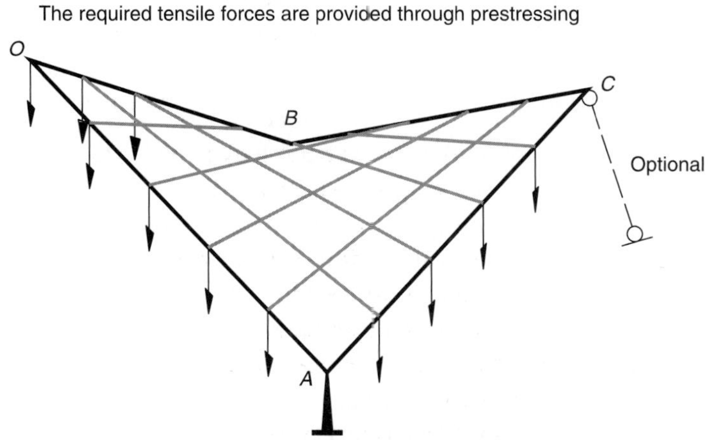

Figure 12.47 Realisation of cable action through prestressing of a shell

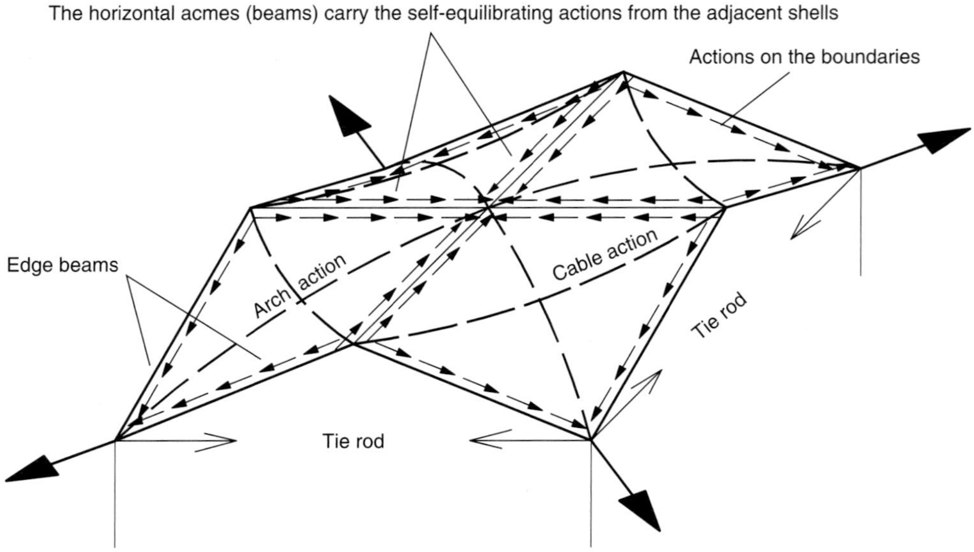

Figure 12.48 Layout of four hypar shells to cover a square ground plan

The covering of a square ground plan using four individual shallow hypar shells is shown in Figure 12.48. Each shell develops a shearing state of stress in the sense of its generatrices, which, as explained above, is caused by the simultaneous presence of the equal tensile and compressive forces of the 'arch' and 'suspension' action, respectively.

It is concluded that the two imaginary cut strips along the top horizontal edges are themselves in equilibrium under the oppositely acting shearing forces, as shown in Figure 12.48. Thus, the required shearing forces along the interior boundary of each shell are provided by the adjacent shell. However, the required shearing forces at the 'external' boundaries have to be provided by edge beams which, being acted on by the opposite forces, have to give up their resultants to the corner columns. These resultants have a horizontal component acting along the diagonals of the ground plan, according to the arch action indicated in Figure 12.48 (see Section 8.1). The equilibrium of the column top necessitates the provision of a horizontal diagonal action in order to avoid an overstressing of both the columns and the edge beams, and this is accomplished by providing diagonal ties that are prestressed accordingly. ■

Despite the fact that the preliminary design of hypar shells may be based on the assumption that the loads are carried by membrane action, it should be noted that various incompatibilities arising from this consideration may also affect the bending action of the shell. Thus, it is noted, among other things, that, for example, non-uniform loadings may cause bending in the shell, or even that the mechanism through which the self-weight of the straight edge beams is carried through membrane action may never be complete.

As has been emphasised elsewhere, in order to determine the final state of stress of the shell, the use of appropriate software is necessary, so that all details of its constructional formation can be taken into account.

12.5.5 *Elastic stability*

The presence of compressive forces in hypar shells constitutes a danger of buckling which should be evaluated sufficiently during their preliminary design.

In the case of shells with parabolic boundaries, it is clear that the presence of tensile forces along the upward-curved parabolas clearly contributes to the stability of the embedded 'arches'. Thus, the use, in this case, of the expression for the critical load of a cylindrical shell having the same dimensions in plan as well as the same curvature will definitely lead to safe results (see Section 12.3.2).

In the case of hypar shells with straight boundaries that are rigidly supported, having ground plan dimensions a and b ($a \geq b$), a height difference f and a thickness d, the critical load p_K may be assessed from the relation (Beles and Soare, 1972)

$$p_K = 0.40 \cdot E \cdot \frac{f^2 \cdot d^2}{a^2 \cdot b^2}$$

If there are also edge beams having a moment of inertia equal to I, the critical load is greater. This can be assessed from the relation

$$p_K = 2 \cdot E \cdot \frac{f^2 \cdot d^2}{a^2 \cdot b^2} \sqrt{\frac{2}{3} \cdot \left(1 + \frac{24 \cdot I}{d^3 \cdot a}\right)}$$

12.6 Conoidal shells

Along with the translational shells are also classified those shells which, in order to cover a rectangular ground plan between two parabolic fronts over two opposite sides of the rectangle, are formed through the segment of the other side as a generatrix by using the two parabolic edges as 'guides'. This generatrix moves in parallel to a vertical plane, passing through the corresponding side (usually the oblong one), inclined towards the two parabolic guides, thus offering obvious advantages regarding the construction of the shell formwork (Figure 12.49). One of the two parabolic fronts is deliberately made higher than the other, which could also simply be a straight. The thus-formed conoidal shell can be supported on all four sides, i.e. at its parabolic edges through appropriately formed arches, and at the straight sides either through free-standing beams or rigid continuous supports.

The preliminary design of such a shell can be based on considering it as an arch having a height f_m equal to the average height of the two front parabolas. If L is the span of the parabolic arch, the compressive force per unit metre of the shell due to the vertical load g is

$$N_d = g \cdot L^2/(8 \cdot f_m)$$

In addition, the critical buckling load can be assessed on the basis of the average radius of curvature $R = L^2/8 \cdot f_m$, from the corresponding relation for cylindrical shells (see Section 12.3.2).

In the case when the shell is not directly supported along its straight edges of length B, it can be assumed that the above compressive 'arch' forces of the shell are carried by the deep beams, which are formed along the edge region with an assumed structural height of $B/4$ and act as simply supported beams (see Figure 12.49).

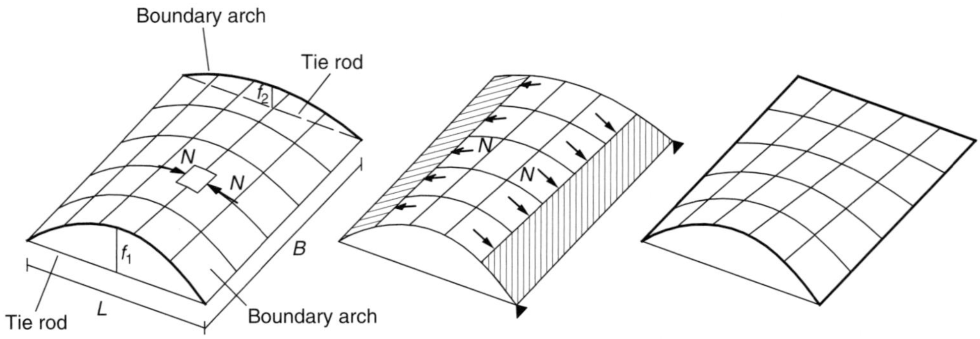

Figure 12.49 Load-bearing action of conoidal shells

A not unusual use of conoidal shells is that of a shelter in the form of a cantilever, fixed at its parabolic edge and having all its three other straight edges free (Figure 12.50).

Its structural function may be conceived as that of a cantilever of length B having a parabolic cross-section, which presents an advantageous increase in its 'moment of inertia' towards the fixed end, where the maximum tensile and compressive stresses are developed.

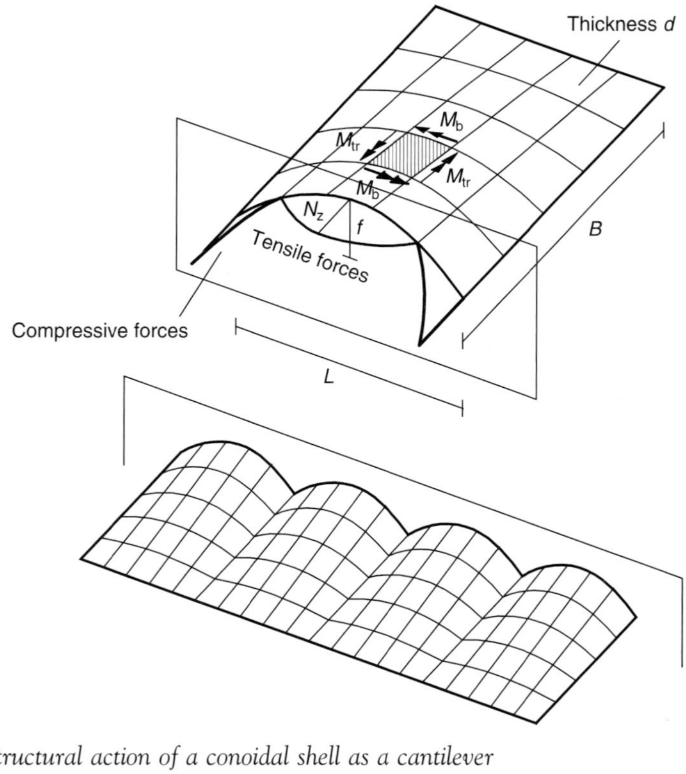

Figure 12.50 Structural action of a conoidal shell as a cantilever

According to the above approach, the maximum tensile force N_z which results under a vertical load g is (see Kollar, 1984)

$$N_z = \frac{B^2 \cdot g/2}{0.265 \cdot f} = 1.90 \cdot \frac{B^2 \cdot g}{f}$$

and is developed at the top of the supported parabolic edge.

Beyond this 'membrane' load-carrying mechanism with regard to the cantilever moment $M_e = g \cdot B^2/2$, a bending moment M_b of much less intensity also has to be considered, being assessed as $M_b = (d/f)^2 \cdot M_e$.

Furthermore, a bending moment M_{tran} in the transverse direction has to be taken into account, which causes tension at the top fibres since it increases the transversal curvature of the shell. Its value at the middle of span L may be assessed as

$$M_{tran} = 0.13 \cdot g \cdot L^2$$

In the case when such shells are arranged monolithically connected in order to shelter larger spaces (see Figure 12.50), the transversal moments are limited to about 25% of the above value, with alternating signs across each field, causing a tension at the bottom fibres over the top of the connected 'arches'.

References

Beles A.A., Soare M.V. (1972) *Berechnung von Schalentragwerken*. Wiesbaden: Bauverlag.

Billington D. (1965) *Thin Shell Concrete Structures*. New York: McGraw-Hill.

Ciesielski R., Mitzel A., Stachurrski W., Suwalski J., Zmudzinski Z. (1970) *Behälter, Bunker, Silos, Schornsteine, Fernsehtürme und Freileitungsmaste*. Berlin: Wilhelm Ernst.

Flügge W. (1960) *Stresses in Shells*. Berlin: Springer-Verlag.

Franz G., Schäfer K. (1988) *Konstruktionslehre des Stahlbetons*, Band II, A. Berlin: Springer-Verlag.

Kollar L. (1984) *Schalenkonstruktionen – Beton – Kalender* 1984 – Teil II. Berlin: Wilhelm Ernst.

Pflüger A. (1963) Zur praktischen Berechnung der axial gedrückten Kreiszylinderschale. *Stahlbau* **32**, 161–164.

Pflüger A. (1966) Zur praktischen Berechnung der axial gedrückten Kreiszylinderschale unter Manteldruck. *Stahlbau* **35**, 249–252.

Salvadori M. and Levy M. (1967) *Structural design in Architecture*. Englewood Cliffs, NJ: Prentice-Hall.

Schlaich J. (1970) Zum Tragverhalten von Hyparschalen mit nicht unterstützten Randträgern. *Beton u. Stahlbetonbau* **3**, 54–63.

Schmidt H. (1961) Ergebnisse von Beulversuchen mit doppeltgekrümmten Schalenmodellen aus Aluminium. *IASS Symposium Delft*, pp. 159–181.

Seide P. (1981) Stability of cylindrical reinforced concrete shells. *American Concrete Institute ACI – SP* **67**, 43–62.

Stavridis L.T., Armenakas A.E. (1988) Analysis of shallow shells with rectangular projection: analysis. *Journal of Engineering Mechanics ASCE*, **114**(6), 943–952.

Timoshenko S. (1956) *Strength of Materials*, Part II. Princeton, NJ: Van Nostrand.

13
Thin-walled beams

13.1 General characteristics

The concept of a thin-walled beam is applied in a wide variety of structural forms. These beams are used mainly in bridges, but they find application in building construction as well. The cross-section of a thin-walled beam is formed by using thin surface elements – mainly plane elements – that are monolithically connected to one other at their common edges. They thus form beam-type oblong members, resulting in a huge saving in material compared with a corresponding solid cross-section of the same constructional height and, simultaneously, have better load-carrying properties. It is for this reason that bridge superstructures, either suspended or not, are formed almost exclusively of thin-walled beams.

As in beams having a solid cross-section, in thin-walled beams the dimension of length L is clearly larger than the maximum transverse dimension B, with a length/width ratio (L/B) of more than 5, while the thickness t of the plane elements is thin enough that $t/B \leq 0.1$ (Figure 13.1). Thus, the beam is formulated as a folded structure according to Section 11.6, and it can basically be analysed using suitable computer software. However, the main goal of the present chapter is the examination of this structural type as a linear beam element so that its load-bearing action can be appraised and understood accordingly.

The main difference between the thin-walled and the solid beam, and which causes the biggest difficulties in analysing the thin-walled beam as a beam element, lies in the way the torsion is taken up. Not only shear but also longitudinal normal stresses contribute, a fact that is not foreseen in the so-called 'classical technical theory' of solid beams. The development of torsional stresses in thin-walled beams occurs both in rectilinear bridges, due to the eccentricity of traffic loads with respect to the cross-sectional axis, and in curved bridges, where only the self-weight leads to a torsional response.

Thin-walled cross-sections are distinguished as 'open' and 'closed' sections, each category having its own load-carrying characteristics. The closed cross-sections are also called 'box sections' due to their closed form. ■

The comparison between a closed and an open double-T section, as they are used in bridge design with respect to the evaluation of their structural behaviour, shows clear advantages for the former, without excluding at all the effectiveness of the latter, in a number of cases.

Copyright Thomas Telford Limited © 2010
All rights reserved

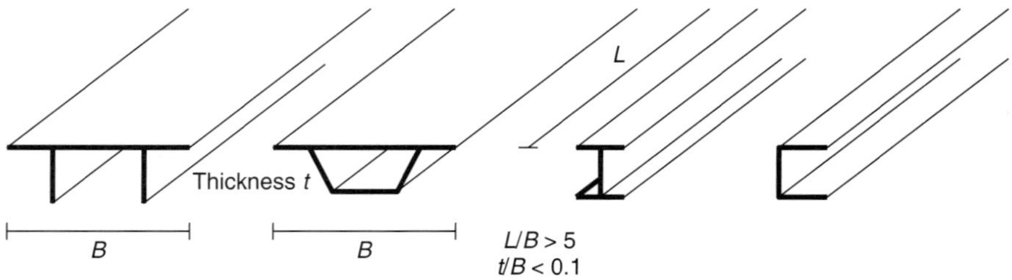

Figure 13.1 The formation of thin-walled beams

In a beam with a double-T section, the possibility of developing compressive forces in the bottom region over the supports, either in a continuous system or in a cantilever (Figure 13.2), is certainly limited in comparison with a box section with the same depth (see Figure 13.1), as the bottom plate of the box can participate fully in taking up the compressive force, something that the double-T section cannot do.

This is automatically translated as a restriction of the bearable tensile force by the top plate (in the case of concrete, through suitable reinforcement, prestressed or not – see Chapters 4 and 5), resulting in a limited corresponding bending resistance of the cross-section. ■

As stated previously, in beams made of concrete, prestressing is necessary for spans longer than 15–20 m. Apart from the fact that the bending resistance of a closed section is clearly greater, another 'unfavourable' characteristic of the open section appears in its use under service conditions, due to the combination of a smaller core depth and a higher positioned centroidal axis, as explained below.

Consider the midspan section of a simply supported prestressed beam. The prestressing compressive force P that is initially applied to the cable trace is finally shifted upwards to a distance $a = M_g/P$, due to the self-weight bending moment M_g, as discussed in Section 4.3.1 (Figure 13.3). However, in an open section this position is closer to the bottom core limit than is the case in a corresponding closed section. Hence the diagram of compressive stresses σ shows a greater difference $\Delta\sigma$ between the extreme fibres in the open cross-section than is the case in the corresponding closed cross-section. Now, as already mentioned in Section 4.1.1, the curvature $1/r$ of

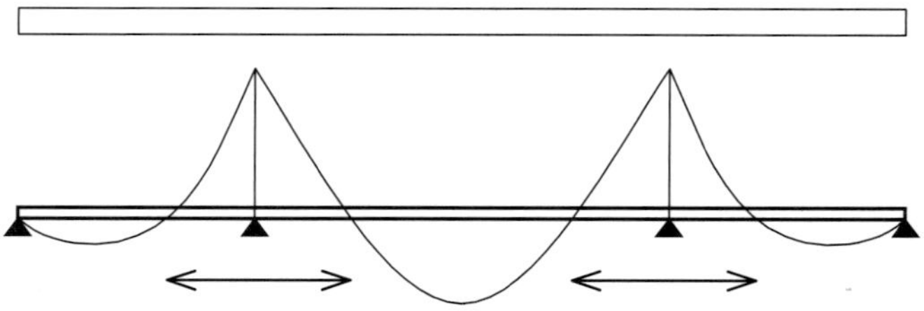

Figure 13.2 The need for a closed cross-section formation

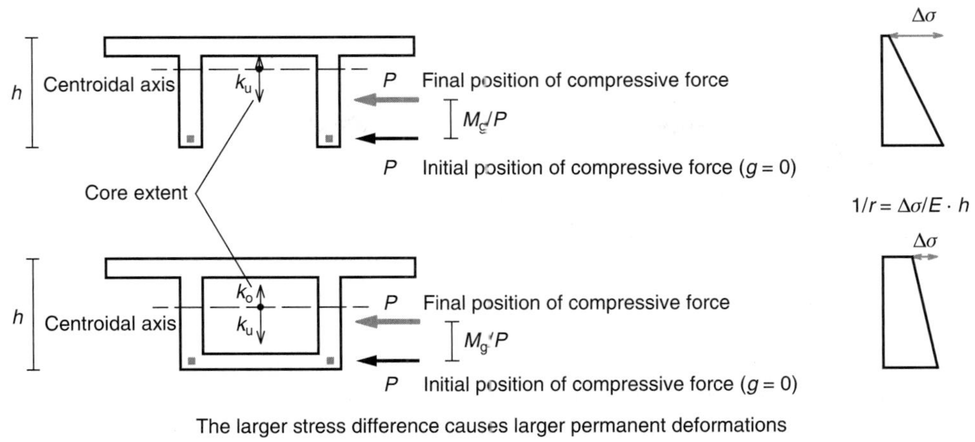

Figure 13.3 The influence of the type of section on the variation in stress due to prestressing across the depth

the beam is expressed as $|\varepsilon_o - \varepsilon_u|/h$, i.e. as $|\Delta\sigma|/(E \cdot h)$. Thus the open section presents larger deformations due to prestressing than does the closed one, and this difference becomes two to three times larger under the influence of creep (see Section 1.2.2.1), which is, optically at least, annoying (see Figure 13.3). ■

The last and perhaps most important difference between the two types of cross-section is related to torsion. A torsional moment M_T applied to the open section is taken up, according to the classic technical theory, by the three oblong orthogonal parts through the development of shear stresses, the maximum values τ_{max} of which occur at the external longitudinal sides (Figure 13.4), as explained in Section 2.5.3. Thus $\tau_{max} = (3 \cdot M_{T,i})/(B \cdot t^2)$, where B is the length of each oblong side and $M_{T,i}$ is that part of M_T which is carried by each part and offered naturally by the couple of the oppositely directed stresses τ. Note that the distribution of M_T to the parts is proportional to their torsional inertia $I_T = B \cdot t^3/3$ (see Section 3.2.10). However, due to the small lever arm (roughly $2 \cdot t/3$) of the above couple, the possibility of taking up a torsional moment through this mechanism is always limited (see Section 2.5.3).

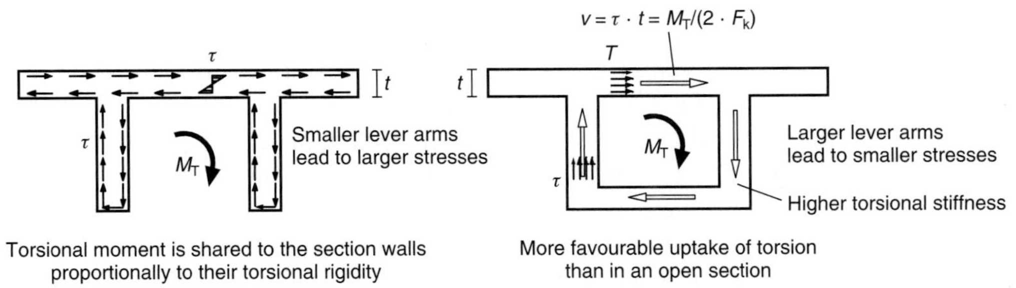

Figure 13.4 The basic torsional behaviour of open and closed cross-sections

Conversely, in a closed section, shear stresses of the same magnitude with a constant flow all over the profile have the ability to take up a much larger torsional moment, due to the relatively large lever arm they can offer. Thus, in each wall of the section, uniformly distributed shear stresses τ_1, τ_2, τ_3 and τ_4 are developed, each of which has the characteristic property of presenting a constant shear flow $v = (\tau \cdot t)$, i.e. a constant shear force per unit length (see Figure 13.4). Thus $v = \tau_1 \cdot t_1 = \tau_2 \cdot t_2 = \tau_3 \cdot t_3 = \tau_4 \cdot t_4$, while, according to the Bredt's formula,

$$v = \frac{M_T}{2 \cdot F_k}$$

where F_k is the area encompassed by the midline of the cross-section (see Section 2.5.3).

Certainly, the torsional rigidity of a closed section is much greater than that of an open section having the same cross-sectional area, as for constant thickness t, the torsional inertia I_T is equal to $4 \cdot t \cdot F_k^2 / L_\Omega$, where L_Ω is the perimeter of the midline of the closed cross-section (see Section 2.5.3). This generally means that a closed section has smaller torsional deformations than a corresponding open section. ■

However, the possibilities of the open section to take up torsional moments are not exhausted by the classical shear stress mechanism mentioned previously, also known as the St Venant mechanism. In addition to this mechanism there is another that is not predicted by classical technical theory and which helps the St Venant mechanism in taking up the torsional moment. However, despite this, the ability of an open section to carry torsional moments remains limited compared to the ability of the closed section.

In a last comparison of the two types of cross-section, the possibility of the participation of the two webs in receiving an eccentrically applied load will be examined, based on a gross estimate of the additional mechanism of torsional resistance mentioned above for open sections, a detailed examination being reserved for later. Both types of cross-section have a width b and a height h (Figure 13.5).

Consider a concentrated load Q applied at the midsection of a simply supported beam, placed precisely on the left web of each section. The load is split into a symmetric and an antisymmetric load, as shown in Figure 13.5. The symmetric loading causes bending and shearing of the beam, whereas the antisymmetric load causes torsion.

Examining first the beam with an open section located to the left of its midpoint, it may be concluded that for the symmetric loading the shear force $Q/2$ acts through the downward forces $Q/4$ at each web, while the plate shear flows have the directions shown in Figure 13.5 and vanish at the midpoint of the plate. In the antisymmetric loading, the torsional moment $(Q \cdot b/4)$ is counterbalanced by a couple of opposing shear forces in the webs, which are equal to $(Q \cdot b/4)/b = Q/4$, and the shear senses in the plate that result are as shown in Figure 13.5. (At this point, it should be noted that the directions of the shearing stresses generally correspond to those followed by an analogous hydraulic flow, regardless of whether they originate from shear or torsion.) Of course, the presence of these shear forces implies bending of the left-hand web downwards and the right-hand web upwards, and clearly this opposite

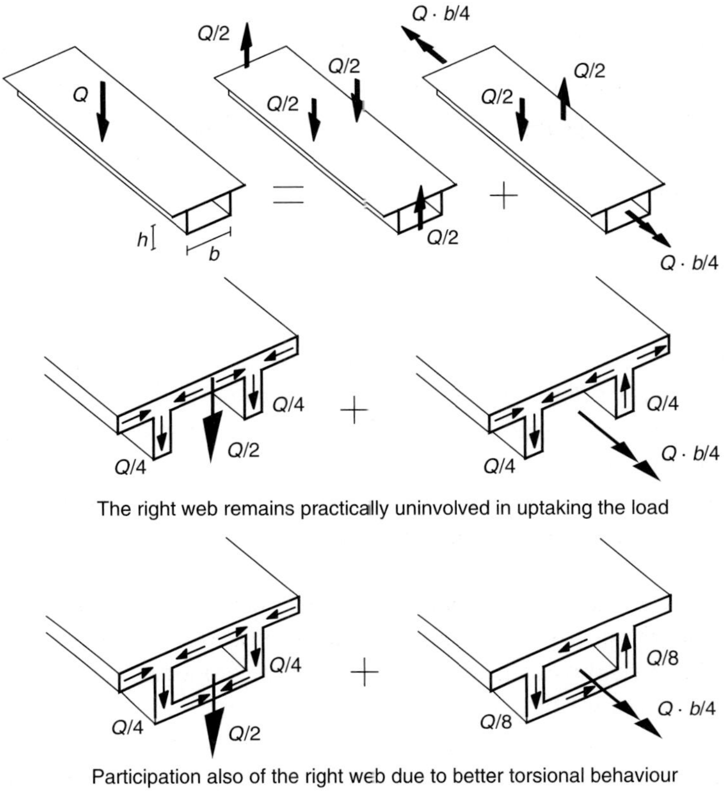

Figure 13.5 The uptake of an eccentric load by beams having an open and a closed cross-section

bending of the two webs will cause such rotations at their ends that Bernoulli's law of plane sections is automatically ruled out. This is a rough description of the mobilised bending mechanism of an open cross-section, which generally takes up a bigger share of the torsional moment than does the St Venant mechanism. This leads to the conclusion that the left-hand web receives all the shear force $Q/2$, while the right-hand web remains unstressed (see Figure 13.5).

The beam with a closed cross-section shows a better behaviour. Again, a cross-section to the left of the beam midpoint is examined. In the symmetric loading the shear force $Q/2$ again causes shear forces of $Q/4$ in the webs, and the shear stresses in the top and bottom slab result according to the above rule of 'hydraulic flow', as shown in Figure 13.5. In the antisymmetric loading, however, the torsional moment $(Q \cdot b/4)$ causes a peripheral shear flow according to Bredt, which develops in each web a shear force (see Figure 13.5)

$$\frac{Q \cdot b}{4 \cdot (2 \cdot b \cdot h)} \cdot h = \frac{Q}{8}$$

instead of $Q/4$, as in the open cross-section. Thus, the left-hand web receives a shear force $(3 \cdot Q/8)$, i.e. 75% of the shear force $Q/2$, while the right-hand web receives a shear force $Q/8$, i.e. the remaining 25%. It is understandable that the superior behaviour of the closed cross-section in the distribution of an eccentrically applied load is owed to a more favourable uptake (with smaller shear forces) of the torsional moment ($Q \cdot b/4$) (see Figure 13.5).

The basic principles governing the behaviour of thin-walled cross-sections will now be examined in a more systematic way.

13.2 The basic assumption of a non-deformable cross-section

A thin-walled beam with an open cross-section that is subjected to a torsional moment presents a perceptible distortion in each cross-sectional plane (Figure 13.6). More specifically, the longitudinal fibres of the beam are deformed along their length so that no cross-section remains planar but undergoes *warping*. The reason for this will be explained later. However, the new outline of the cross-section, mapped over its initially single plane, gives a form that is rotated with respect to its initial position, and does not present any deformation relative to its initial form (see Figure 13.6). This is the so-called assumption of a 'non-deformable cross-section'. Despite the fact that this assumption can be satisfied implicitly, the insertion of transverse diaphragms at certain reasonable distances over the length of the beam may be additionally required, which, given the negligible rigidity transversely to their plane, does not affect the deformability of the longitudinal fibres of the beam, i.e. its warping.

The reason why a distortion of the cross-section occurs is that a torsional moment applied to an open section is taken up not only by the local shear flow in each of its constituent thin-walled elements (as mentioned in Sections 2.5.3 and 13.1) but also, and mainly, by their bending within their plane, as shown for example in Figure 13.6. This involves, of course, a bending rotation of the sections of the elements within their plane, and the 'synthesis' of these bending rotations of the elements leads to warping of the cross-section. It is important to note that the presence of warping should not be confused with the requirement of non-deformability of the cross-section.

For beams having a closed cross-section, the magnitude of warping is very limited. It should be noted, however, that the assumption of the non-deformability of closed cross-sections that is adopted throughout this chapter may potentially be abandoned for

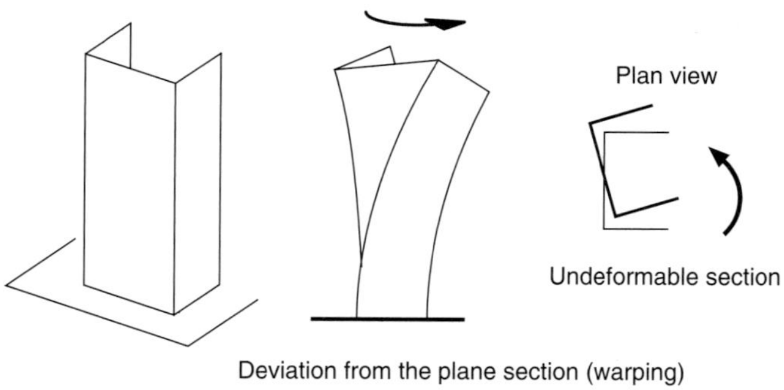

Figure 13.6 Warping and non-deformability of the cross-section

reasons of constructional practice reasons, with corresponding consequences on the structural behaviour, as will be examined in Chapter 14.

13.3 Shear centre

The concept of a shear centre is of fundamental importance in describing the behaviour of thin-walled cross-sections in torsion. In the example shown in Figure 13.7, the thin-walled beam with a channel section is fixed at its left-hand end and is subjected to a vertical distributed load, positioned over the web. At the fixed end, besides the bending moment, this load causes a shear flow over the cross-section of the beam.

In this case, the determination of the senses of shear forces in the flanges requires more than the analogy with a hydraulic flow (see above). To establish the longitudinal

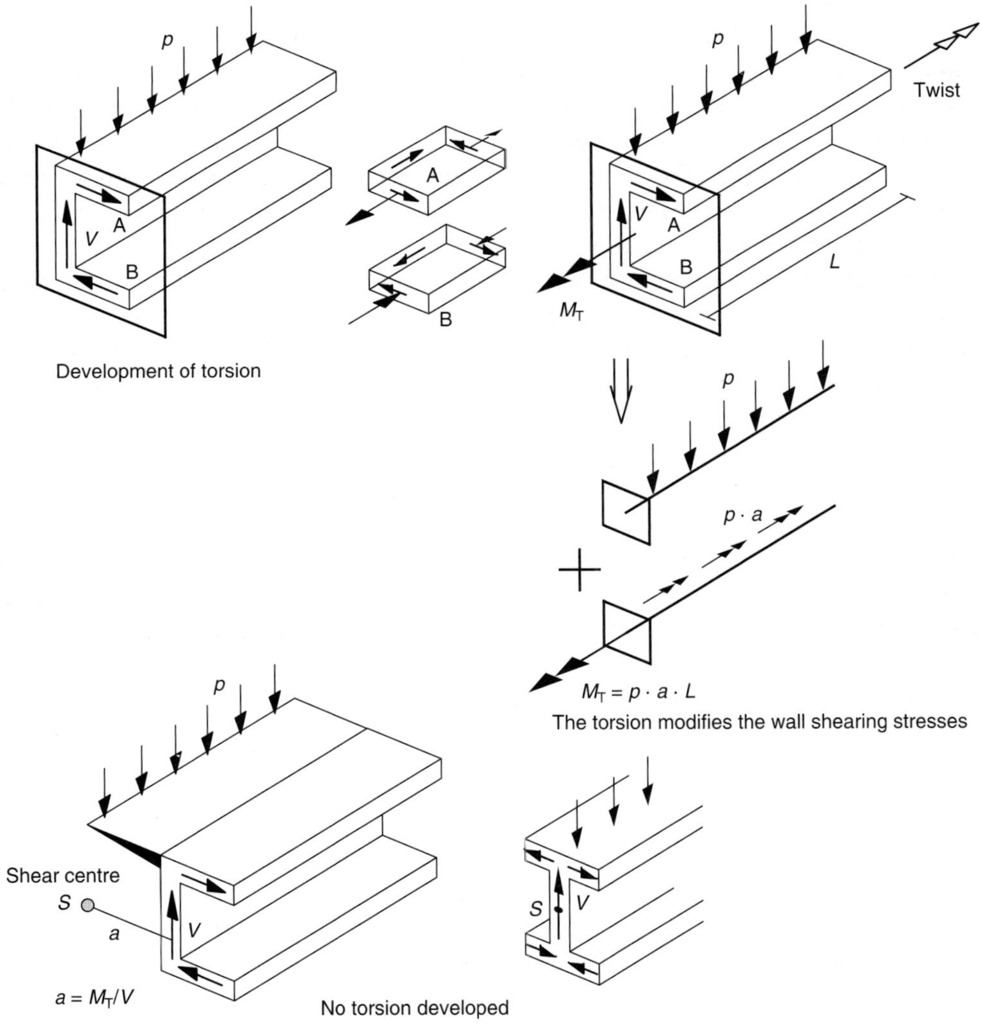

Figure 13.7 The shear centre in a thin-walled beam

equilibrium of a cut-out part of the top and bottom flange that is acted on by different longitudinal normal forces requires a shear force on its internal face. The sense of the shear stresses on the plane of the cross-section is obtained, based on the above, according to Cauchy's theorem (see Section 4.1.1).

If, now, the beam is considered as a free body under the distributed uniform load, the shear flow forces and the normal stresses on the fixed-end section, then the equilibrium is not satisfied because the shear forces of the two flanges produce an unbalanced moment. It is clear that to establish the equilibrium a torsional moment must be applied from the fixed end of the beam, with a vector directed towards the fixed end (see Figure 13.7). The twisting angle of each cross-section will be denoted by a vector with opposite sign.

In order to exclude the above torsion, the plane of the distributed load must be shifted to a suitable distance a on the outside, thus passing from the point S as it appears in Figure 13.7. With this loading arrangement, the three applied shear forces on the walls at the fixed cross-section will be in equilibrium with the loads, both in the vertical direction as well as with the developing moments with respect to point S, so that no torsional moment need be imposed by the fixed end to maintain equilibrium.

Thus, it is obvious that the beam does not twist. The point S is called the *shear centre* of the cross-section, and it is the point to which torsion is always referred (see Figure 13.7). Any load that passes through the shear centre does not cause torsion but only a bending deformation, whereas the action of each load, acting at a distance e from it, is equivalent to that of the same load shifted to the shear centre plus a torsional moment. This moment, when the load p is a distributed one, constitutes a distributed torsional moment $m_D = p \cdot e$ along the beam axis (see Figure 13.7), whereas when the force P is concentrated it constitutes a concentrated torsional moment $P \cdot e$. It is obvious that the development of a torsional reaction on the beam will cause an additional shear flow in the walls of the cross-section.

Clearly, if a thin-walled section has one axis of symmetry, then its shear centre must lie on this axis. It should also be pointed out that the shear centre is the point about which the cross-section rotates. In other words, it is the point of the cross-section that under pure torsion remains immovable.

The above means of 'qualitatively' searching for the shear centre (i.e. through the development of shear flows) can, in principle, be applied to any cross-section. Of course, the analytical determination of the shear centre (Zbirohowski-Koscia, 1967) is cumbersome and must be performed using suitable computer software. Software programs of similar nature may also be employed to determine the shear flow distribution in the walls of a closed cross-section.

13.4 Warping of thin-walled beams and the stress state due to its prevention

The uptake of a torsional moment by a thin-walled beam causes warping of its cross-section, as mentioned in Section 13.2. However, the different mechanisms of taking up a torsional moment in open and closed sections involve some characteristic differences in the behaviour of thin-walled beams, depending on their type, as shown below.

13.4.1 Open sections

The beam shown in Figure 13.8 having an I cross-section is examined. This is considered in free body equilibrium under two equal and opposite torsional moments, as shown below.

It is understandable that the constant torsional moment at each cross-section of the beam is taken up by the total of the shear stresses developed in each wall of the cross-section. These shear stresses vanish, of course, at the middle of each wall thickness, so that the midplane of each wall is free from shear stresses (see Section 2.5.3).

Under the influence of the torsional moments, each cross-section of the beam, including the extreme ones, will rotate about its centre of symmetry (shear centre), which means that the section flanges will move in opposite directions. As shown in Figure 13.8, the assumption that the extreme cross-sections will retain their initial plane leads to a certain shear deformation of the beam flanges, which naturally is not

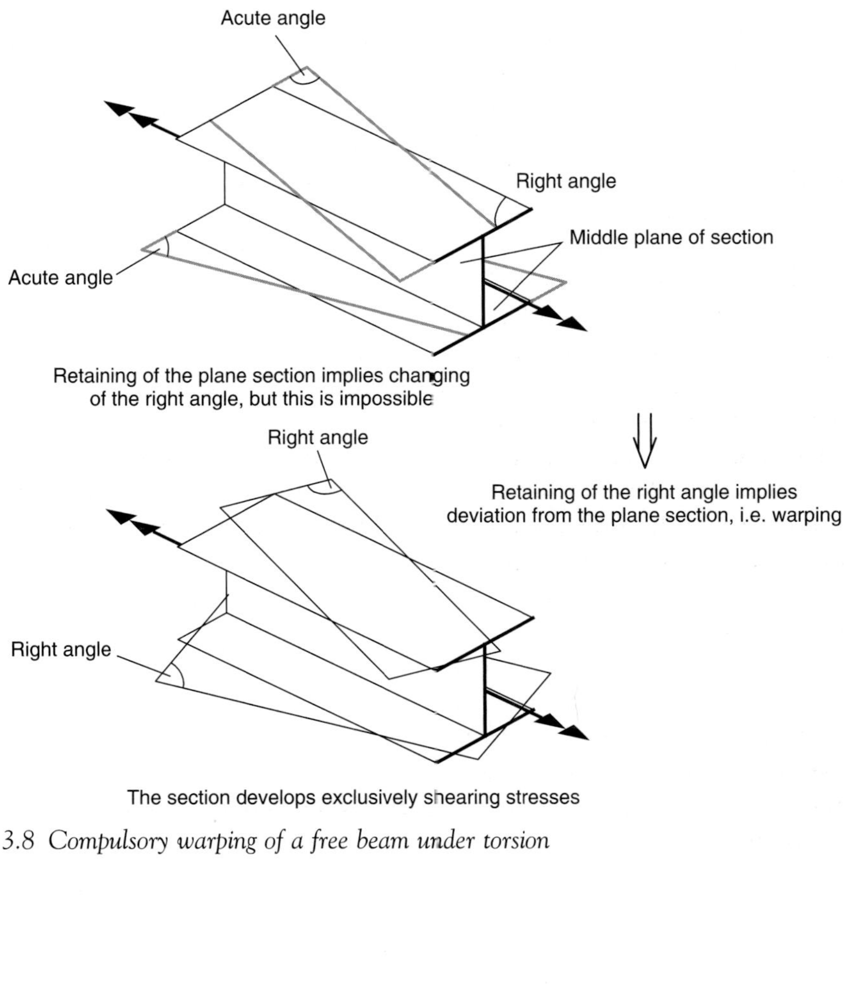

Figure 13.8 Compulsory warping of a free beam under torsion

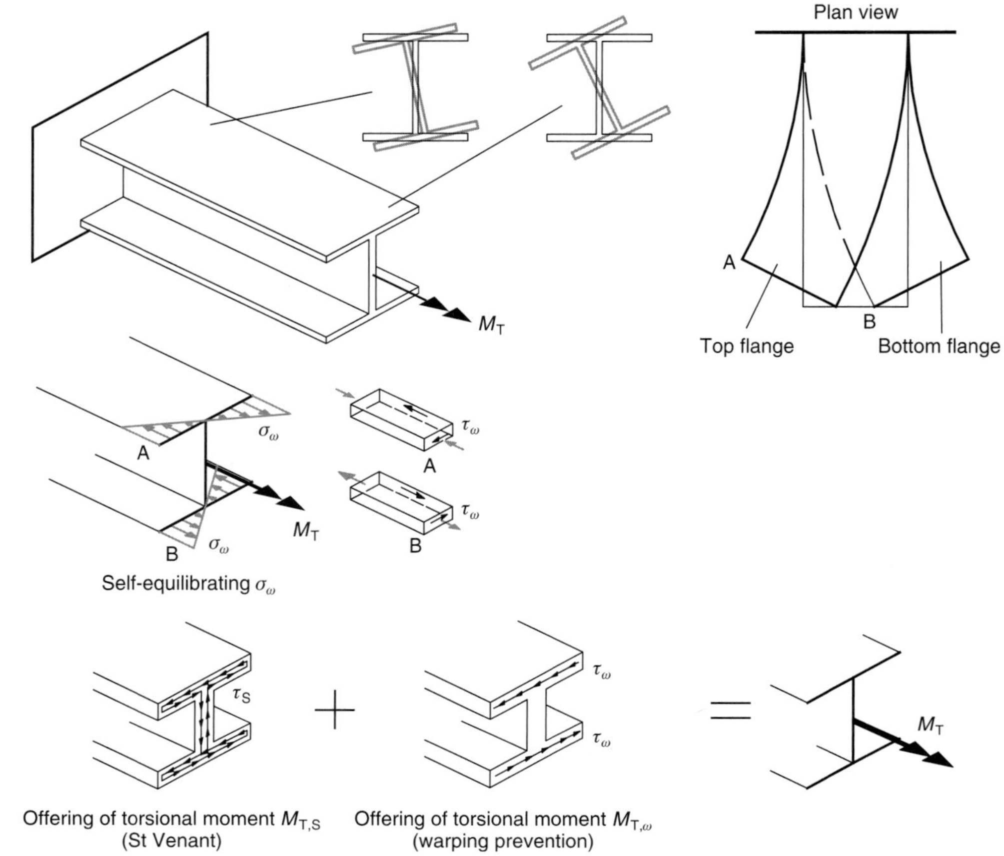

Figure 13.9 The development of the longitudinal stress state due to restrained warping

possible, given the stressless state of the midplane of the section. Hence it is concluded that the occurrence of the above shear stresses in the section walls for the uptake of torsion involves the warping of the cross-section. It is clear that no other stress needs to develop in the element examined. ■

The above stress state changes drastically if one end of the beam, say the left-hand one, is fixed while at the free end the previous torsional loading is maintained. It is understood that at the fixed end the cross-section cannot warp, and obviously remains planar (Figure 13.9).

For the stress state prevailing at the extreme free cross-section, all that has been mentioned previously is valid. The applied torsional moment M_T causes a rotating shear flow in each wall. The total developed torsional contribution of the walls according to St Venant is $M_{T,S}$. Obviously, at the extreme cross-section $M_T = M_{T,S}$, but this is not the case for any other point in the beam.

Because of the developing rotation of the cross-sections by θ, which naturally varies along the length of the beam, the two section flanges are bent within their plane in opposite directions. For the particular sense of M_T the top flange is bent to the left and the bottom flange to the right. This bending of the flanges, i.e. the developing curvature, is naturally accompanied by longitudinal stresses σ_ω that vary along the length of beam, as the curvature also varies, being maximum at the fixed end and zero at the free end, where obviously normal stresses cannot be developed. It is clear that this variation in longitudinal normal stresses σ_ω leads, for equilibrium reasons, to the development of shear stresses τ_ω in both the top and the bottom flange (see Figure 13.9).

Thus, at each position of the top or the bottom flange, a bending moment is developed corresponding to a linearly varying diagram of longitudinal normal stresses, as well as a transverse shear force in each flange for each position. It can be seen from Figure 13.9 that, while the developing longitudinal normal stresses σ_ω in the two flanges are self-equilibrating, i.e. they do not create an axial force in the cross-section, the shear forces in the two flanges offer a torsional moment $M_{T,\omega}$. This moment, together with the moment $M_{T,S}$, constitutes the imposed torsional moment M_T:

$$M_T = M_{T,S} + M_{T,\omega}$$

It is obvious that the contribution of $M_{T,S}$ to the total moment M_T of the cross-section decreases as one moves away from the free end, where of course it constitutes 100%.

It becomes clear that both the stresses σ_ω and the developing τ_ω along the length of the beam, which finally give rise to the contribution of $M_{T,\omega}$ in taking up the torsional moment M_T, owe their existence to the prevention of the cross-sectional warping at the fixed end. In general, any prevention of warping of the cross-section in places where torsion develops causes the above stresses. It is clear that the higher this distortion, the bigger the stresses, and vice versa. A detailed examination of warping is, therefore, absolutely necessary, because the consequences of its prevention constitute the only reason why the technical theory of torsion (St Venant) needs to be revised. As a matter of fact, the latter predicts the sectional distortion, but not the stresses σ_ω and τ_ω that appear due to its prevention.

13.4.2 Closed sections

Before examining warping in detail, it is important to point out that the above contribution of shear stresses τ_ω to the total torsional moment M_T applies only to open sections. The reason for this is that in closed sections the developing τ_ω constitute a self-equilibrating system of stresses (in the same manner as the stresses σ_ω) and, consequently, do not contribute to the uptake of the torsional moment of the corresponding section. Thus, in closed sections the developed shear flow, according to Bredt, is actually the only mechanism of taking up the applied torsional moment.

This is now shown (Figure 13.10) for the same cantilever beam examined previously. The applied torsional moment is considered to be introduced through a transverse diaphragm that is assumed to be very stiff in its own plane. However, the diaphragm raises no resistance to transverse displacements, such as those that accompany the warping of section. The diaphragm transmits (according to Bredt) a peripheral shear

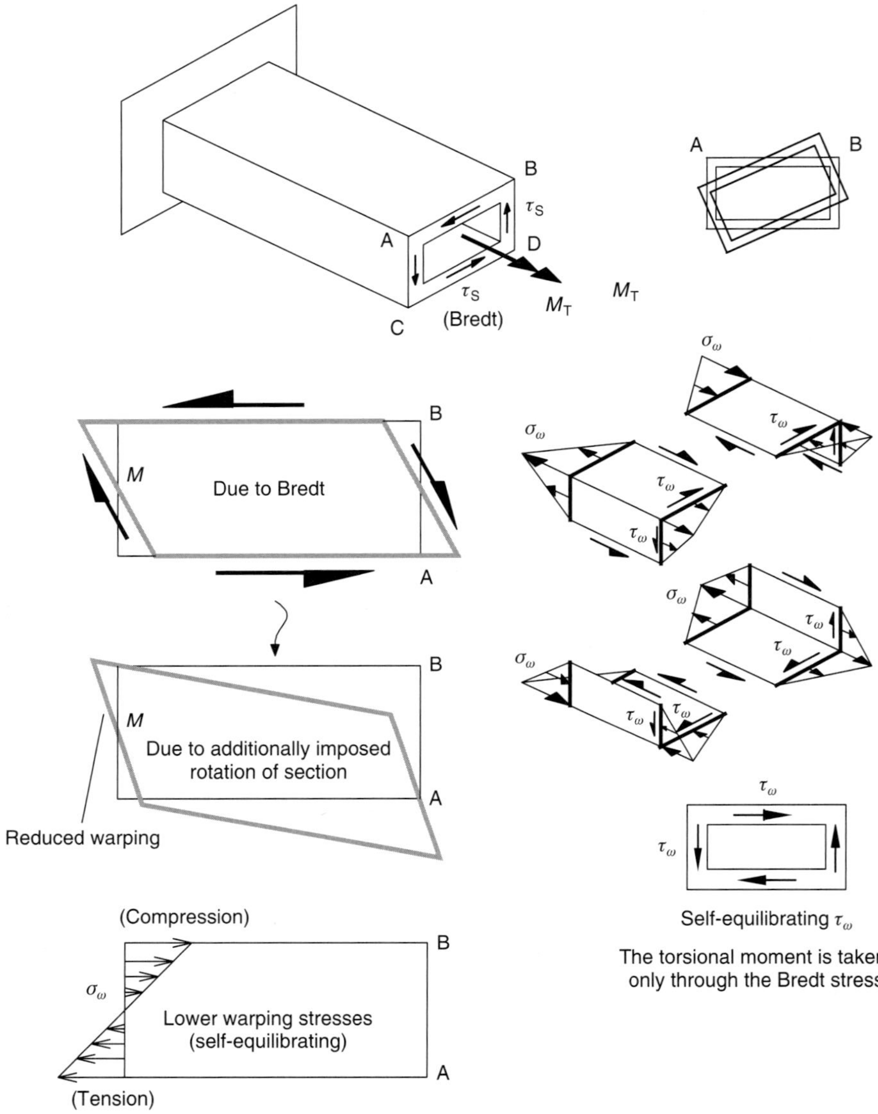

Figure 13.10 Restrained warping and the longitudinal stress state in a closed cross-section

flow to the walls of the closed section (see Section 13.1). Thus the cross-section rotates about the shear centre due to the applied torsional moment by an angle that reduces to zero at the fixed support.

To elucidate the deformation of the beam, the top plate is first considered cut out from the webs, with the midpoint M of its left side simply pinned against the horizontal displacement (see Figure 13.10). The plate is in equilibrium under the shear actions (according to Bredt) at both ends and the shear forces (according to Cauchy) on its longitudinal sides. Under the above forces the plate is deformed in shear and the two end sides acquire a slope with respect to their initial (undeformed) positions, thus causing a deviation from the cross-sectional plane and, therefore, warping. Certainly, this deviation (warping) is reduced by the imposed movement of the right-hand side of plate 'downwards', with point M being fixed, due to the anti-clockwise rotation of the cross-section of the loaded end (see Figure 13.10). Thus it is clear that the developed warping in a closed cross-section is definitely smaller than that in an open one.

It is clear that the existence of fixity does not allow the left-hand side to develop this slope (i.e. to warp), and therefore in order for the plane section to maintain planar, longitudinal stresses σ_ω must develop (see Figure 13.10) over the entire perimeter of the section, and the same applies for each individual wall of the cross-section. It is thus clear that each wall is subjected to bending. The developing longitudinal stresses are smaller than those of an analogous open section, because the warping of the closed section is reduced. Here also the stresses σ_ω constitute a self-equilibrating system, and the fact that they obviously vanish at the free end means that they keep reducing all along the length of the beam, thus giving rise to the development of shear stresses τ_ω.

Indeed, by cutting the top and bottom plates, as well as the two webs at their mid-width along some length of the beam, the need to develop longitudinal shear stresses is apparent from the longitudinal equilibrium of the free part, which in turn (according to Cauchy's theorem) produces corresponding shear stresses τ_ω on the cross-section itself. However, as can be seen from Figure 13.10, the total of τ_ω on the cross-section is self-equilibrating, and thus it does not contribute to the uptake of the torsional moment of the corresponding section, although it modifies to a small degree the existing shear stresses according to Bredt. Hence, in closed sections,

$$M_T = M_{T,S}$$

As to the length over which the stresses σ_ω and τ_ω extend, this is of the order of the cross-sectional width, given the self-equilibrating character of σ_ω and the non-deformablity of the cross-section, according to the corresponding conclusion of elasticity theory (the St Venant theorem).

As the additional stresses σ_ω and τ_ω do not particularly modify the stress state obtained by means of the technical theory, they may generally be omitted in the preliminary design stage, provided that the 'non-deformability' of the cross-section is ascertained.

13.4.3 *Analysis of warping*

If each point of the cross-section outline is characterised by its distance s on it, measured from a certain constant point, then the longitudinal displacement of a point s of the section that is found on the abscissa x of the beam may be represented by the function $u(x, s)$, while $\theta(x)$ is the twisting angle at the specific position (x) of the beam (Figure 13.11).

As is already clear from the above, the warping of a cross-section at position x results from the new positions $u(x, s)$ of its points (x, s), after the longitudinal deformation of the

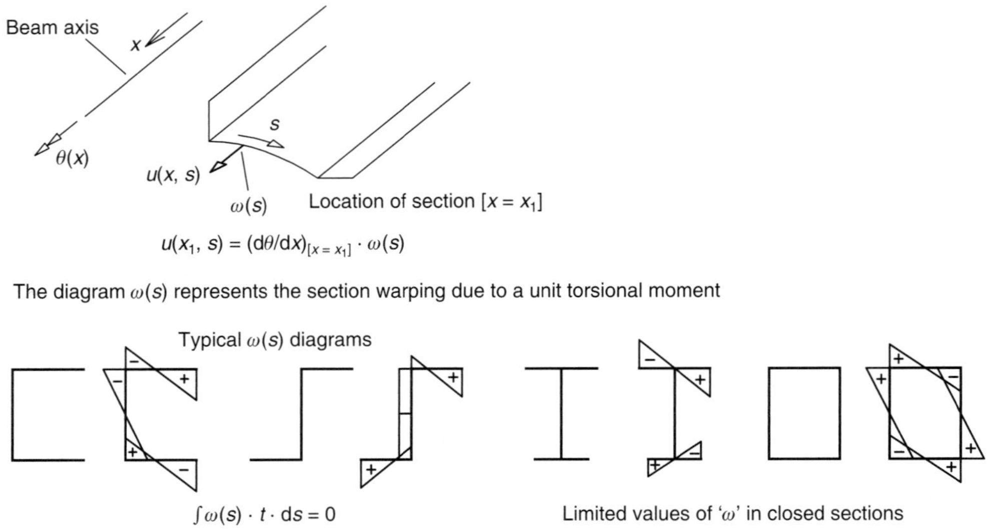

Figure 13.11 Warping behaviour of cross-sections

beam fibres. It is found that (Zbirohowski-Koscia, 1967)

$$u(x, s) = \frac{d\theta}{dx} \cdot \omega(s)$$

where the 'warping function' $\omega(s)$ is referenced at each point of cross-section outline and is determined on the basis of the position of its shear centre. Thus each thin-walled cross-section has its own characteristic diagram of the warping function $\omega(s)$. This diagram shows for each point of the cross-section how much this point will be shifted along the beam axis due to a unit change $d\theta/dx$ in the twisting angle of the beam. The diagram $\omega(s)$ (dimensions m^2) always satisfies the equation

$$\int \omega(s) \cdot t \cdot ds = 0$$

The determination of the warping diagram $\omega(s)$ for a cross-section is a cumbersome process and should be done using a specialised software program. The qualitative warping diagrams of some characteristic cross-sections are shown in Figure 13.11. ■

It should be noted that there are some thin-walled cross-sections that do not develop any warping under the torsional response, and consequently they are never subjected to the stresses σ_ω and τ_ω that arise due to the restraint of warping. The following cross-sections belong to this category (Figure 13.12):

- cross-sections having a centre of symmetry – the shear centre coincides with this point
- open cross-sections having walls passing through a common point – the shear centre coincides with this point
- all the closed cross-sections that can be circumscribed to a circle having a constant wall thickness (triangular, etc.)

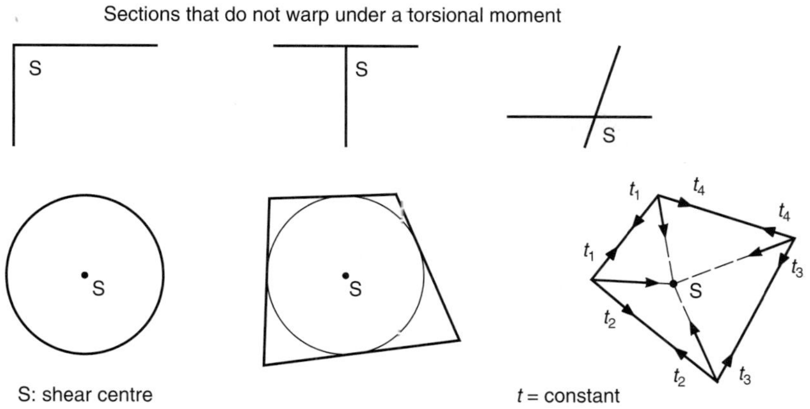

Figure 13.12 Cross-sections that do not develop warping

- more generally, if in a closed cross-section the wall thicknesses are considered as vectors and the resultants of these vectors at each corner pass through a common point, then the cross-section does not undergo warping.

13.4.4 Longitudinal stresses due to prevention of warping

It is reiterated that the longitudinal normal stresses σ_ω due to the restraint of warping are not predicted by the classical technical theory of beams, as the only normal stresses that appear therein refer either to bending moments or to axial forces. In each case the developed stresses are statically equivalent to the occurring bending moment or axial force. As the stresses σ_ω are self-equilibrating, they do not have any static equivalence to any force or moment, i.e. they do not contribute to the global equilibrium of the structure. However, it has been shown that they may be expressed in the same way as the normal stresses in technical theory, through a sectional magnitude M_ω called the *bimoment* (Figure 13.13).

In absolute analogy with the technical theory of bending, it is found that for each point of the cross-section the longitudinal warping stress can be expressed as (Vlasov, 1961)

$$\sigma_\omega = \frac{M_\omega}{I_\omega} \cdot \omega(s)$$

where

$$I_\omega = \int_A \omega^2 \cdot dA$$

The magnitude I_ω (dimensions m^6) is called the *warping constant*, and is obtained by integrating over each elementary part of the surface dA of the cross-section. It is understood that $\omega(s)$ plays the role of 'distances from the neutral axis', as is the case in pure bending. The similarity with pure bending is obvious. It is observed that, for a given value of the bimoment, the distribution of σ_ω is identical to the distribution of $\omega(s)$.

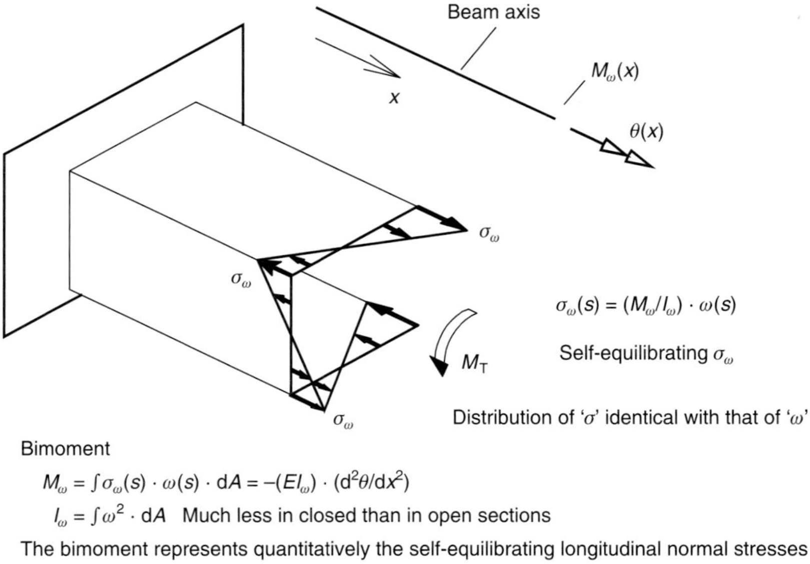

Figure 13.13 Size and significance of the bimoment

The analytical determination of I_ω is very cumbersome and should be performed using specialised computer software.

The bimoment itself (dimensions kN m^2), the physical meaning of which is covered in the following two sections, is defined, analogously to the bending moment in technical theory, through the expression

$$M_\omega = \int_A \sigma_\omega(s) \cdot \omega(s) \cdot dA$$

If $\theta(x)$ is the function of the twisting angle along the length of the beam, then

$$M_\omega = -EI_\omega \cdot \frac{d^2\theta}{dx^2}$$

As closed cross-sections exhibit a clearly smaller warping compared to open cross-sections, they clearly also have a much more 'moderate' warping function $\omega(s)$, and consequently their warping constant I_ω is comparatively smaller. Thus in closed cross-sections the normal stresses σ_ω are limited.

13.5 The bimoment concept

In a thin-walled beam the bimoment concept is directly related to the self-equilibrating system of the normal stresses that arise from the prevention of warping of the cross-section under a torsional moment. It is clear that preventing a constructional warping of even a single section (e.g. the end of one beam) is sufficient to exert a restraining effect on the warping of all other sections of the beam. It should also be understood that this prevention of warping activates the bending rigidity of the section walls, and

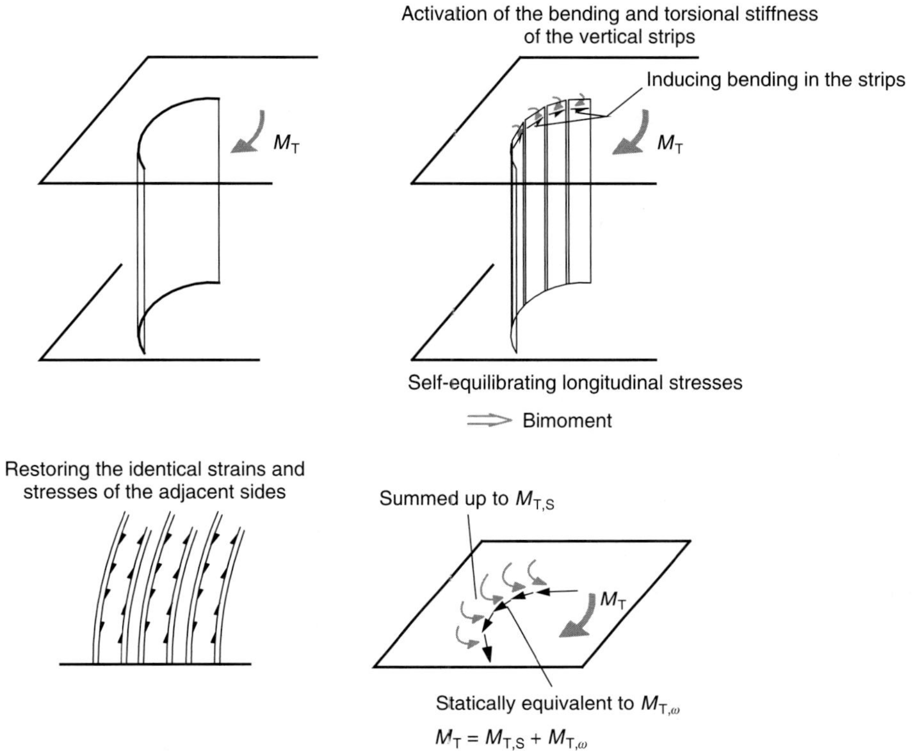

Figure 13.14 An example of the physical meaning of the bimoment

automatically causes longitudinal normal stresses that are not negligible, being in any case obligatory in the inherent self-equilibrium along the longitudinal sense, given that no external axial force is present. It may be considered that these normal stresses represent the bimoment itself.

The fact that a torsional moment generally causes bending, thus giving rise to the concept of the bimoment, besides what was examined in Section 13.4.1, is illustrated in the model shown in Figure 13.14. The vertical beam with an open profile is fixed at its bottom end, while the top end is rigidly connected to a horizontal diaphragm that is non-deformable within its own plane and has no transverse bending rigidity. A torsional moment M_T is applied on the diaphragm.

The beam may initially be considered as being composed of mutually independent vertical plane strips that exhibit a bending rigidity only within their plane and have a torsional rigidity according to St Venant. The forced movement of the horizontal diaphragm under the torsional moment, consisting of a displacement and a rotation, causes a shear force and a torsional moment on each wall segment according to its own bending and torsional rigidity, respectively. The horizontal diaphragm is in equilibrium under the externally applied torque, as well as the reacting shear forces and torsional moments of the walls, which are acting in opposite senses. It is clear that these shear forces will amount to a total moment $M_{T,\omega}$ because, for equilibrium reasons, it is impossible to offer any other force. $M_{T,\omega}$, together with the summation of the St Venant moments $M_{T,S}$, will then balance the moment M_T:

$$M_T = M_{T,\omega} + M_{T,S}$$

The development of a shear force at the top of each wall segment causes bending, leading to axial normal stresses. These stresses are obviously self-equilibrating, i.e. they have no resultant along the longitudinal direction, as no axial force is acting on the segment. This situation will not change even when a self-equilibrating system of longitudinal shear forces (see Figure 13.14) is imposed along the free sides of the wall segments in order to restore the incompatible deformations occurring there. This set of self-equilibrating axial stresses is actually equivalent to the existence of a bimoment.

Returning now to the last equation, it is again pointed out that a torsional moment M_T is composed of a St Venant moment $M_{T,S}$ and a warping moment $M_{T,\omega}$. It is understandable that the relationship between these two contributions depends not only on the cross-section itself but also on the length of the beam. For example, as the length increases, the ratio of the bending to the torsional rigidity for each wall segment decreases, and consequently the percentage contribution of $M_{T,\omega}$ with respect to $M_{T,S}$ in taking up the torsional moment M_T decreases too.

13.6 Two theorems of the bimoment

Theorem 1

A force P parallel to the beam axis and applied to a point on the cross-section having a warping value ω causes a bimoment equal to $M_\omega = P \cdot \omega$ (Figure 13.15). (The compressive loads are considered to be negative.)

In the beam with an I cross-section, where three self-equilibrating forces are applied ($N = 0$), a bimoment equal to $M_\omega = 2 \cdot (P/2) \cdot \omega_1$ is developed. This bimoment obviously causes axial stresses according to the basic equation given in Section 13.4.4, contrary to what would be predicted using the technical theory.

The equal and opposite compressive forces P on the beam in Figure 13.15, which can be considered as prestressing forces, cause a positive bimoment $M_\omega = P \cdot \omega_1$, ($P < 0$, $\omega_1 < 0$) and, consequently, tensile normal stresses at the ends of the cross-section ($\omega > 0$), contrary to what would be concluded according to the technical theory. However, a uniform compressive stress p_0 applied on the entire the cross-section does not cause a bimoment, because (see Section 13.4.3)

$$M_\omega = \int p_0 \cdot dA \cdot \omega = p_0 \cdot \int \omega \cdot dA = 0$$

and, therefore, the whole section is acted on by the above compressive stress.

Theorem 2

A moment M applied on a plane parallel to the beam axis at a distance e from the shear centre causes a bimoment $M_\omega = M \cdot e$. (The bimoment is considered positive if the vector of the applied moment M is directed towards the shear centre.)

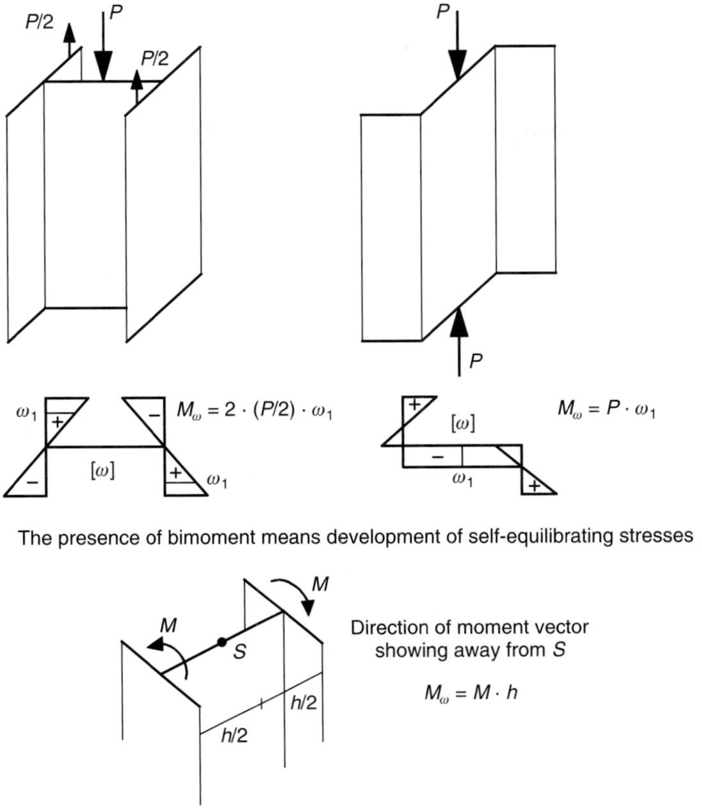

Figure 13.15 The development of a bimoment in the absence of a torsional load

According to the example shown in Figure 13.15, even though the two externally applied moments M on the beam are in equilibrium, they cause a bimoment

$$M_\omega = -M \cdot (h/2) - M \cdot (h/2) = -M \cdot h$$

and accordingly they give rise to a longitudinal stress response.

13.7 Warping shear stresses

As explained previously, the warping shear stresses τ_ω are required in order to ensure the equilibrium, given the variation in the normal stresses σ_ω along the length of the beam. As stated, their distribution over the cross-section is not self-equilibrating, as is the case for σ_ω, but leads to the warping torsional moment $M_{T,\omega}$. It should be noted, however, that this particular effect occurs only in open sections. As shown in Section 13.4.2, the stresses τ_ω in closed sections constitute a self-equilibrating system, which affects the distribution of shear stresses according to Bredt. However, it does not contribute to the uptake of the torsional moment M_T, which is taken up exclusively by the mechanism of Bredt's peripheral shear flow.

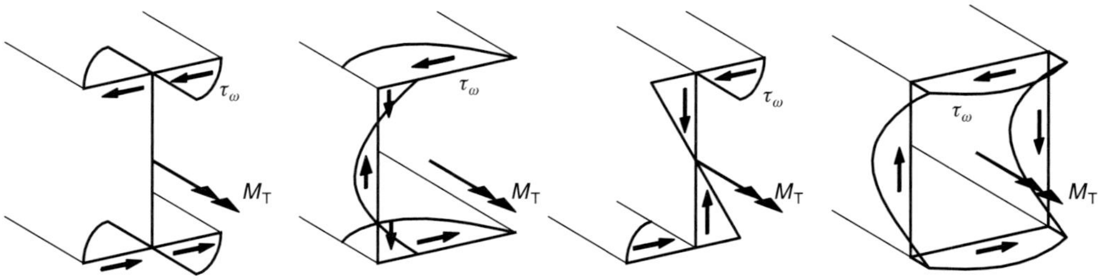

Figure 13.16 Shear stresses due to the prevention of distortion

It should be mentioned here that a warping moment $M_{T,\omega}$ is also present in closed cross-sections, and this leads to self-equilibrating shear stresses τ_ω, as explained previously.

The analytical determination of the distribution of stresses τ_ω in the cross-section (open or closed) based on the moment $M_{T,\omega}$ is generally cumbersome, and for this reason specialised computer software must be used. The distribution of the warping shear stresses τ_ω in some typical cross-sections is shown in Figure 13.16.

13.8 The governing equation for torsion and its practical treatment

Consider a suitably supported rectilinear thin-walled beam that is subjected to a distributed torsional load m_D, and, possibly, to a concentrated torque T as well. The beam is referenced to axes with an abscissa x, as shown in Figure 13.13.

The characteristic deformation magnitude on the basis of which all the torsional magnitudes of the beam are expressed is the rotation $\theta(x)$. Thus it is found that the warping torsional moment $M_{T,\omega}$ is expressed as

$$M_{T,\omega} = -EI_\omega \cdot \frac{d^3\theta}{dx^3}$$

while the torsional moment according to St Venant $M_{T,S}$ is

$$M_{T,S} = G \cdot I_D \cdot \frac{d\theta}{dx}$$

Hence the equation established in Section 13.4 ($M_T = M_{T,\omega} + M_{T,S}$) may be written as

$$M_T = G \cdot I_D \cdot \frac{d\theta}{dx} - EI_\omega \cdot \frac{d^3\theta}{dx^3}$$

and, given that the equilibrium relation is, in effect,

$$\frac{dM_T}{dx} = -m_D$$

the last equation takes the form (Vlasov, 1961)

$$EI_\omega \cdot \frac{d^4\theta}{dx^4} - G \cdot I_D \cdot \frac{d^2\theta}{dx^2} = m_D$$

It should be pointed out that this equation is valid for both open and closed cross-sections, despite the fact that in closed cross-sections the size of $M_{T,\omega}$ leads, as previously mentioned, to self-equilibrating shear stresses τ_ω. It is, however, possible that the last result may be also obtained from the equation

$$\tau = \tau_{T,\omega} + \tau_{T,S}$$

which is valid for both open and closed cross-sections.

Moreover, as the bimoment M_ω in Section 13.4 is expressed as

$$M_\omega = -EI_\omega \cdot (d^2\theta/dx^2)$$

the following relationship between the bimoment and the warping torsional moment is obtained directly from the above:

$$M_{T,\omega} = \frac{dM_\omega}{dx}$$

This above equation is the fundamental relationship that governs the torsional behaviour of thin-walled cross-sections and, of course, in order to determine all the above torsional magnitudes in terms of the function $\theta(x)$, the satisfaction of the pertinent respective boundary conditions is required. Thus, the following boundary conditions are distinguished:

(1) In the case of a fully fixed cross-section at position x

$$\theta = 0$$

$$d\theta/dx = 0$$

(2) In the case where twisting of the cross-section, but not its warping, is prevented, which means that in the considered cross-section no longitudinal warping stresses σ_ω can be developed, then $\theta = 0$ and $M_\omega = 0$, and on the basis of the previous equations

$$\theta = 0$$

$$d^2\theta/dx^2 = 0$$

(3) In the case of a completely free cross-section $M_\omega = 0$ and $M_T = 0$, and therefore on the basis of previous equations

$$d^2\theta/dx^2 = 0$$

$$G \cdot I_D \cdot \frac{d\theta}{dx} - EI_\omega \cdot \frac{d^3\theta}{dx^3} = 0$$

The boundary conditions for other cases can be worked out based on the above. Thus, for example, when in the place of the internal support of a continuous beam instead

Table 13.1 Analogy between the magnitudes in the bending and the torsional response

Beam in bending		$\longleftrightarrow$	Beam in torsion	
Displacement	w	$\longleftrightarrow$	Rotation angle	θ
Tensile force	H	$\longleftrightarrow$	Torsional rigidity	$G \cdot I_T$
Moment of inertia	I	$\longleftrightarrow$	Warping constant	I_ω
Uniform load	p	$\longleftrightarrow$	Uniform torsional moment	m_D
Concentrated transverse load	P	$\longleftrightarrow$	Concentrated torque	T
Vertical component of the tensile force	H_v	$\longleftrightarrow$	Torsional moment according to St Venant	$M_{T,S}$
Bending moment	M	$\longleftrightarrow$	Bimoment	M_ω
Shear force (vertical)	Q	$\longleftrightarrow$	Torsional moment	M_T

$\overrightarrow{M_{T,\omega}} = \overrightarrow{V} - \overrightarrow{H_v}$ (see Figure 13.17).

two beams are linked, the arranged bearings, with or without a diaphragm placed above the support, prohibit the end rotation, and thus $\theta_{(1)} = \theta_{(2)} = 0$, while due to the continuity of normal stresses σ_ω the equality of bimoments, $M_{\omega(1)} = M_{\omega(2)}$ must also be required at the same position. ■

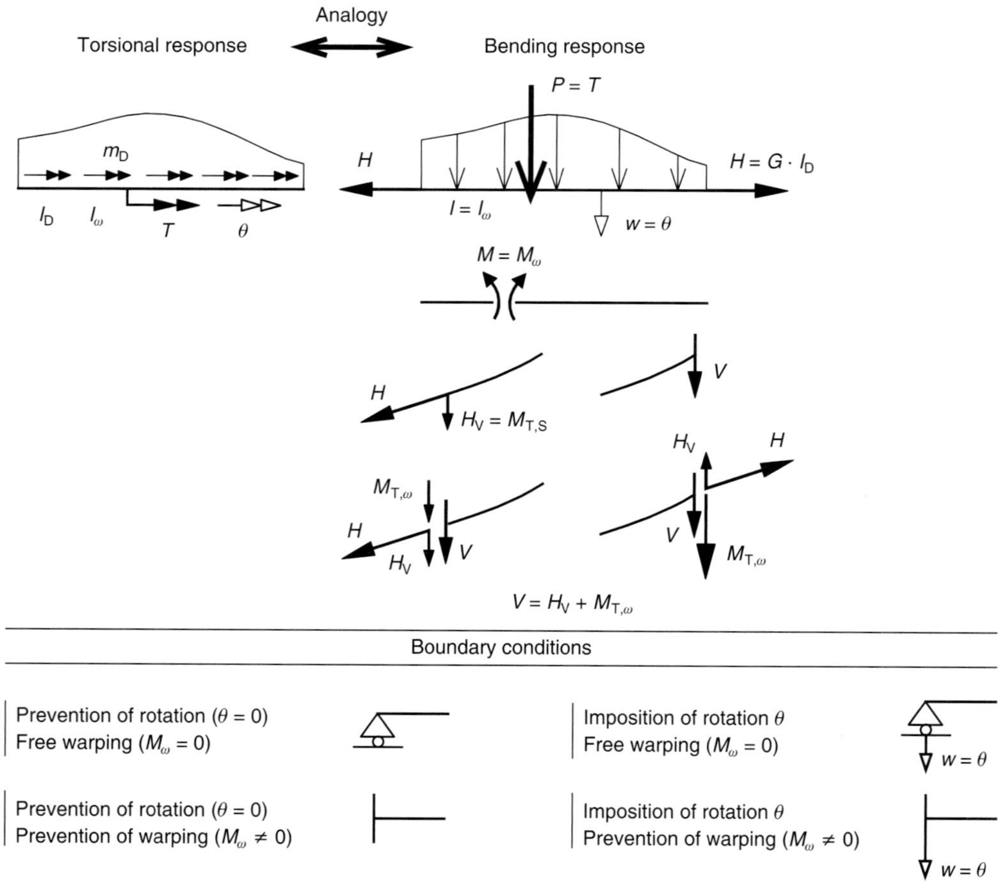

Figure 13.17 The analogy between the torsional and the bending stress states of a tensioned beam

Although the solution of the above basic equation governing the torsional stress state of a beam through the classical analytical treatment of differential equations is, in principle, feasible, it is very awkward and impractical, especially for more complex cases. However, it should be noted that the form of the equation in question is precisely the same as the equation for a tension beam under transverse loading according to the second-order theory, as already examined in Section 7.2 (Roik, 1983). This equation is

$$EI \cdot \frac{dw^4}{dx^4} - H \cdot \frac{dw^2}{dx^2} = p(x)$$

This observation is of crucial importance because it establishes a complete equivalence between the two problems, and this allows the torsional problem to be confronted using the conceptually much simpler problem of beam bending, for which many accessible software programs exist. This equivalence, and the corresponding dimensions, is illustrated in Table 13.1 and Figure 13.17. It is clear that all the boundary conditions of the problem that are expressed through the specific requirements for the rotation angle θ, the bimoment M_ω or the torsional moment M_T, may be attributed to the corresponding terms for the beam in bending (see Figure 13.17). This result provides an overall and naturally more direct illustration of the torsional problem.

13.9 Examples

Example 1

The aim is to determine the torsional stiffness of a cantilever of length 10.0 m, i.e. to determine the torsional moment required at its free end that will cause a rotation $\theta = 1$ rad. The behaviour of both an open and a closed cross-section, having the same external dimensions and the same cross-sectional area, is examined, considering in each case the possibility of preventing or not the warping of the free end. It is assumed that the profile of the cross-section remains undeformable.

Figure 13.18 shows the details of approaching the torsional behaviour through the established analogy with the beam in bending based on the second-order theory (see Section 13.8). The torsional stiffness is determined as the shear force that develops due to an imposed support settlement on the bending model of the beam, after the assumed twisting angle at the free end.

It can be seen that in the open section the prevention of warping at the free end increases considerably the torsional rigidity, which is significantly higher than the value predicted according to the classic expression after St Venant ($G \cdot I_D/L$). In the closed cross-section these differences are much smaller and thus have a much lesser practical importance.

Example 2

Consider the stress state of a simply supported concrete bridge, with a span of 40 m and an open cross-section. The bridge is loaded at the midspan with an eccentric load of 1000 kN applied directly on the top of one web. Assume that the profile of the cross-section remains undeformable due to a suitable insertion of transverse diaphragms (Figure 13.19).

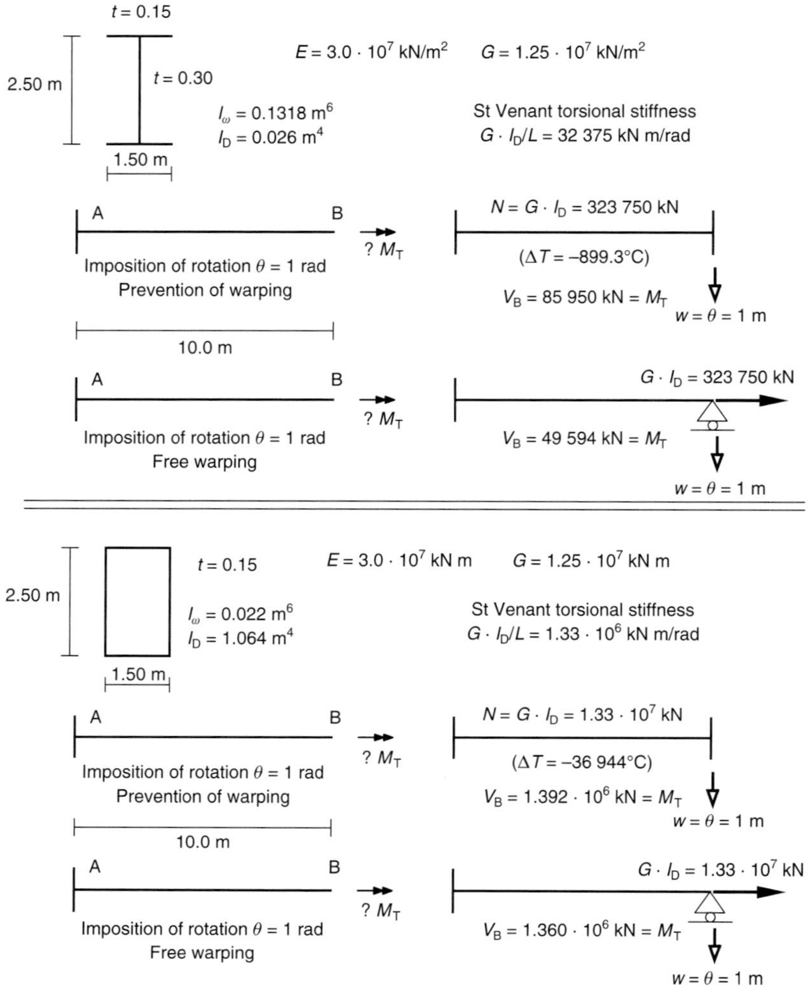

Figure 13.18 Determination of the torsional stiffness of a thin-walled beam

The load is analysed exactly as described in Section 13.1, i.e. by considering a symmetric and an antisymmetric load. For the sake of later comparison, it is noted that the symmetric load, which consists of the two loads of 500 kN, causes at the midspan a maximum tensile stress of 6300 kN/m^2. In this respect the interest is, of course, focused on the antisymmetric loads of 500 kN (see Figure 13.19), a loading that causes a corresponding torsion. It is supposed that the layout of the bearings can offer to the beam the required reactions for the development of the torsional moment of 1425 kN m at its ends. The analysis is done according to the analogy established in Section 13.8, and is shown in Figure 13.19.

The resulting bending moment of 18 946 kN m at the midspan represents the bimoment, and this is used as the basis for determining the longitudinal warping stresses, as detailed in Section 13.4.4. Based on the corresponding diagram [ω], the greatest tensile stress is 4121 kN/m^2, i.e. 65% of the tensile stress caused by bending. The contribution of $M_{T,S}$,

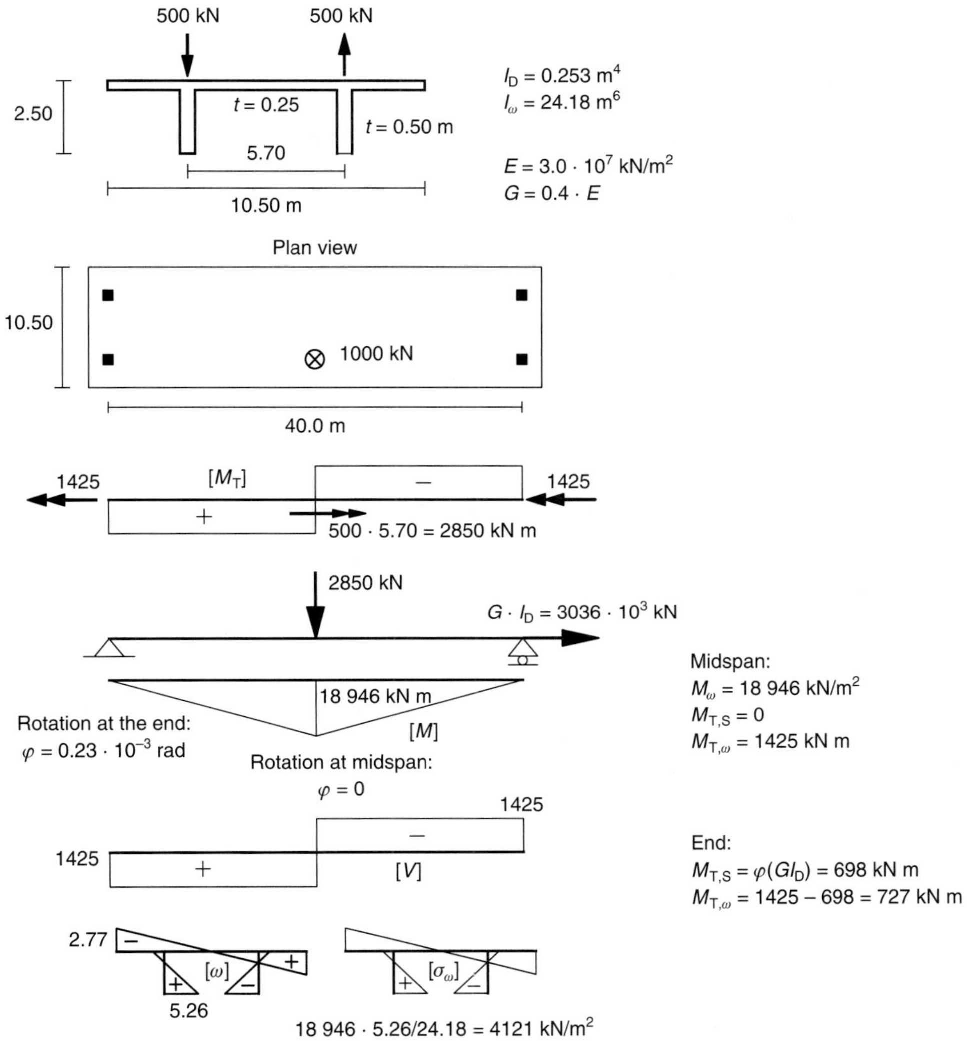

Figure 13.19 Determination of the stress state in a thin-walled girder due to an eccentric load

i.e. of the St Venant shear flow, in taking up the torsional moment M_T is determined according to Section 13.8. As shown in Figure 13.19, this contribution is zero at the midspan, while at the ends it amounts to 49%. The corresponding percentages for the warping torsional moment $M_{T,\omega}$ are 100% and 51%, respectively.

References

Roik K. (1983) *Vorlesungen über Stahlbau. Grundlagen*. Berlin: Wilhelm Ernst.

Vlasov V.Z. (1961) *Thin Walled Elastic Beams*. London: Oldbourne Press.

Zbirohowski-Koscia K. (1967) *Thin Walled Beams – From Theory to Practice*. London: Crosby Lockwood.

14
Box girders

14.1 General

Box girders, used mainly for bridges, are thin-walled beams having a closed section, and, as such, they obey everything in the previous chapter, on one condition: in order for the basic equation as well as for all the related concepts (bimoment, etc.) concerning torsional response to be valid, the undeformability of the box section must be ensured. This condition, for beams with relatively small-section dimensions, can be guaranteed by a certain wall thickness, but for larger dimensions, stiff diaphragms are required over each support, as well as at two or three places along the span. If these diaphragms are not desirable for constructional reasons, then the uptake of torsional loads induces deformation of the section, and, consequently, an additional bending response in the longitudinal as well as in the transverse sense is developed. In this chapter, rectilinear beams are first examined, followed by curved beams.

14.2 Rectilinear girders

14.2.1 General loading case

In the typical box section shown in Figure 14.1, an eccentric layout of the traffic load is considered. This is specified in each case by the load standards followed.

It is appropriate to consider those external actions applied to the nodes A and B of the upper plate, which eliminate both their displacement and rotation. These 'fixing actions' (see Section 3.3.1) consist of longitudinally distributed vertical loads q_A and q_B, as well as longitudinally distributed moments m_A and m_B, which are determined directly from the fixed-end actions on the assumed clamped nodes A and B.

Thus, following Section 3.3.1, it can be considered that the state of stress of the girder results from the superposition of the fixed state I containing the loads with the appropriate nodal actions and of the state II containing only these nodal actions q_A and m_A, together with q_B and m_B, acting, however, in the opposite sense.

The fixed state I, as a trivial state, is directly determinable, so that state II is predominantly considered. This state – provided the section is symmetric – can always be split into symmetric and antisymmetric parts. The symmetric part consists of the symmetrically distributed loads $q_R/2$ and moments $m_R/2$ applied to the edges A and B, respectively. The antisymmetric part consists of the mutually opposite loads $q_S/2$, as well as the moments in the same sense, $m_S/2$ (see Figure 14.1).

Copyright Thomas Telford Limited © 2010
All rights reserved

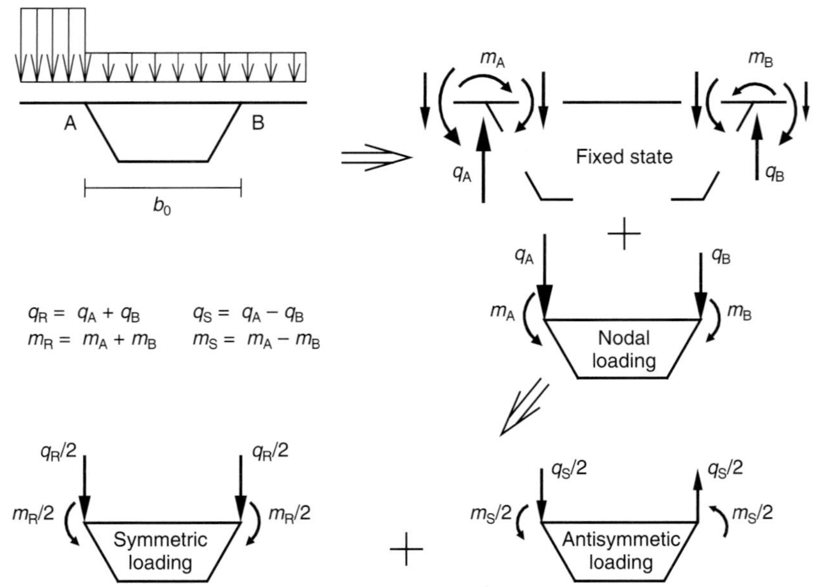

Figure 14.1 Derivation of nodal actions on a beam profile from the acting eccentric loading

The symmetric loading activates the longitudinal bending stiffness of the girder, and results in the well-known sectional force diagrams for bending moments and shear. Moreover, it causes a transverse response, if the equilibrium of a segmental part of unit length is considered. This girder strip is acted on by the resultant of the shear flows applied to each of its two faces (Figure 14.2). Given that the shear flow v is expressed as $v = V \cdot S/I$, then $dv/dx = (S/I) \cdot (dV/dx) = (S/I) \cdot q_R$ so that the above resultant is exactly the differential shear flow dv acting on the strip, equal to $dv = (S/I) \cdot q_R$. It should be noted that S represents the static moment of inertia of the cut-out part of the section being considered,with respect to its centroidal axis.

Thus, the girder strip is in equilibrium under the symmetric loads $q_R/2$, as well as under $m_R/2$, and the differential shear flow dv. The loads $q_R/2$ and dv do not induce

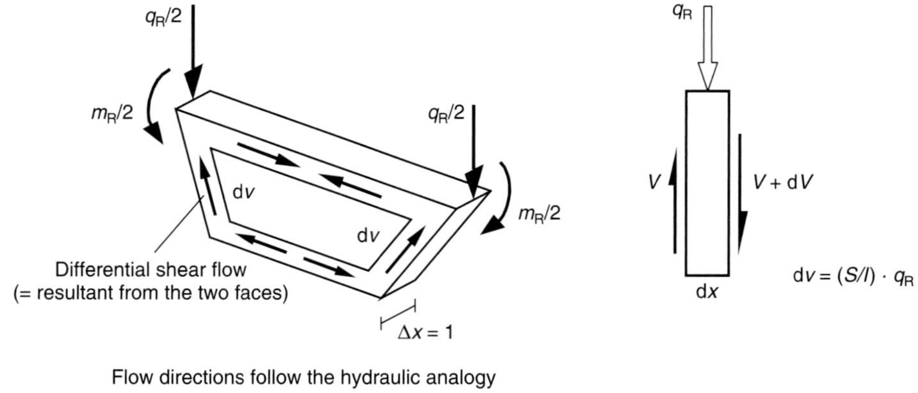

Figure 14.2 Transverse response under symmetric loading

any bending at all, but only axial stresses. The loads $m_R/2$ cause both bending and axial stresses (see Figure 14.2). This state of stress may be determined using classic computer software for plane frames. ■

When speaking of the symmetric loading of a box girder, two points should be made regarding haunched bridge girders either in a cantilevered system or, analogously, in a continuous one over its internal supports – in addition to what has already been mentioned in Sections 4.4 and 5.1 – mainly with respect to the free cantilevered method of construction.

The first remark concerns the variation in the tensile axial forces Z in the top slab of a cantilever, which is associated with the longitudinal shearing flow v, as shown in Figure 14.3. As explained above, the inclined compressive force of the bottom slab helps with its vertical component, in offering to the cross-section the required resistance to the induced vertical shear force, thus achieving, for a parabolic curving of the bottom slab, an approximately constant shear flow throughout the cantilever length. By cutting out from the web an elementary segment of the top slab subjected to the axial tensile forces, as well as to the acting shear flow, and considering its horizontal equilibrium,

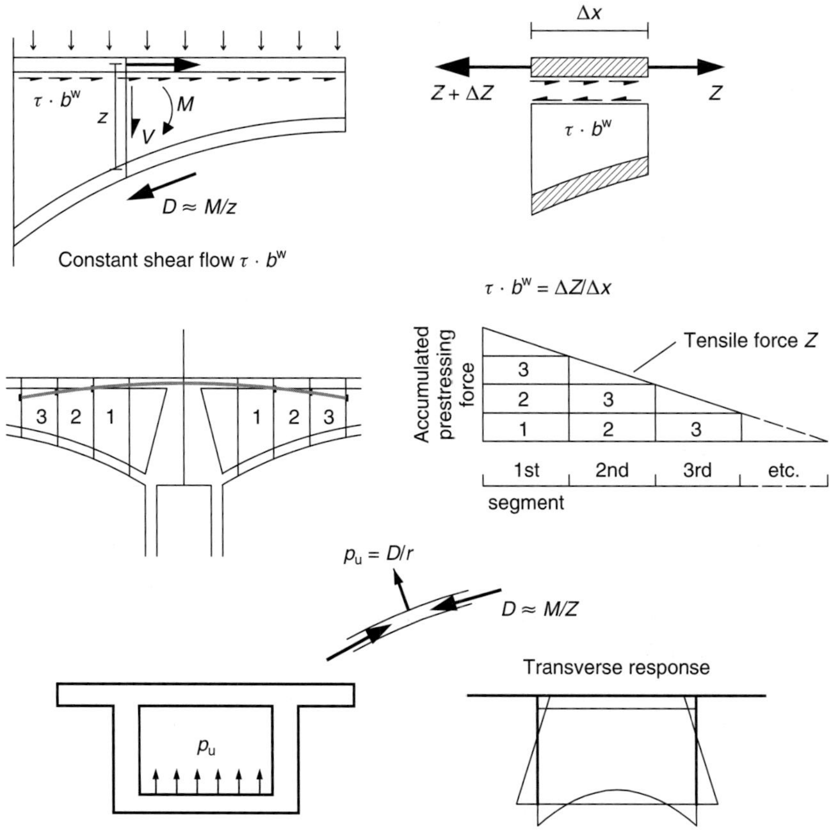

Figure 14.3 Structural behaviour of a cantilevered haunched girder

it is concluded that

$$v = \tau \cdot b^{w} = \frac{\Delta Z}{\Delta x}$$

where

$$\tau \cdot b^{w} = \frac{1}{z} \cdot \left(V - \frac{M}{z} \cdot \frac{\Delta z}{\Delta x} \right) \quad \text{(see Section 4.4)}$$

The above relations are quite generally valid. In the case of a constant depth ($\Delta z/\Delta x = 0$), the tensile force Z exhibits a parabolic variation along the girder ($\Delta z/\Delta x = V/z$). However, in the case of a variable depth leading to a constant flow v, the tensile force Z is linearly varied, which, with regard to the balanced cantilevered construction for continuous bridges, mentioned in Section 5.1 and suggested in Figure 14.3, means that an essentially constant amount of prestressing 'dosage' needs to be added at each consecutive construction stage, thus representing a constructional advantage. It should be pointed out that the prestressing reinforcement for the free cantilevered process (see also Section 5.5.2) is placed principally on the top slab of the girder section.

The second remark regards the fact that the ever-changing inclination of the compression force of the bottom slab, due to bending, creates an inward deviation pressure $p_{u} = D/r$ (see Figure 14.3), which stresses the bottom slab in the transverse sense, obviously counteracting its self-weight, but necessarily affecting the whole closed transverse segmental strip in bending, as shown in Figure 14.3. ■

As the symmetric loading does not present any other peculiarities, the whole problem is shifted to the antisymmetric loading part. This acts as a distributed torsional load, $m_{D} = (q_{S}/2) \cdot b_{O} + m_{S}$ (see Figure 14.1), resulting in a particular torsional moment diagram (Figure 14.4). At each position of the beam, the torsional moment developed is taken up by the corresponding Bredt shear flow. Thus, for each wall of the section considered in its longitudinal sense, a shear force diagram can be assigned.

If, now, the non-deformability of sections can be ensured through the layout of transverse diaphragms, as mentioned previously, an additional longitudinal response consisting of axial and shearing stresses will take place, according to Chapter 13. The fact that the section is closed leads to a limited response arising from the restraining of the limited warping of the section.

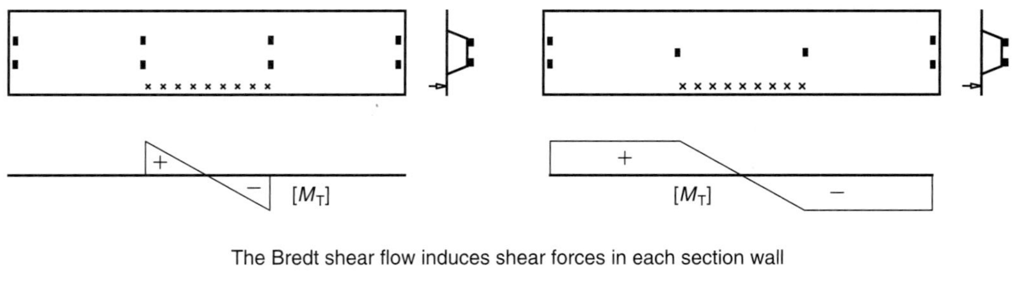

Figure 14.4 Influence of the support layout on the distribution of the torsional response

However, the presence of the transverse diaphragms is generally not desirable for constructional reasons. Thus, the girder section must be considered as deformable, the development of torsional response is no longer governed by the basic equation of Section 13.8, and, consequently, the structure has to be considered, accordingly, as a folded system. In this case, as will be examined in the following section, the above-mentioned additional response to the shear flow, according to Bredt, consists of a longitudinal bending for each wall, as well as of a transverse response for the section profile.

14.2.2 Response due to the deformability of the profile section

The imposed torsional loading ($m_D = (q_S/2) \cdot b_0 + m_S$), due to the antisymmetric couple, causes deformation of the closed section, causing longitudinal bending of the section walls (Figure 14.5), which is coupled with the resulting transverse bending of the section profile itself (Schlaich and Scheef, 1982). It is again pointed out that this response is in addition to the initial Bredt shear flow in the section walls.

In order to now investigate the influence of the deformability of the section on the response, the equilibrium of a cut-out girder strip of unit length is first considered with the antisymmetric loads $q_S/2$ and $m_S/2$ acting at the section edges A and B (see Figure 14.5). It is clear that the segment is in equilibrium under the above external

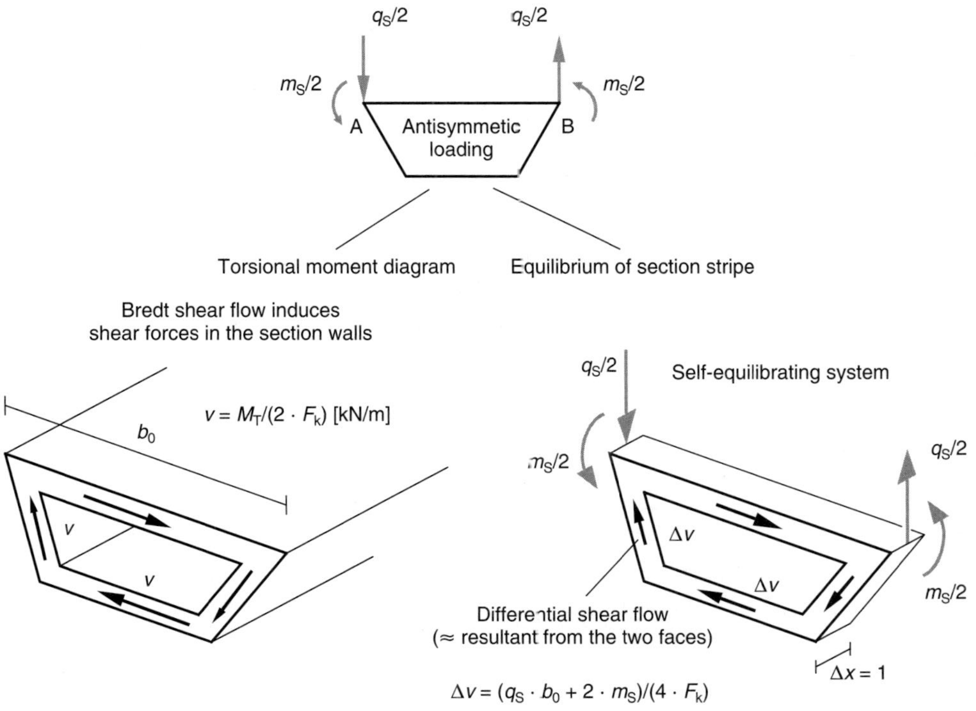

Figure 14.5 Analysis of antisymmetric loading

forces and the differential shear flow Δv, which is obtained as the resultant of the Bredt shear flows on the two faces of the strip considered.

Given that

$$v = \frac{M_T}{2 \cdot F_k}$$

then

$$\frac{\Delta v}{\Delta x} = \frac{\Delta M_T}{\Delta x} \cdot \frac{1}{2 \cdot F_k} = \frac{m_D}{2 \cdot F_k}$$

and, consequently,

$$\Delta v = \frac{q_S \cdot b_0 + 2 \cdot m_S}{4 \cdot F_k}$$

According to the section dimensions given in Figure 14.5, $F_k = d \cdot (b_0 + b_u)/2$.

Obviously, the examined strip tends to deform under the forces $q_S/2$, $m_S/2$ and Δv. This deformation consists essentially of a change in length of its diagonals, and induces an additional state of stress for each section wall. So, it is considered appropriate – following Section 3.3 – to insert a hinged strut of infinite axial stiffness in the sense of the augmenting diagonal, i.e. a non-deformable one (Figure 14.6). With the insertion of this element throughout the length of the girder, the deformability of the section is eliminated, whereas the section itself develops a bending and axial state of stress, the former being induced by the acting moments $m_S/2$. It is clear that this diagonal element develops a tensile axial force R (kN/m) distributed along the length of the

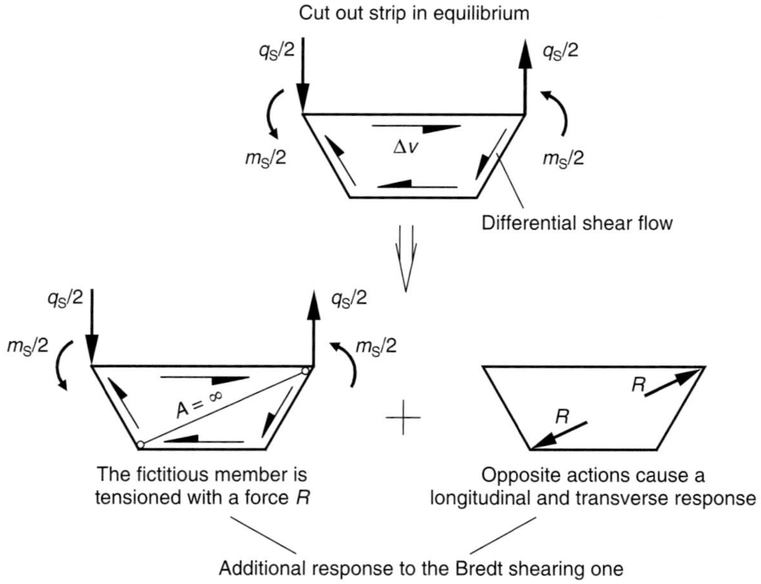

Figure 14.6 Additional girder response due to antisymmetric loading

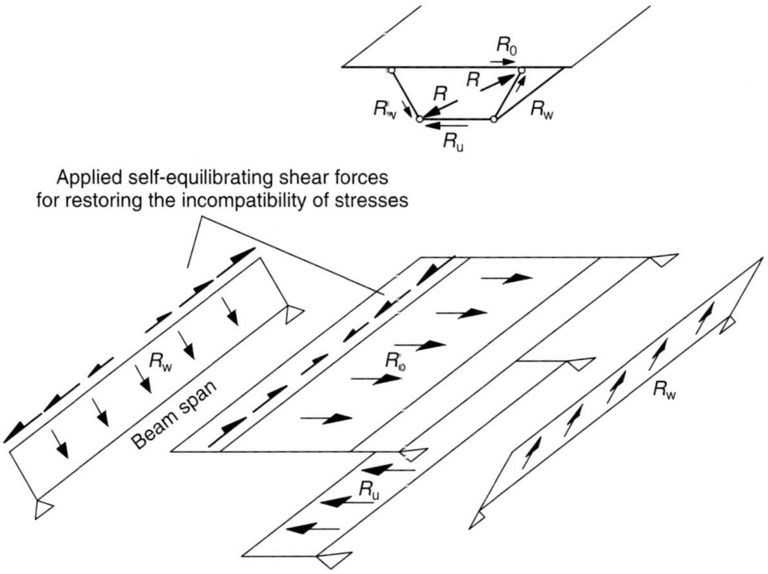

Figure 14.7 Taking up the diagonal loads through the longitudinal wall beams

girder. Applying now the opposite of the acting forces R on the corresponding longitudinal edges of the girder in absence of the diagonal strut, it is obvious that the superposition of the thus resulting state of stress with that of the blocked strip will give the final response (see Figure 14.6). ■

For the examination of this last 'diagonal loading' it is first considered that the forces R act on a girder having hinged connections at the section edges instead of monolithic ones (Figure 14.7). The forces R may be resolved equivalently at each edge into the two concurring walls; therefore, each of them can be considered as a longitudinal beam loaded by the corresponding component, developing bending moments M_0 and corresponding normal stresses σ, according to the classical theory of bending. However, the resulting strains σ/E at the common edges of the walls are not equal – as they should be – and for this reason some distributed longitudinal forces have to be additionally introduced along the edges of each wall, in order to establish the strain compatibility at each edge (see Figure 14.7). It is clear that in this way the initially determined normal stresses σ will be changed.

By following the above analysis, the determination of the longitudinal axial stresses is possible, e.g. for the left web of the 'hinged' box section in Figure 14.8 they can be determined through the classical bending formula on the basis of the aforementioned moment M_0. In this case, the moment of inertia I^* used is slightly larger than the normal value I_w for the web by a factor k^w, and the new 'neutral axis' lies at a distance y_0 from the top fibre, which is somewhat less than its 'normal value', i.e. half of the web height (Schlaich and Scheef, 1982).

The moment M_0 results from the loading of the left web with the corresponding component R_w of the 'diagonal force' R (see Figure 14.7). It is found that $R_w = R \cdot b_w/s$.

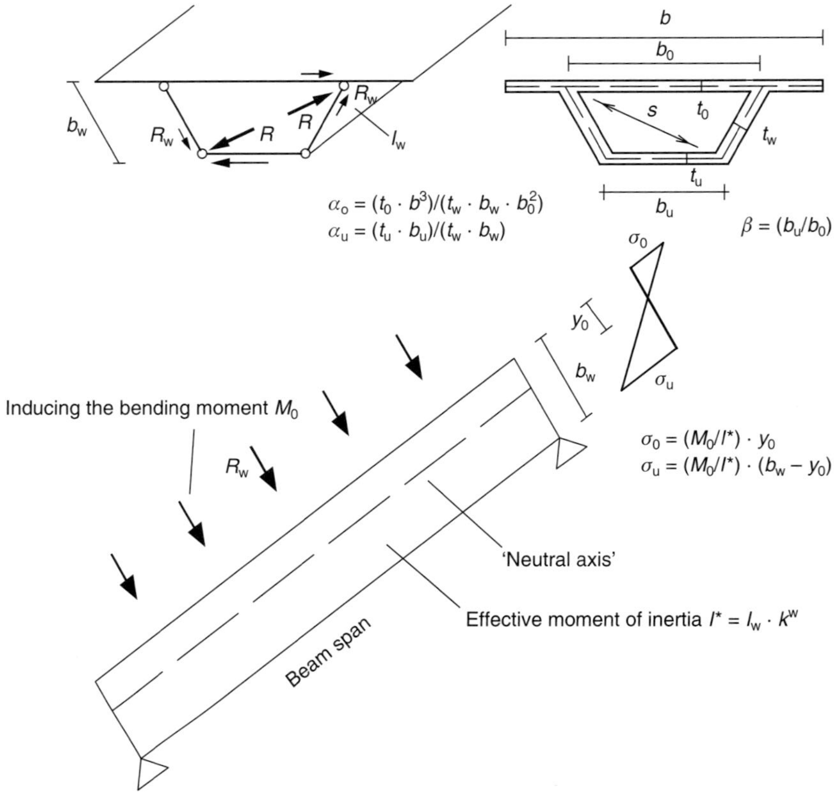

Figure 14.8 Bending response of the longitudinal section walls in a 'hinged' system

Thus,

$$\sigma_0 = -\frac{M_0}{I^*} \cdot y_0$$

$$\sigma_u = \frac{M_0}{I^*} \cdot (b_w - y_0)$$

where

$$I^* = I_w \cdot k^w$$

Furthermore (see Figure 14.8),

$$k^w = \frac{2 \cdot \beta \cdot [(\alpha_0 + 2) \cdot (\alpha_u + 2) - 1]}{(1 + \beta) \cdot (3 + 3 \cdot \beta + \alpha_0 + \alpha_u \cdot \beta)}$$

$$y_0 = \frac{1 + 2 \cdot \beta + \alpha_u \cdot \beta}{3 + 3 \cdot \beta + \alpha_0 + \alpha_u \cdot \beta} \cdot b_w$$

Thus, the longitudinal web beam obeys the following typical differential equation (see Section 2.3.6):

$$EI^* \cdot \frac{d^4 w}{dx^4} = R_w$$

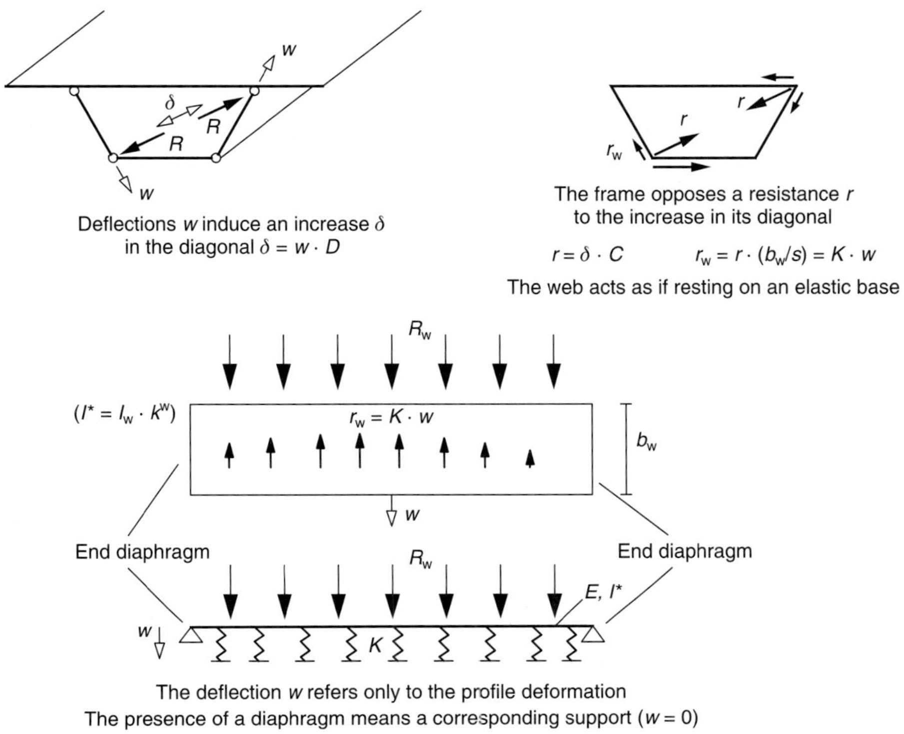

Figure 14.9 Structural action of a section wall in the monolithic system

where R_w and I^* represent the initial distributed web loading and the equivalent moment of inertia, respectively, as explained above. It is noted that w represents the in-plane deflection of the web, and it is due merely to the assumed deformability of the section (see Figure 14.8).

Only now may the monolithic connection of the section walls be taken into account.

The bending deformation of the hinged section profile, under the action of diagonal forces R, has as a consequence an increase of its diagonal length by δ (Figure 14.9). However, this change cannot be realised without resistance, given that the transverse stiffness C of the closed monolithic section profile is automatically mobilised. This stiffness is expressed through the relation $r = \delta \cdot C$, as that diagonal pair of forces r required to produce the deformation $\delta = 1$. The stiffness C clearly depends on the geometric and elastic data of the section, and may be determined using typical computer software for plane frames. Thus, the tendency of the hinged section profile to be deformed by δ is counteracted by the resistance r of the monolithic closed frame, and in this way the web is subjected to, in addition to its 'initial' loading R_w, the loading of the component r_w of the force r, obviously with opposite sense. Thus, $r_w = r \cdot (b_w/s)$ (see Figure 14.9).

Taking into account that the bending deformation w of the web can be expressed through the diagonal deformation δ by the relation $\delta = w \cdot D$ (see Figure 14.9), where

$$D = \frac{2 \cdot b_w}{\beta \cdot s} \cdot \frac{(1+\beta) \cdot (2 + 2 \cdot \beta + 2 \cdot \beta^2 + \alpha_0 + \alpha_u \cdot \beta^2)}{3 + 3 \cdot \beta + \alpha_0 + \alpha_u \cdot \beta}$$

it is found that the component r_w is also expressed through w by the relation

$$r_w = w \cdot D \cdot C \cdot (b_w/s) = w \cdot K$$

where

$$K = D \cdot C \cdot (b_w/s)$$

The differential equation for the web beam can now be adjusted to the restored rigid section profile, and can be written as

$$EI^* \cdot \frac{d^4w}{dx^4} = R_w - r_w = R_w - w \cdot K$$

and, finally, takes the form

$$EI^* \cdot \frac{d^4w}{dx^4} + K \cdot w = R_w$$

This equation is recognised as the typical equation of a beam on an elastic foundation with a modulus of subgrade reaction K (see Section 17.3.3.1). Indeed, the web is carried by the elastic support offered by the profile resistance when undergoing diagonal deformation (see Figure 14.9). ■

It is, of course, clear that the presence of a single concentrated load on the web will lead to an applied concentrated moment and a corresponding shearing flow (Bredt) and, finally, to a concentrated load applied to the model of the elastically supported beam (Figure 14.10). If this eccentric load is the only one acting on the girder, then, in the above equation, the distributed load R_w on the right-hand side does not exist. In this case, it is clear that the diagonal actions 'R' on the girder consist of a single pair of equal and opposite forces. These forces, if a transverse diaphragm exists at this location, are taken up directly by it, without causing any 'secondary' stresses in the girder, as would happen in an 'undeformable section'. If such a transverse diaphragm does not exist, then the diagonal forces are taken up by the deformable section of the girder itself, as has been previously examined in detail (see Figure 14.10) (Menn, 1990).

Recapitulating, the deformation of the web under the component R_w of the diagonal loading R induces the transverse stiffness of the closed profile. This stiffness arises as a load r_w acting on the web in the opposite sense, being linearly related to its deflection, i.e. as occurs in a beam resting on an elastic foundation. It is clear that the thus-resulting longitudinal bending response of the web will lead, through the developed axial stresses according to the above relations, to the edge stresses of the remaining section walls, as at each edge the strains ε and, consequently, the stresses $\varepsilon \cdot E$ are common. ■

Regarding the boundary conditions that have to be fulfilled by the deflection w, it should be pointed out that w specifically concerns only that deformation which arises from the deformability of the section profile and not the overall deflection of the web beam, to which the torsional twist angle of the section also contributes as a rigid body. Thus, at each position, where the profile remains undeformed because of the presence of a transverse diaphragm, the condition $w = 0$ has to be satisfied. This means that, at the

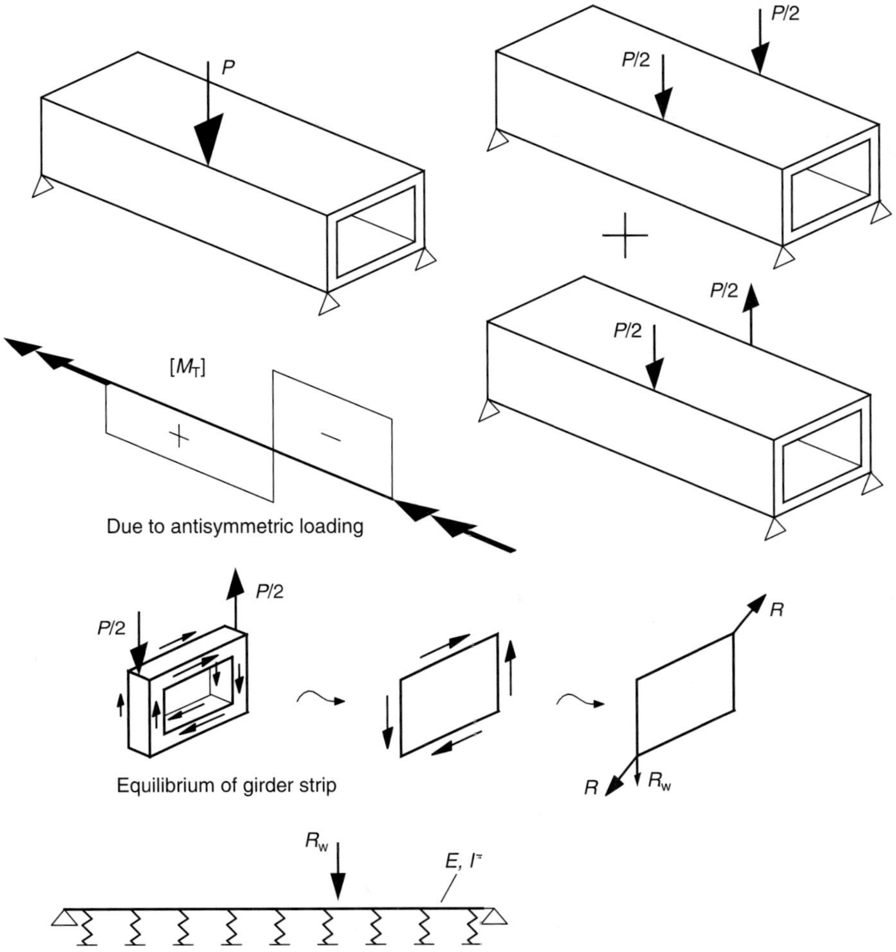

Figure 14.10 Behaviour of a box girder under an eccentric concentrated load

corresponding point of the elastically supported beam model, an unyielding support must be inserted (see Figure 14.10). Moreover, at a fixed position of the girder, the following condition is imposed:

$$\frac{dw}{dx} = 0$$

Finally, the treatment of a girder support with no transverse diaphragm under the developed support reactions arising from torsional response is examined (Figure 14.11).

It should be noted that, in this case, no support needs to be provided for the model of the elastically supported beam but, instead, the developed vertical reactions should be considered as antisymmetric loads. These exert a concentrated torsional moment and, according to the above, they lead to a concentrated force on the elastically supported web model. ■

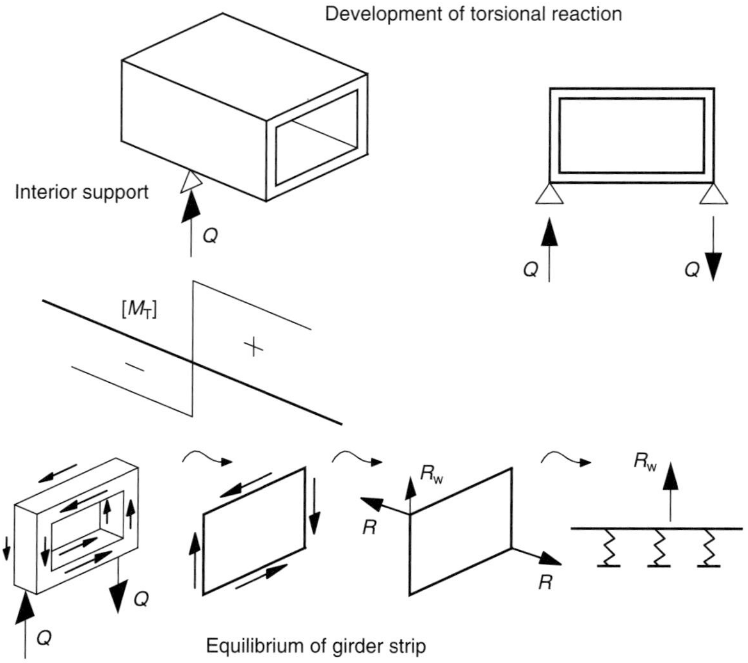

Figure 14.11 Behaviour at a support without a transverse diaphragm

Of course, the simultaneous transverse response of the section profile should not be overlooked.

Based on the deformation w of the web, the diagonal loading r of the closed rigid profile, as shown in Figure 14.12, can be readily determined, thus allowing the evaluation of the transverse response of the section. According to the above,

$$r = w \cdot K \cdot \frac{s}{b_w}$$

14.2.3 Numerical example

A simply supported orthogonal box girder with a 40.0 m span, shown in Figure 14.13, is examined. At the midspan and over the left web a concentrated load of 1000 kN is

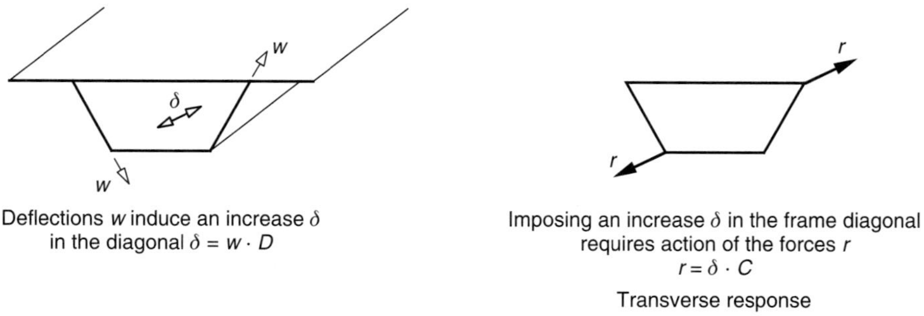

Figure 14.12 Transverse response due to profile deformation

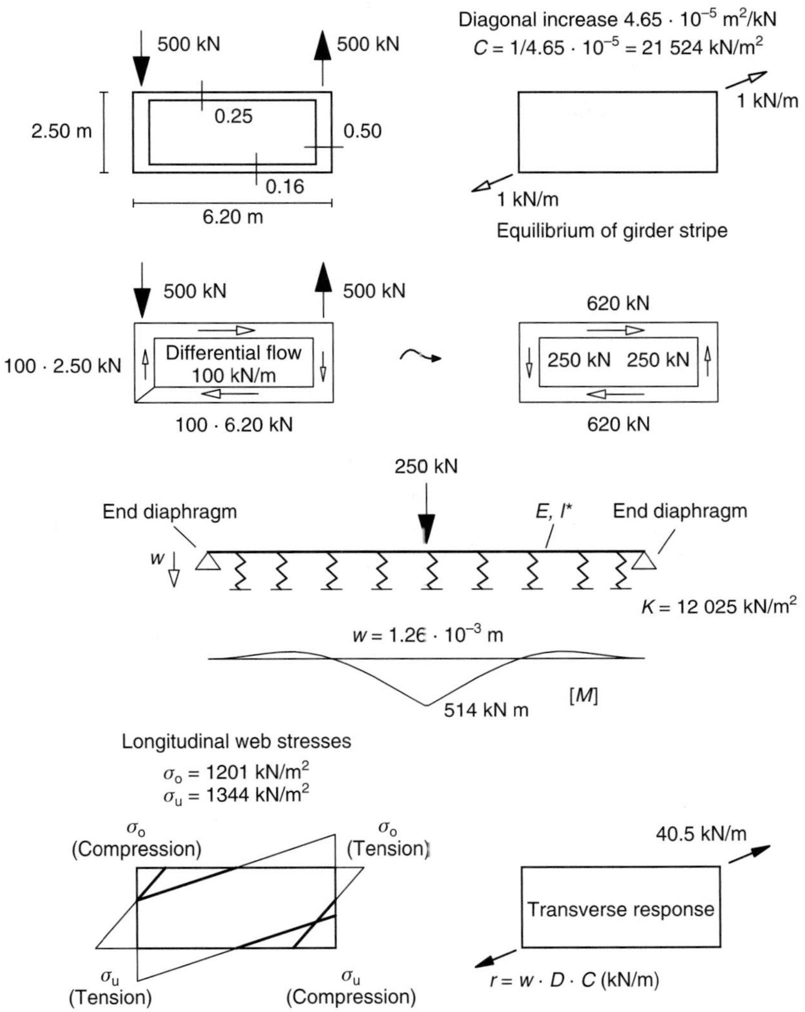

Figure 14.13 Antisymmetric loading of a girder caused by an eccentric load

considered, which, according to Figure 14.10, induces torsion through the antisymmetric couple of 500 kN.

On the basis of the expressions in Section 14.2.2, the following cross-section data are determined:

$$\beta = 1,\ \alpha_0 = 1.24,\ \alpha_u = 0.79,\ k^w = 1.002,\ y_0 = 1.18,\ D = 1.495,\ s = 6.69\,\text{m}$$

as well as

$$I_w = 0.504\,\text{m}^4$$

Thus obtaining

$$I^* = 0.504 \cdot 1.002 = 0.505\,\text{m}^4$$

Moreover, the stiffness C (kN/m^2) of the unit length of the girder strip with respect to a unit increase (1.0 m) of the diagonal distance of its edges is determined as $C = 21\,524.0\,kN/m^2$.

From the equilibrium of the girder strip at midspan, the web is acted on by a concentrated force of 250.0 kN. More specifically, the strip is in equilibrium under the antisymmetric couple of 500 kN as well as the peripheral differential shear flow v, which, according to Section 14.2.2, is

$$v = \frac{500 \cdot 2 \cdot 6.20}{4 \cdot 2.50 \cdot 6.20} = 100\,kN/m$$

The force R_w acting on the left web is

$$R_w = 500.0 - v \cdot 2.50 = 250.0\,kN \quad \text{(see Figure 14.13)}$$

The web, as explained previously, acts as a beam on an elastic foundation with a modulus of subgrade reaction equal to $K = 1.495 \cdot 21\,524.0 \cdot 2.50/6.69 = 12\,025\,kN/m^2$.

The corresponding bending moment diagram is obtained, as can be seen in Figure 14.13, exhibiting a maximum value at midspan equal to 514 kN m. Moreover, the corresponding deflection is $1.26 \cdot 10^{-3}$ m.

Thus, the normal stresses at the top and bottom fibres are obtained, from the relations in the previous section, as

$$\sigma_O = 514.0 \cdot 1.18/0.505 = 1201\,kN/m^2 \quad \text{(compression)}$$

$$\sigma_u = 514.0 \cdot (2.50 - 1.18)/0.505 = 1344\,kN/m^2 \quad \text{(tension)}$$

The above longitudinal stresses are solely due to the torsional response of the girder, by taking into account the deformability of its cross-section. The complement of the stress image for the whole section is shown in Figure 14.13.

The development of the deflection w causes an increase in the diagonal length of the profile, which requires the action of the corresponding diagonal forces r. These are, according to the above,

$$r = 1.26 \cdot 10^{-3} \cdot 1.495 \cdot 21\,524.0 = 40.5\,kN/m$$

On the basis of these forces, the corresponding bending, axial and shearing responses of the closed girder section can be readily deduced (see Figure 14.13).

14.2.4 *Summary*

According to the analysis above, the examination of the behaviour of rectilinear box girders under an eccentric loading consists of the following:

- The eccentric loading is reduced through the 'trivial fixed state', represented by the transverse bending state $[m^0]$, into a non-symmetric nodal loading.
- The non-symmetric nodal loading is split into a symmetric and an antisymmetric part.
- The symmetric part causes a longitudinal bending and shearing response determined by the corresponding sectional force diagrams [M] and [V], as well as a transverse

response concerning a transverse girder strip of unit length arising from a self-equilibrated loading consisting of the acting loads and the differential shear flow. The transverse response is represented by the corresponding bending moment, shear force and axial force diagrams $[m_{sym}]$, $[v_{sym}]$ and $[n_{sym}]$, respectively, plotted along the section profile.

- The antisymmetric part causes a torsional response along the girder, represented by a torsional diagram $[M_T]$, which enables the determination of the Bredt shear flow at each position. This shear flow causes, in turn, on the one hand, shear of the section walls, depicted by a the diagram $[V^T]$ for each wall, and, on the other hand, a longitudinal 'diagonal loading' $\{R\}$, arising from the resulting differential shear flow of a strip. To this last response, the transverse loading arising from fixation of the section edges has to be added, represented by the diagrams $[m_f]$, $[v_f]$ and $[n_f]$.
- The 'diagonal loading' $\{R\}$ causes a longitudinal bending and shear response, represented by the sectional force diagrams $[M^R]$ and $[V^R]$, respectively, arising from the consideration of a characteristic section wall (usually the web), as a beam resting on an elastic foundation. This beam is supported only at places where the deformation of the section profile is prevented.
- The above response is always accompanied by the transverse response of the section profile, determined on the basis of its mobilised 'diagonal stiffness', and is represented by the corresponding sectional diagrams $[m^R]$, $[v^R]$ and $[n^R]$ along the profile perimeter.

Thus, for the response of each wall, the following diagrams must be taken into account:

- Longitudinal bending: $[M] + [M^R]$.
- Longitudinal shear: $[V] + [V^T] + [V^R]$.
- Transverse bending: $[m^0] + [m_{sym}] + [m_f] + [m^R]$.
- Transverse axial response: $[n_{sym}] + [n_f] + [n^R]$.
- Transverse shear: $[v_{sym}] + [v_f] + [v^R]$.

The transverse shear is usually of secondary importance.

14.3 Curved girders

14.3.1 General

The box girder is particularly suitable for bridges curved in plan, as these are subjected to a permanent torsional response caused even by non-eccentric loads, e.g. the self-weight of the girder.

The skeletal model used for a curved girder is of the grid type (see Chapter 10), i.e. a plane structure loaded transversely to its plane, and, as such, it develops three sectional forces at each position, namely a vertical shear force V, a bending moment M_B and a torsional moment M_T (Figure 14.14).

As will be shown immediately below, an equilibrium interrelation exists between bending and torsional moments, and thus any redistribution of the bending response

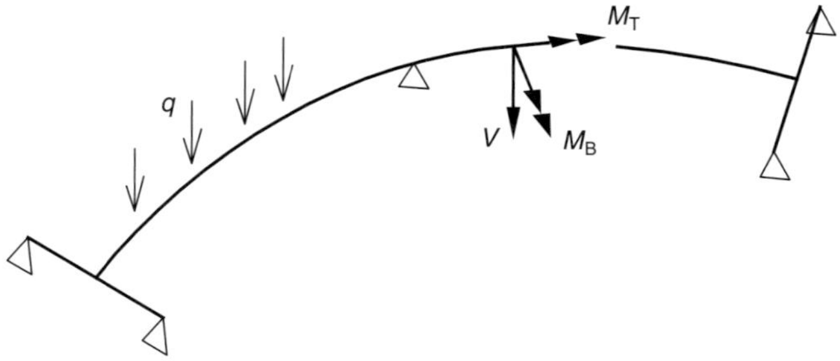

Figure 14.14 Curved beam model with its developed sectional forces

of the girder, for reasons associated with its plastic analysis (see Section 6.6.2), has, necessarily, to be accompanied by an adjustment of the corresponding torsional moments.

14.3.2 Determination of the bending and torsional response

14.3.2.1 Evaluation of the equilibrium equations

The girder is assumed to have a constant radius of curvature equal to R. The loading consists of a vertical distributed load q passing through the shear centre of the cross-section, as well as of a distributed torsional moment m_D (Figure 14.15). It should be recalled that, exactly as in the case for grids, any part of the beam must satisfy three

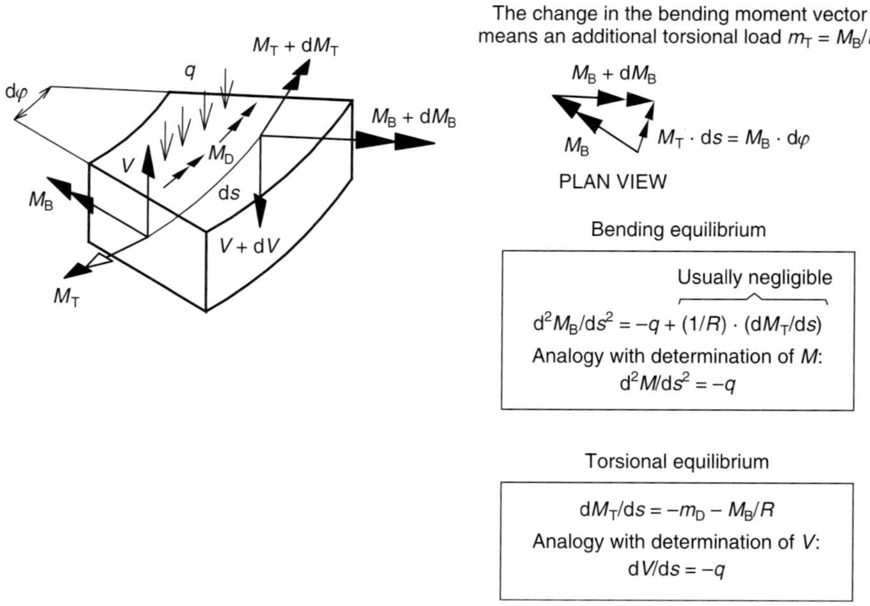

Figure 14.15 Evaluation of equations of equilibrium

conditions of equilibrium, i.e. those with respect to vertical forces, and those with respect to the projections of moment vectors on two arbitrary horizontal axes.

The equilibrium equations of an elementary segment of length ds forming an angle $d\varphi$ ($1/R = d\varphi/ds$), as shown in Figure 14.15, can be written in the form:

$$\frac{dV}{ds} = -q \quad \text{(equilibrium of vertical forces)}$$

$$\frac{dM_T}{ds} = -m_D - \frac{M_B}{R} \quad \text{(equilibrium of moment vectors, on a tangential axis)}$$

$$\frac{dM_B}{ds} = V + \frac{M_T}{R} \quad \text{(equilibrium of moment vectors, perpendicular to the axis)}$$

It should be understood here, according to Figure 14.15, that the vectorial change in a bending moment between two adjacent sections is equivalent to the action of a distributed torsional load. This can also be recognised from the presence of the term M_B/R on the right-hand side of the second equation above. In other words, the presence of a bending moment at the two edges of a curved beam segment means that, apart from m_D, an additional distributed torsional load M_B/R is acting on the beam. In an analogous way, the vectorial change in the torsional moment M_T is related to bending development, as can be seen from the third equation (Menn, 1990).

The elimination of the shear force V from the above equations provides the two final equations, which clearly describe the interdependence of the bending and torsional moments of the girder:

$$\frac{d^2M_B}{ds^2} = -\left(q - \frac{1}{R}\cdot\frac{dM_T}{ds}\right) \quad \text{(a)}$$

$$\frac{dM_T}{ds} = -\left(\frac{M_B}{R} + m_D\right) \quad \text{(b)}$$

If the arc span length L is much smaller than the radius of curvature R (i.e. $L/R < 0.3$) and, moreover, the distributed torsional load m_D is much smaller than $q \cdot R$, then, by expressing M_B as $q \cdot L^2/c$ – where c is of the order of about 10 – from equation (b) it may be concluded that

$$\frac{-\frac{1}{R}\cdot\frac{dM_T}{ds}}{q} = \frac{L^2}{R^2\cdot c} + \frac{m_D}{R\cdot q}$$

which means that the term $-(1/R)\cdot(dM_T/ds)$ on the right-hand side of equation (a) is much smaller than q.

Thus, the first equation may be more simply written as

$$\frac{d^2M_B}{ds^2} = -q$$

i.e. resembling the equilibrium relation of a rectilinear beam between the bending moment and the load (see Section 2.2.3). This means that, under the above conditions, the bending moments of the curved girder may be approximated by the bending moments M_B of a straight beam of span L equal to the arch length of the girder.

Once the moments M_B are determined in this way, then the right-hand side of equation (b) takes a concrete value, and the equation itself refers directly to the equilibrium relation of a beam, $dQ/ds = -p$, between shear forces and the distributed load. This means that the torsional moments M_T may result as the shear forces of a straight beam of length L carrying the load $M_B/R + m_D$ (see Figure 14.15).

If the above conditions regarding the ratios L/R and $m_D/q \cdot R$ are not met and a better approximation is needed, then, on the basis of the above initial values of M_B, one can return to equation (a) and evaluate the term

$$q - \frac{1}{R} \cdot \frac{dM_T}{ds} = q - \frac{1}{R} \cdot \left(\frac{M_B}{R} + m_D \right)$$

as a new loading of the beam, in order to obtain 'improved' values of the bending moments M_B, and afterwards, through equation (b), to determine anew the torsional moments M_T as shear forces for the loading

$$\frac{M_B}{R} + m_D$$

The repetition of the above procedure converges rapidly.

14.3.2.2 *Summary*

- First, the 'stretched' girder is considered.
- On the basis of vertical loads, the bending moment diagram M_B is considered.
- The 'straight' girder is loaded with the torsional load M_B/R, according to the sense dictated by the composition of the vectors M_B, as applied to the corresponding elementary curved girder segment considered in plan view. The distribution of this torsional loading is obviously identical to the corresponding bending diagram $[M_B]$. Any existing load m_D must be added.
- The torsional loads M_B/R are considered as distributed vertical downward loads, as long as the bending moments M_B cause tension at the bottom fibres; otherwise, they are directed upwards. The possibly present m_D may also be considered as a distributed vertical load with the same direction as M_B/R, if its sense coincides with that of the torsional load M_B/R; otherwise, it has the opposite direction.
- The torsional moment diagram $[M_T]$ is identical to the diagram of shear forces for the above fictitious loading, but not with respect to its sign. ■

It is clear that the above equations concern only the equilibrium, and, as they do not take into account the compatibility of deformations in the case of redundant supports, they do not lead to a strictly accurate solution, unless the ratio $\kappa = EI/G \cdot I_T$ is considered equal to zero. Nevertheless, their use for purposes of preliminary design is quite appropriate. It is noted that for a simply supported curved girder and for a ratio L/R less than 0.3, which means a central angle not greater than 20°, the discrepancies from the 'exact' solution are less than 1%. Moreover, for a fixed-ended girder, and thus for the approximately equal internal spans of a continuous girder with $L/R \leq 0.3$ and $\kappa \leq 5$, the discrepancies are less than 1.5%. Furthermore, as mentioned initially, the above equations may have a direct utility when, for the purposes of plastic design, the bending moments obtained from a computer solution have to be altered, in which case the redistribution of torsional moments must necessarily satisfy the equilibrium criterion.

14.3.2.3 Application examples

The above procedure is applied to the following examples.

In the *first example* (Figure 14.16) a simply supported girder allowing the take up of torsional moments at its supports and a fixed-end girder are considered. Figure 14.16 shows that the fixed-end girder develops a clearly lower torsional response, while its torsional reactions are zero. This may be understood on the basis of the remark made in Section 2.3.6 that, as the beam is loaded by its bending moment diagram, its reactions

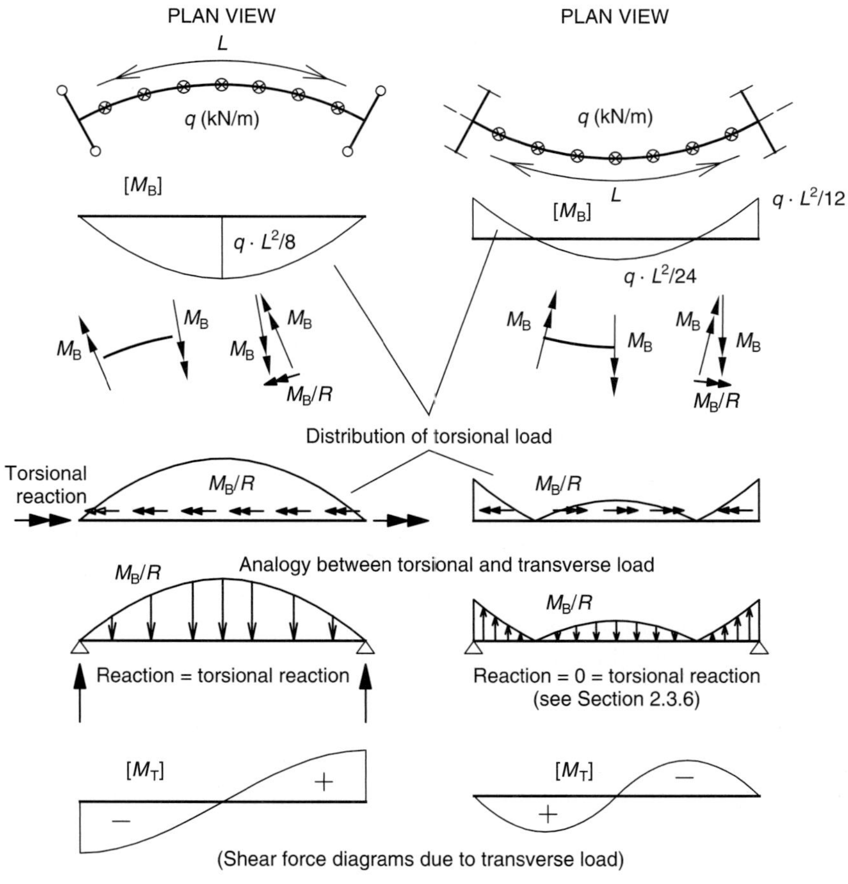

Figure 14.16 Determination of the bending and torsional response of a curved girder

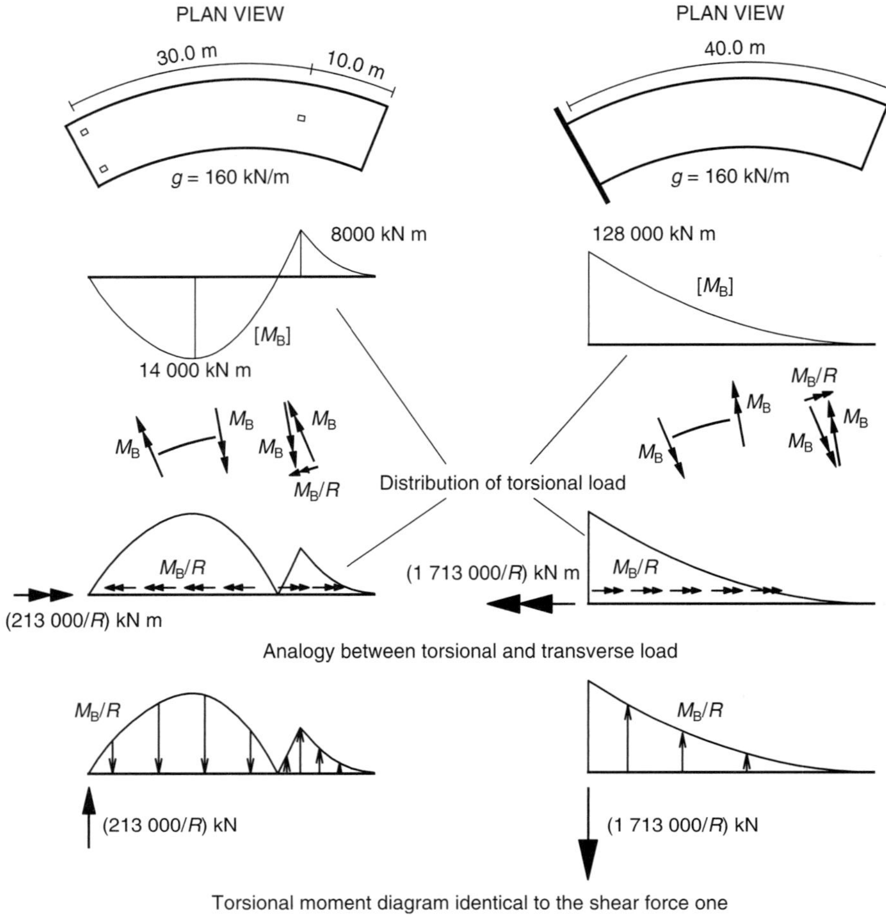

Figure 14.17 Evaluation of bending and torsional response with cantilever action

(specifically the torsional ones) must correspond to the developed end rotation angles, which are obviously zero. In both cases examined, attention must be focused on the correct determination of the loading sense of the torsional loading M_B/R, as shown in Figure 14.16.

Regarding the internal span of a continuous beam, the consideration of a fixed-end beam instead does not essentially alter the torsional response.

In the *second example* (Figure 14.17), a beam is examined which can develop a torsional reaction only at its left support, being in the first case supported on an intermediate point, whereas in the second case it acts as a cantilever.

Despite the fact that the intermediate support cannot develop a torsional reaction, it relieves the beam significantly with respect to the torsional response, since the torsional reaction in the case of a cantilever is about eight times greater. Again, due attention is paid to the direction of the torsional loading M_B/R, according to Figure 14.17. ■

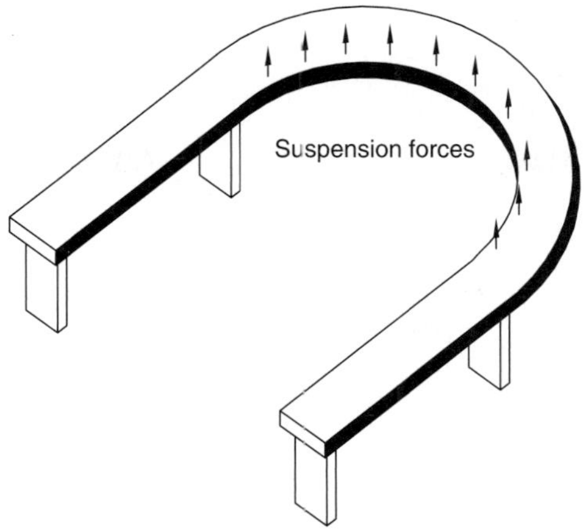

Figure 14.18 Layout of a semicircular cable-stayed bridge

It is now appropriate to examine how a girder behaves when subjected to a distributed torsional moment m_D along its whole length. First, it is obvious that a straight girder, where the ratio L/R is equal to zero, develops only torsional response and no bending at all.

However, as the arc length L over a certain chord, and consequently the ratio L/R, increase, the torsional response is progressively reduced, being accompanied by ever-increasing bending. This can be confirmed using equations (a) and (b) in Section 14.3.2, which are, of course, valid without making any assumption regarding the geometry of the curved girder. However, if a girder in the form of a ring, i.e. exhibiting the maximum possible value of the ratio L/R, namely equal to $2 \cdot \pi$, is subjected to a uniformly distributed torsional self-equilibrating loading moment m_D, it develops no torsion at all but only constant bending, as was ascertained in Section 12.4.1 (see Figure 12.28). Such a ring develops a constant twisting angle all along its perimeter.

The above property of a ring makes possible the bridging of a span length $2 \cdot R$ through a girder in the form of semicircle of radius R, suspended along its internal edge in order to exclusively develop a bending response without any torsion. This goal, however, is achieved under the condition that the girder is being subjected to an appropriate bending moment at both its supported ends, and this is in turn directly accomplished if the girder is connected to two rectilinear beams of appropriate length, as shown in Figure 14.18 (Schlaich and Seidel, 1988).

More specifically, the eccentrically suspended semicircular girder, under its self-weight g, is subjected to a constant torsional load $m_D = g \cdot e$, where e is the distance of the centroid to the suspension line (Figure 14.19). According to Section 12.4.1 (see Figure 12.28), the equilibrium condition of the curved girder leads to the exclusive development of the bending response, if a bending moment $M_B = m_D \cdot R$ can be provided at both its ends. This specific bending moment, which will be acting throughout the curved girder, can be provided in practice by the fixed-support moment $g_S \cdot L_s^2/8$ of a fixed simply supported beam. This is rigidly connected at the

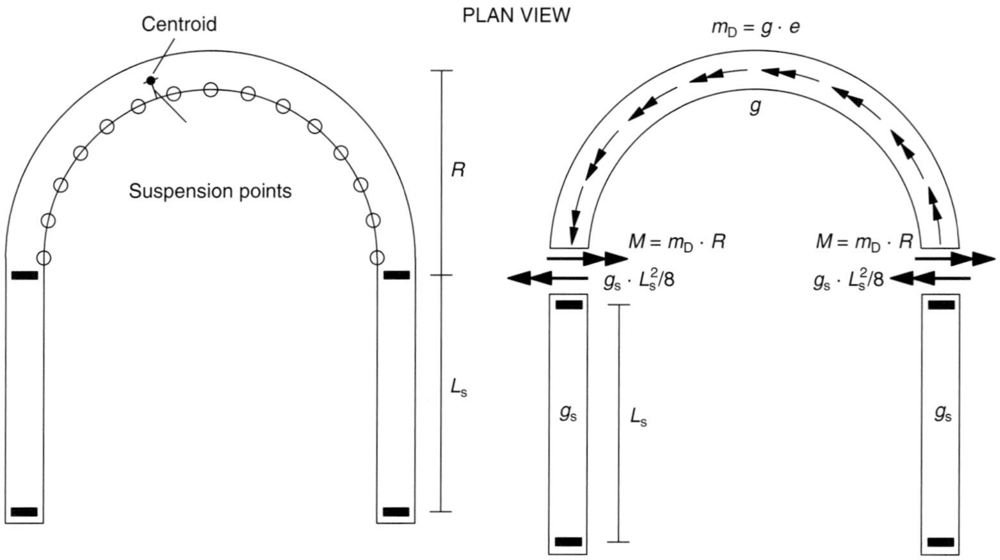

Figure 14.19 Structural action of the system shown in Figure 14.18

two girder ends, having an appropriate self-weight g_S and length L_s, both determined through the obvious requirement that $m_D \cdot R = g_S \cdot L_s^2/8$ (see Figure 14.19).

14.3.3 The response of cross-section walls

The torsional response referring to the skeletal model of the girder as examined in Section 14.3.2 implies, beyond the Bredt peripheral flow, an additional response for the box section walls. This response results from the way that gravity loads are introduced to the girder.

First, it should be noted that the acting compressive forces D and tensile forces Z in the curved top and bottom flanges, respectively, cause distributed deviation forces q, according to the well-known relation

(distributed deviation force) = (axial force)/(radius of curvature)

It is clear from Figure 14.20 that the transversely distributed equal and opposite forces q_D and q_Z, which the top and bottom slabs are obliged to carry, respectively, create a torsional load per unit of curved length, which is simply that resulting from the vectorial variation of bending moments, as examined in the previous section (Menn, 1990):

$$q_D = \frac{D}{R} = \frac{M_B}{h \cdot R}$$

$$q_Z = \frac{Z}{R} = \frac{M_B}{h \cdot R}$$

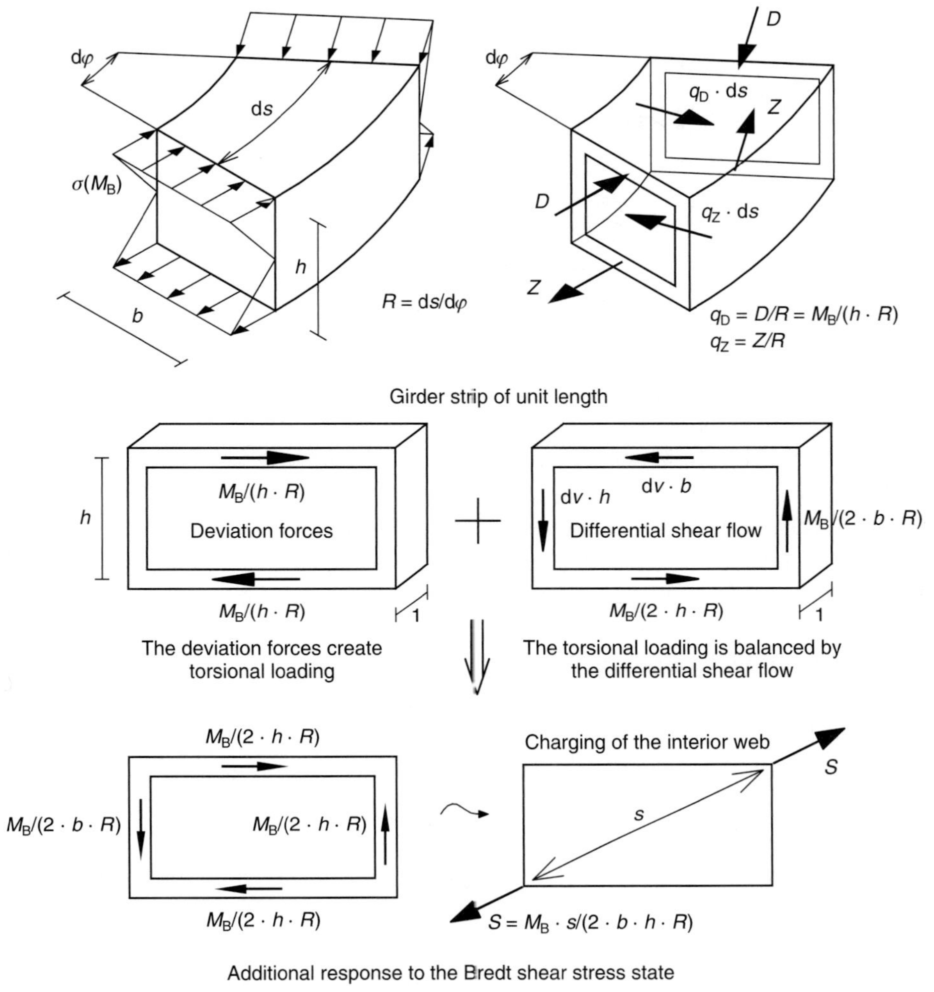

Figure 14.20 Introduction of a torsional response in a box girder walls

and, given that $D = Z$,

$$q_D \cdot h = \frac{M_B}{R}$$

Thus, it can be seen that, even without the action of an externally applied torsional moment m_D, merely the existence of bending along a curved axis implies – as previously found – the imposition of a distributed torsional load M_B/R, which causes, of course, a torsional response, according to equation (b) in Section 14.3.2:

$$\frac{dM_T}{ds} = -\frac{M_B}{R}$$

Of course, how the loads q_D and q_Z are introduced to the girder has to be examined, and, in particular, how and with what consequences a transverse sectional strip is self-equilibrated under the above loads. The cut-out strip of unit length, receiving the forces q_D and q_Z at its top and bottom sides, respectively, is in equilibrium with the developed Bredt shear flow at both its faces (see Figure 14.20). The resultant of these two flows is the so-called *differential shear* dv referred to the unit of transverse length (see Section 14.2.1). According to Bredt's formula (see Section 2.5.2),

$$\frac{dv}{ds} = \frac{dM_T}{ds}\frac{1}{2 \cdot b \cdot h} = -\frac{M_B}{2 \cdot b \cdot h \cdot R}$$

Thus, the strip, being in equilibrium as a plane structure, under the loads q_D, q_Z and dv, gives rise to the self-equilibrated diagonal loading of the profile

$$S = \frac{M_B}{2 \cdot R} \cdot \sqrt{\frac{1}{b^2} + \frac{1}{h^2}}$$

as shown in Figure 14.20, which causes longitudinal bending of the walls as well as transverse bending of the section profile, as examined in detail for the rectilinear girder.

Although the analogy is not quite accurate, for preliminary design needs and for limited curvatures, it can be considered that the left web wall takes the downward uniform load $S_w = M_B/2 \cdot R \cdot b$, acting, as in the case of a rectilinear beam, like a beam resting on an elastic foundation with a 'subgrade modulus' K, on the basis of the stiffness of the section profile against a diagonal length change. Of course, if the resulting response is high, transverse diaphragms should be provided. Then, the deviation forces of the top and bottom slabs that trigger the whole response will be transferred through corresponding horizontal bending to each diaphragm. This, in turn, will transfer a concentrated torque to the section, by excluding at the same time its deformation that would otherwise take place.

The above diagonal loading of the girder cross-section, when transverse diaphragms are not present, makes it clear that – as shown in Figure 14.20 for the longitudinal response – the 'internal' web is burdened whereas the 'external' one is relieved. However, it is interesting to note that in the case of an open section, as shown in Figure 14.21, quite the opposite happens, as explained below.

More specifically, the consideration of an elementary curved segment, as depicted in Figure 14.21, shows that the action of bending moments M_B is equivalent – as previously explained – to the application of a distributed torsional load $m_D = M_B/R$. It can be seen that this torsional load can be taken up by the 'opposite bending' of the two webs, through the resolution in the continuously distributed vertical loads

$$q^* = M_B/R \cdot b$$

which obviously give an additional load to the external web while decreasing the load on the internal one. If it is assumed that the existing vertical load q is equally distributed between the two webs, then their bending response will be obtained as

$$M_{B,ext} = (1/2) \cdot M_B(q) + M_B(q^*)$$

$$M_{B,int} = (1/2) \cdot M_B(q) - M_B(q^*)$$

for the external and the internal web, respectively.

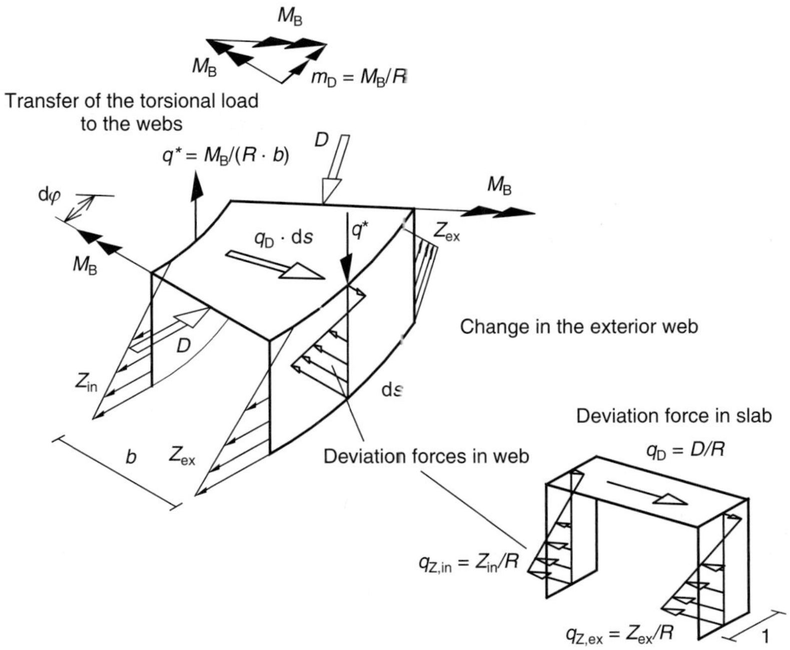

Figure 14.21 Uptake of a torsional load by the webs of an open section

The compressive forces in the slab due to the bending moments M_B compel it to take up 'outward' deviation forces, whereas the existence of compressive and – especially – tensile stresses in the webs create transverse distributed deviation forces which equilibrate those of the slab, causing at the same time transverse bending of the open frame of the section, as shown in Figure 14.21 (Menn, 1990).

14.3.4 The influence of prestressing in curved girders

As was made clear in Chapter 4 on beams, the layout of prestressing tendons aims to provide deviation forces that counteract the gravity loads for which the structure has to be designed. In the case of curved girders, gravity loads indirectly create torsional loads through the existing curvature, eventually being increased by their eccentric layout on the bridge deck. Thus, it is useful to examine how the presence of prestressing in a curved box girder influences its torsional response and, moreover, to consider the possibility of torsional relief through an appropriate tendon layout.

Given that the tendons are arranged within the thickness of each web by using all of its available depth, it is first pointed out that a cable, inclined with respect to the girder axis, implies torsion (Figure 14.22). This is explained by the fact that the inclined compressive force on the section has a component in its own plane, which causes a torsional moment equal to the product of this component by its distance to the shear centre of the section. The fact that, in the usual prestressing layouts, the prestressing does not induce torsion arises because the different torsional contributions of the

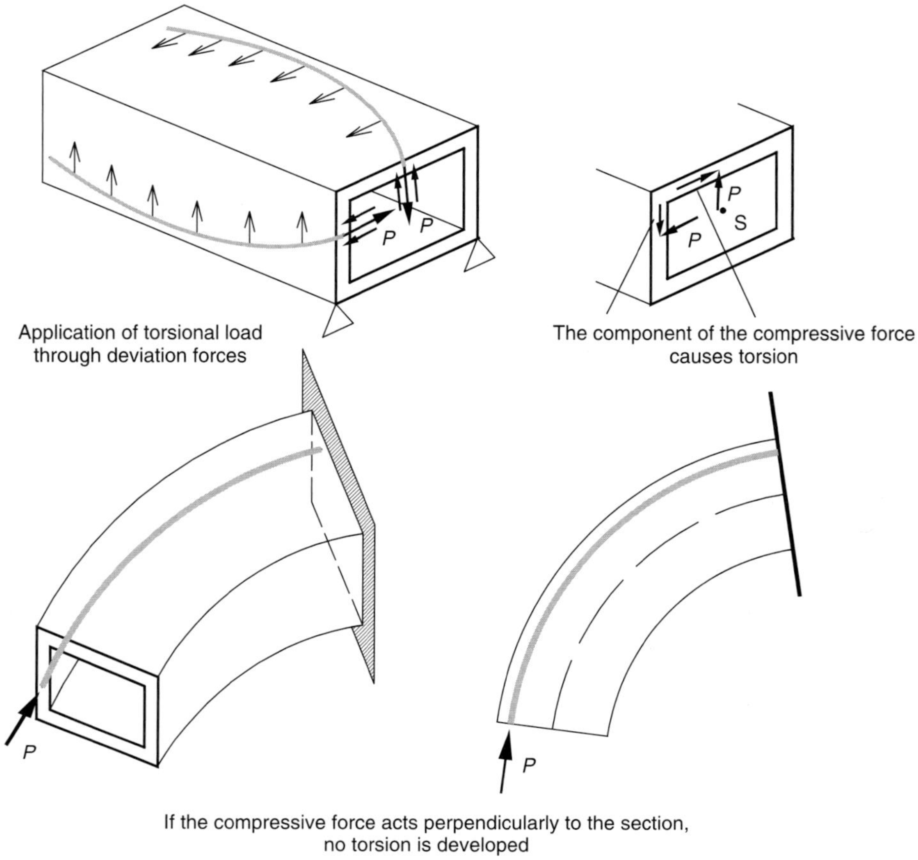

Figure 14.22 Torsional response due to tendon inclination

cables cancel each other. However, an appropriate change in the cable inclinations may induce a torsional response that could relieve that arising from the loads.

Of course, a cable lying in the top or bottom slab with an inclined position to the girder axis may also cause a torsional moment with respect to the shear centre through the component of its compressive action on the section (see Figure 14.22).

Moreover, in a curved beam, where the prestressing cable runs parallel to the shear centre axis and is anchored at its two ends, no torsional loads are imposed, as the prestressing force does not offer anywhere a component on the corresponding section plane (see Figure 14.22).

On the basis of the foregoing, it can be seen that in a curved statically determinate girder an arbitrary tendon layout with a given prestressing force causes exactly the same response (i.e. diagrams of sectional forces M_B, Q and M_T) as that developed in the corresponding rectilinear (i.e. 'stretched out') girder. Why this happens is now explained.

A statically determinate structure containing a prestressed tendon appropriately anchored, if subjected only to the prestressing forces (anchor and deviation forces),

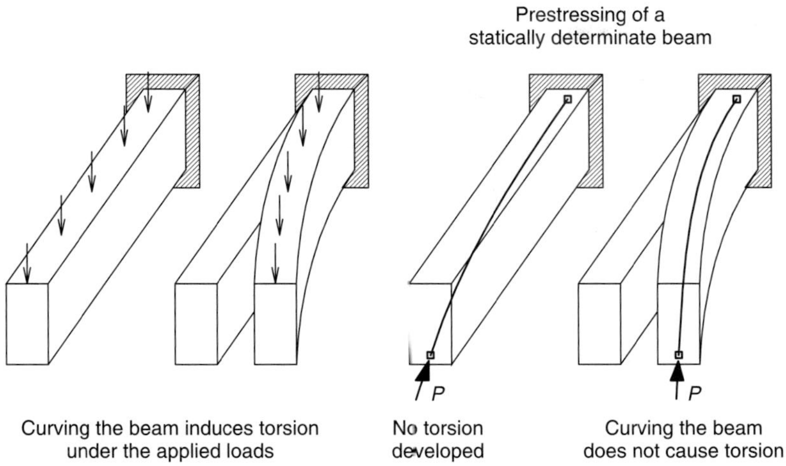

Figure 14.23 External loads and prestressing in a curved statically determinate beam

does not develop any reaction at the points of support. Hence, any bending or torsional moment, or even shear force of a section, results from the prestressing force acting on it. The point of application and the direction of this force always, as explained in Section 4.3.1.1, correspond to the appropriate trace and local tangent of the tendon. If the component of this action on the section plane does not pass from the shear centre, then a torsional moment is caused, as previously mentioned. Otherwise, no torsion is imposed. This situation remains unchanged if the initially considered straight girder becomes curved (Figure 14.23).

Thus, while a rectilinear statically determinate girder, under, for example, its self-weight, will develop torsion if it becomes curved, a prestressing tendon not producing torsion in the rectilinear girder will continue to do so, even if the girder becomes curved. Accordingly, it may be concluded that in a curved statically determinate girder, if the prestressing deviation forces completely counterbalance the external loads (e.g. self-weight), bending will be eliminated but torsion will not.

It should, however, be noted at this point that the 'bending moment' M_P due to prestressing should not be perceived as creating (see Section 14.3.2) a torsional load M_P/R, just because the bending moment M_P is not 'genuine', i.e. it is not directly required for the equilibrium, as recalled in Figure 14.24.

Of course, all the above is changed once the girder becomes statically indeterminate. A curved continuous girder, for example, with a prestressing tendon that in the assumed 'stretched' structure does not induce torsional moments develops a torsional response, as explained below, and depicted in Figure 14.25.

The continuous girder considered in Figure 14.25 has simple supports at both its ends that are capable of developing torsional reactions, whereas it is simply supported in the middle. The primary – statically determinate – structure obtained through removal of the intermediate support under the existing deviation and anchor forces does not develop any torsion, as already explained. However, the redundant vertical force that must be applied in order to eliminate the deflection occurring at the midpoint of the

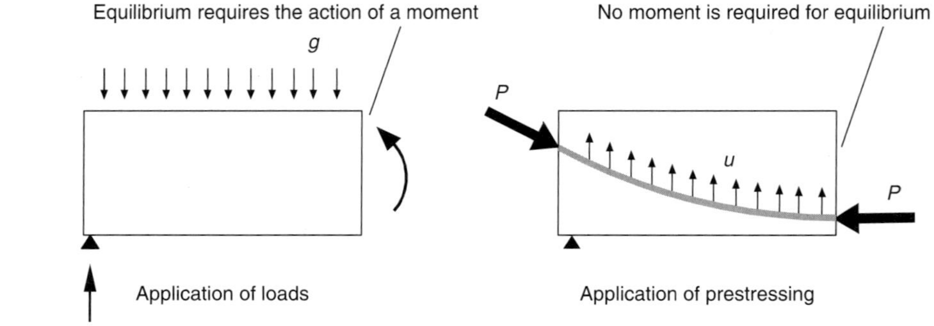

Figure 14.24 Prestressing in a statically determinate beam implies an internal moment

primary structure does induce a torsional response. This externally imposed redundant force, which differs very little from the corresponding reaction of the rectilinear continuous beam, provided that the values L/R as well as κ are small enough (see Section 14.3.2.1), will induce a triangular bending moment diagram in the primary

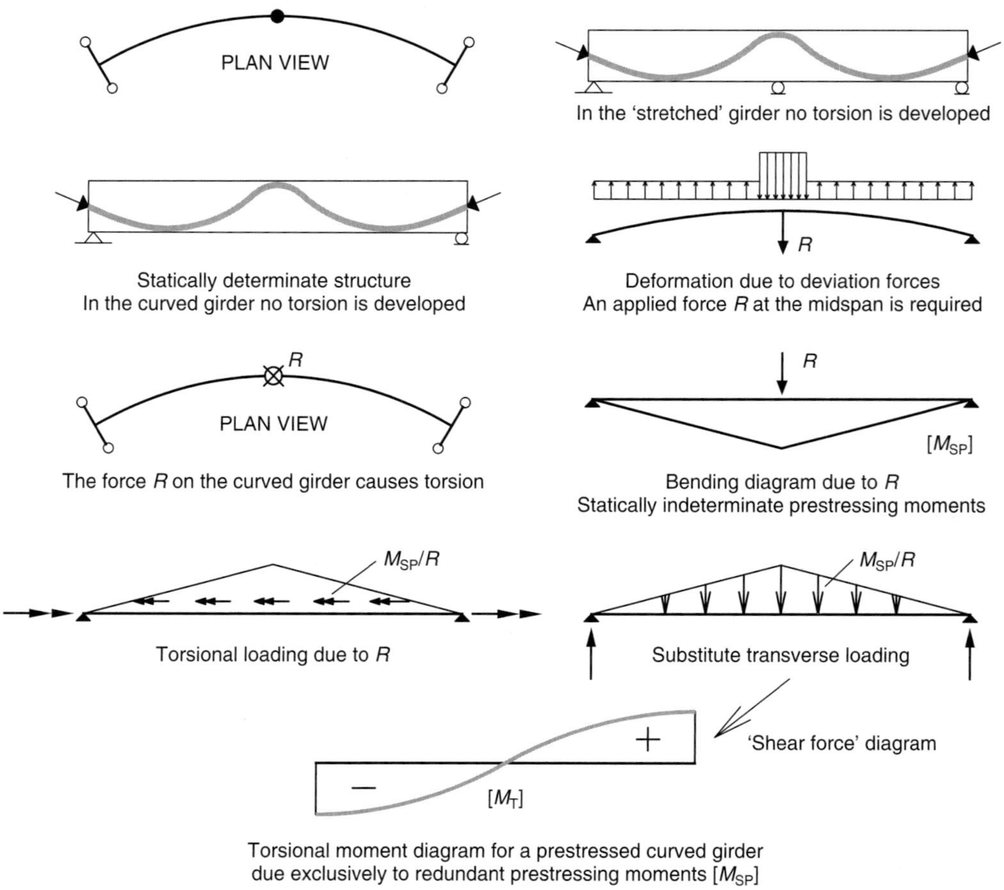

Figure 14.25 Influence of prestressing on a curved redundant beam

structure. This represents the statically indeterminate prestressing moments [M_{SP}], as explained in Section 5.4.1. The bending diagram [M_{SP}] applied to the curved statically determinate girder will automatically produce the torsional loading [M_{SP}/R]. The implied torsional response will then be obtained as the shear force diagram of the corresponding rectilinear beam, due to the transverse triangular loading [M_{SP}/R] (see Figure 14.25). It is clear, then, that a prestressing tendon in a curved continuous girder causes a torsional response solely due to the statically indeterminate prestressing moments M_{SP} of the corresponding rectilinear beam.

Thus, according to the above and referring to Figure 14.26, the torsional moment diagram [M_T] of a curved statically indeterminate prestressed girder can be obtained

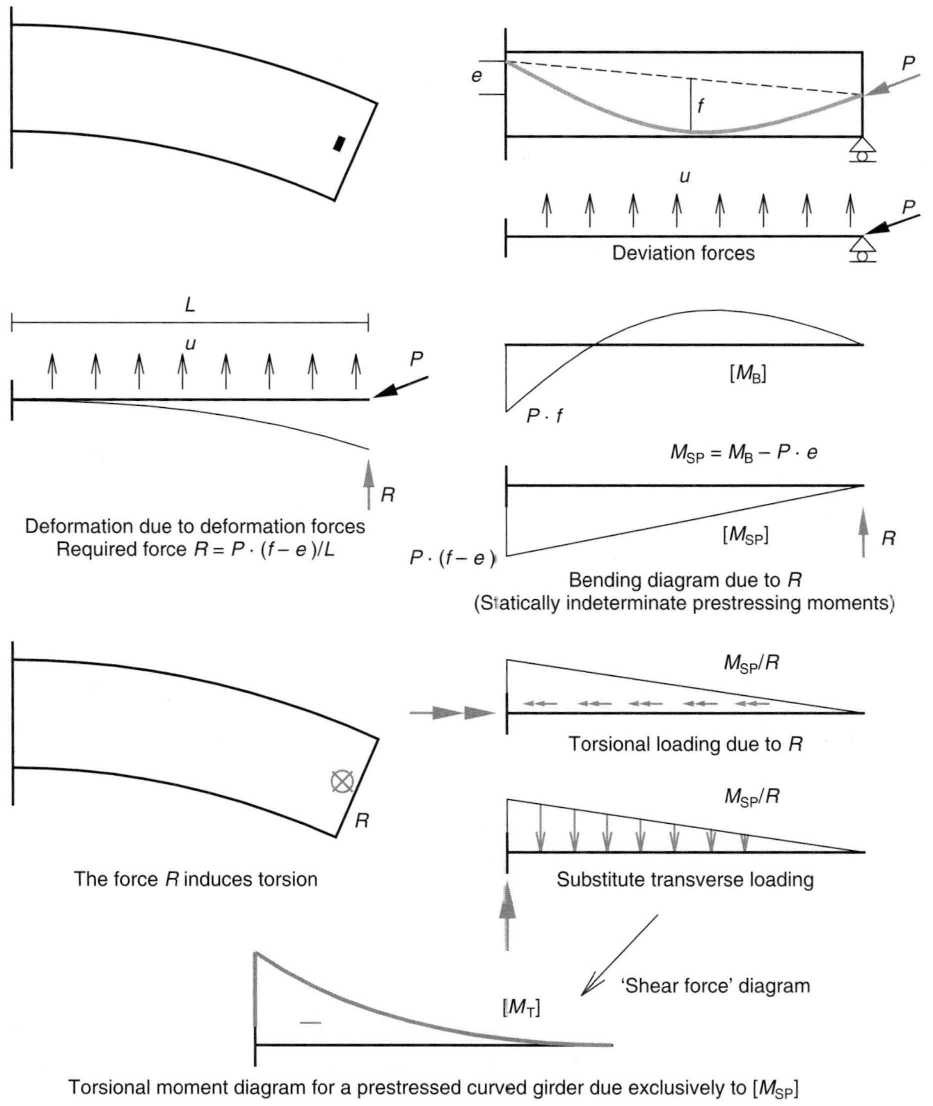

Figure 14.26 Assessment of the influence of prestressing on a statically indeterminate curved girder

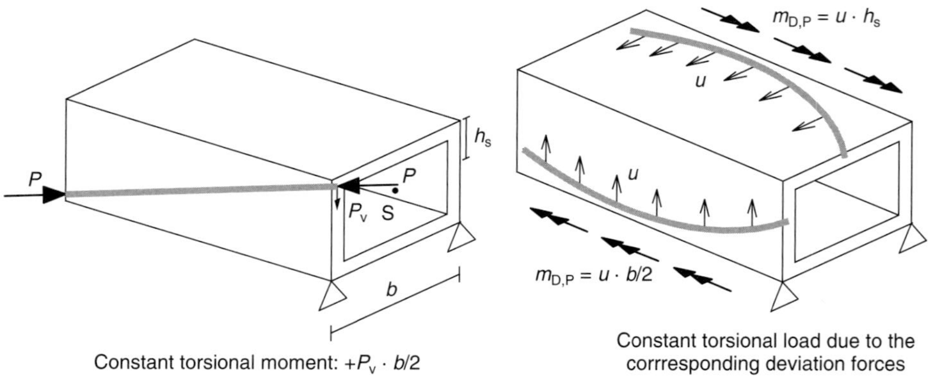

Figure 14.27 Creating torsional loading through separate prestressing cables

on the basis of the bending moment diagram [M_P] due to deviation and anchor forces (see Section 5.4.1), as follows. After the statically indeterminate moments ($M_{SP} = M_P - P \cdot e$) have been determined, where P and e are the prestressing force and the tendon eccentricity at each position, respectively, the torsional load [M_{SP}/R] is applied as a distributed vertical load on the statically determinate rectilinear girder. The resulting shear force diagram represents, then, the torsional moment diagram [M_T] of the statically indeterminate curved girder.

14.3.5 Reducing the torsional response through prestressing

The distributed torsional action m_D, or, more generally, the development of a torsional moment diagram [M_T], in a box girder, may be reduced to a lesser degree through prestressing (Menn, 1990).

As has been stated previously, the presence of an inclination at some point on a prestressing cable in a girder web principally implies a torsional action. If the cable is straight, the torsional response developed is constant. If the cable is parabolic, the constant deviation forces u induce a constant distributed torsional load: $m_{D,P} = u \cdot b/2$ (Figure 14.27).

The arrangement of a straight cable in the top or bottom slab of the girder, in a different direction to the girder axis, will also cause a constant torsional moment, while a parabolic cable, through its deviation forces, will impose a constant torsional load $m_{D,P} = u \cdot h_s$, where h_s is the distance from the shear centre (see Figure 14.27).

It is clear, however, that arranging a parabolic prestressing cable in one web only implies the imposition not only of a distributed torsional load but also of a distributed vertical load, which may be undesirable if only a torsional load $m_{D,P}$ is to be imposed. For this reason, another parabolic cable with opposite curvature should be arranged in the other web (Figure 14.28). Analogously, in a horizontal cable layout, the girder slabs should be provided with cables of opposite curvature.

The prestressing tendons can be specified separately in the first stage of design, to counteract the bending and the torsional response of the girder. In a second stage,

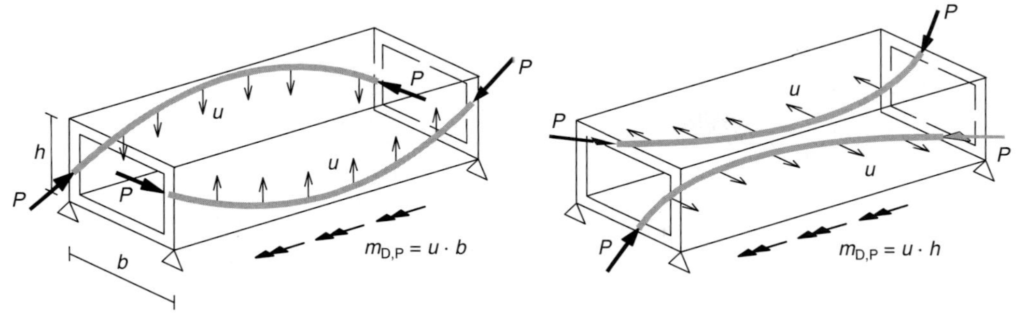

Figure 14.28 Basic layouts of prestressing cables for inducing torsional action

the synthesis of the two separate tracings into a single, final, one can be attempted, by retaining, if possible, the number of cables (i.e. the total prestressing force) required for bending. In this way, a differentiated tendon curvature can be obtained for both webs, which must ensure the required upward deviation forces, on the one hand, and offer the required torsional counter-load, on the other.

The above is illustrated by the example of the simply supported curved girder in Figure 14.29. It becomes clear that the arrangement of a convex cable in the 'outer'

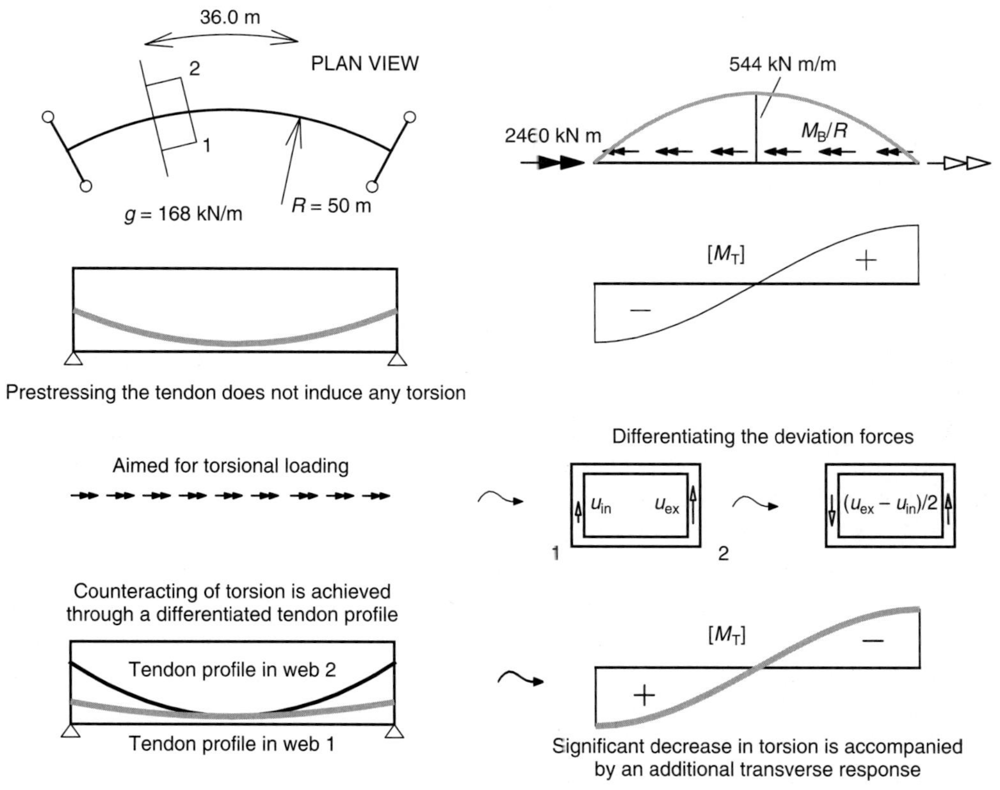

Figure 14.29 Relieving influence of the tendon layout on the torsional response

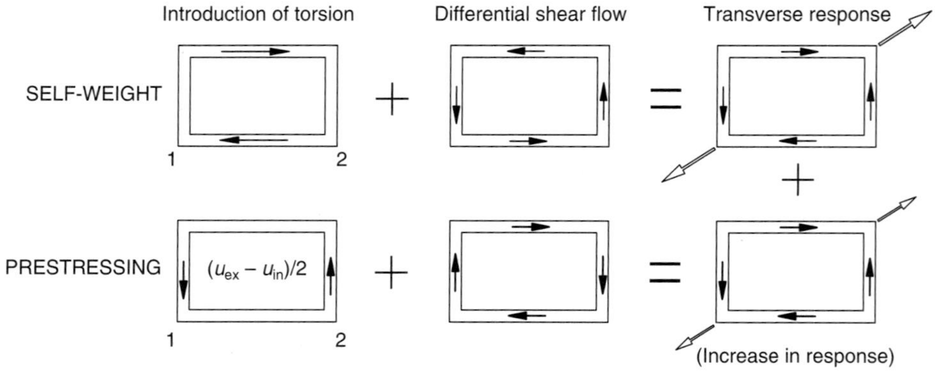

Figure 14.30 Influence of the prestressing layout on the transverse response of a girder

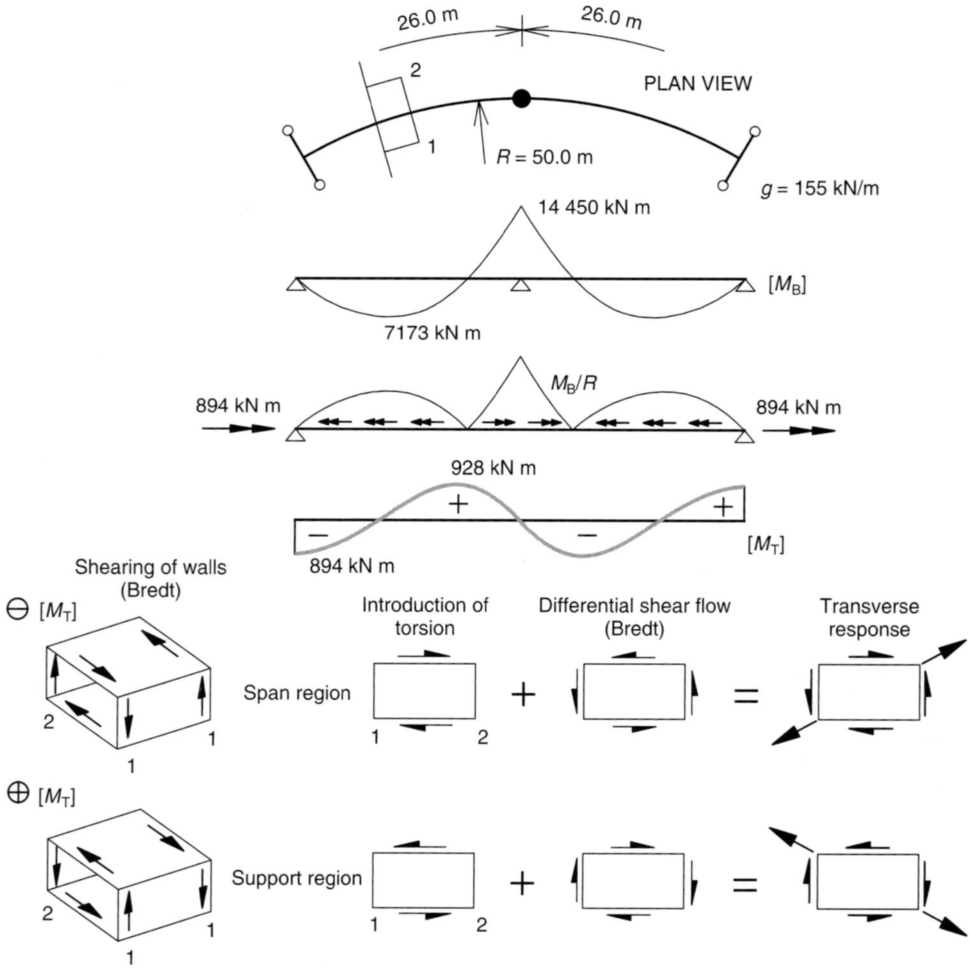

Figure 14.31 Longitudinal and transverse response due to the self-weight of a girder

web and a concave cable in the 'inner' web will produce a torsional load counteracting the torsional effect of the self-weight. Consequently, the final tendon profiles present a stronger curvature for the outer web than for the inner one, both of them being, of course, convex.

If in the above case the deviation forces of the outer and inner web are u_{ex} and u_{in}, respectively ($u_{ex} > u_{in}$), it can be seen that, while the symmetric total deviation force $(u_{ex} + u_{in})/2$ counteracts the gravity loads, the resulting torsional load is due to the antisymmetric loading of each web with $(u_{ex} - u_{in})/2$, thus being equal to $m_{D,P} = (u_{ex} - u_{in}) \cdot b/2$. However, this torsional load will cause a corresponding peripheral shear flow, leading to an additional longitudinal and transverse response of the girder, according to the examination in Section 14.3.3 (Figure 14.30).

14.3.6 Summary examples

It is appropriate at this point to recapitulate the complete torsional response of a single-span prestressed curved girder, according to the following four states of stress:

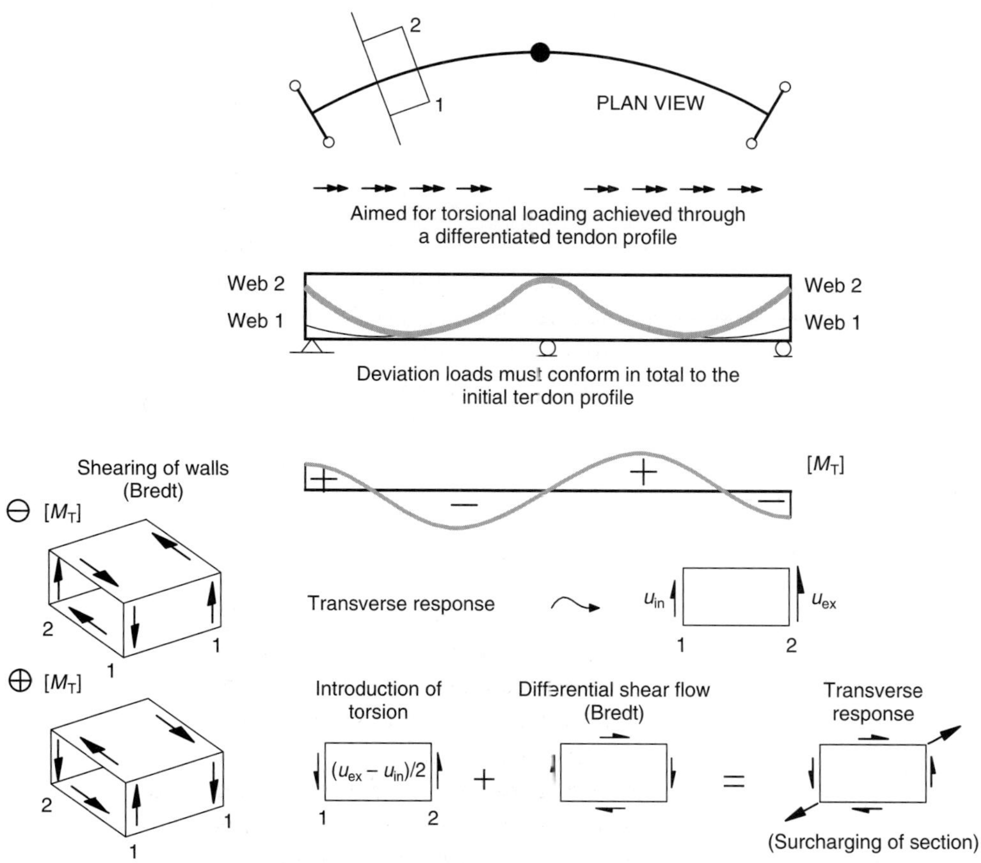

Figure 14.32 Influence of the differentiated tendon profile on the torsional response

(a) The shear response of the section walls according to Bredt for gravity loads, due to the existing curvature of the girder.
(b) The additional response due to the deformation of the section profile, on the basis of the differential shear flow in (a) and the deviation forces on the top and bottom slabs.
(c) The shear response of section walls according to Bredt, due to the torsional loads initiated by the eventually differentiated tendon profile in the webs.
(d) The additional response due to the deformation of the section profile, on the basis of the differentiated shear flow in (c) and the antisymmetric loading of the two webs by the deviation forces of the tendons.

The application of the above in the case of the single-span girder examined, as in Figures 14.29 and 14.30, shows that the transverse response of the section due to the differentiated tendon profile in the webs increases that arising from the self-weight, resulting in greater longitudinal and transverse bending.

It can be seen that in the case of a statically indeterminate girder the following states of stress must, furthermore, be taken into account:

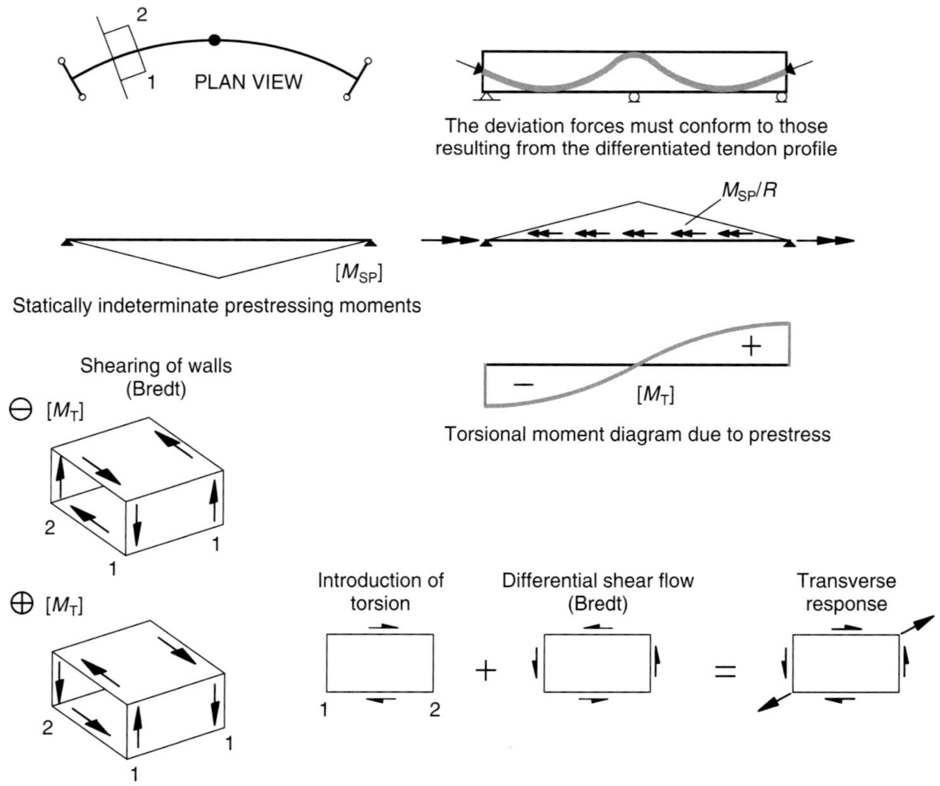

Figure 14.33 Influence of prestressing on the torsional response due to the static redundancy

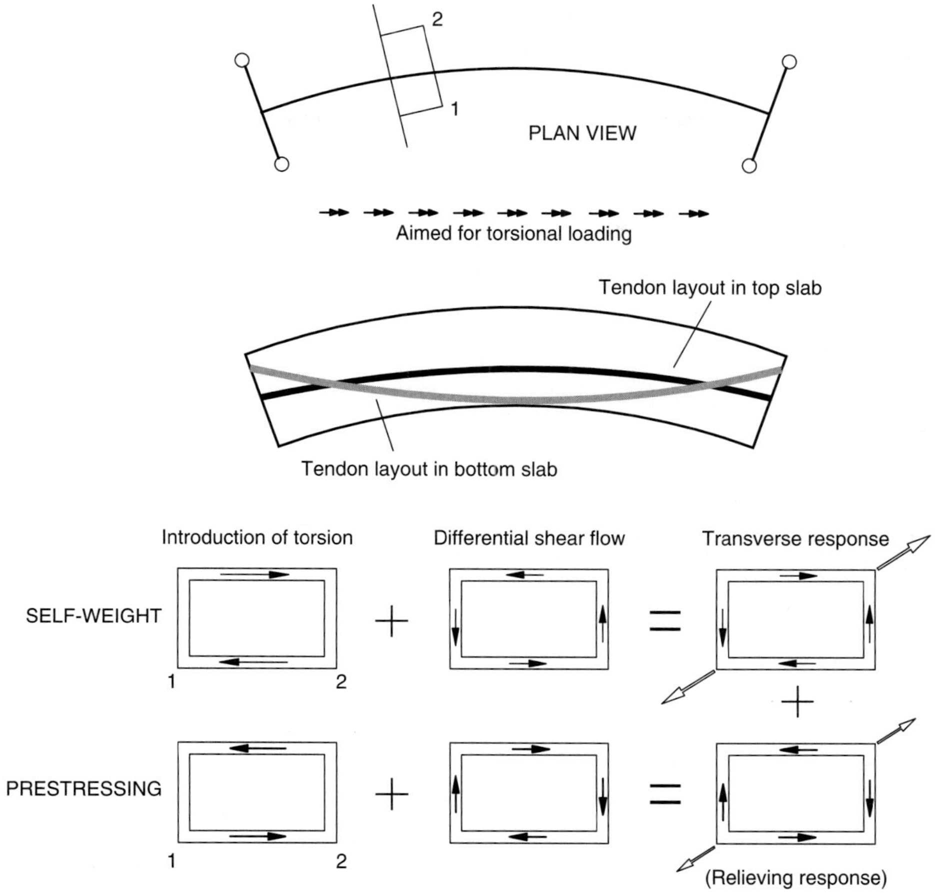

Figure 14.34 Consequences of the arrangement of additional tendons in the top and bottom slabs

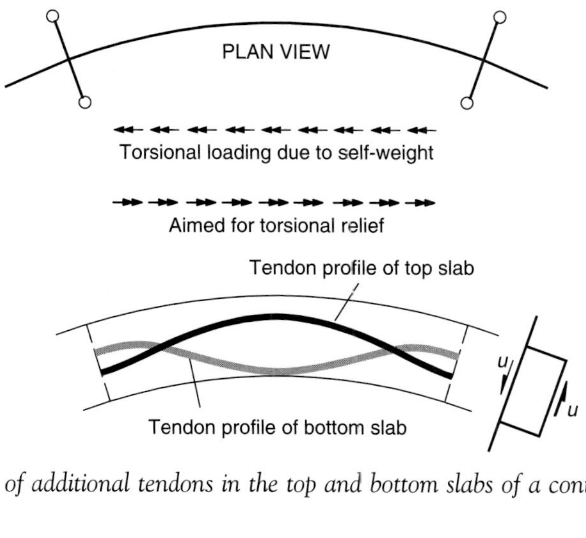

Figure 14.35 Layout of additional tendons in the top and bottom slabs of a continuous girder

(e) The shearing response of the section walls (Bredt), due to the torsional response on the basis of the statically indeterminate prestressing moments M_{SP}, according to Section 14.3.4.
(f) The additional response due to the deformation of the section profile on the basis of the differential shear flow and the deviation forces of the top and bottom slabs occurring in (e), according to Section 14.3.3, in the same way as in (b).

Figures 14.31–14.33 illustrate the treatment of a prestressed continuous curved girder, in order to reduce the torsional response due to the action of a permanent load as the girder self-weight, following the above sequence:

- steps (a) and (b) are illustrated in Figure 14.31
- steps (c) and (d) are illustrated in Figure 14.32
- steps (e) and (f) are illustrated in Figure 14.33. ■

As previously explained, the torsional response due to permanent loads may clearly be reduced through the arrangement of additional tendons in the top and bottom slabs, which, by means of their deviation forces u, offer a relieving torsional moment $m_{D,P} = u \cdot h$, as shown in Figure 14.34.

The introduction of these horizontal deviation forces, together with the differential shear flow they cause, lead, according to the foregoing, to a 'diagonal' state of stress, which drastically reduces that due to the permanent load (see Figure 14.34). It is clear that the above layout is much more effective for torsional reduction in a continuous girder than the previously considered differentiation of the profiles of the tendons within the webs (Figure 14.35). However, this can be achieved only through the extra cost of additional tendons.

The task of restricting the torsional response through prestressing must always be undertaken for each case with regard to its technological–economic consequences, and taking account of the cost of the difficulties that will be encountered in the realisation of the prestressing layout.

References

Menn C. (1990) *Prestressed Concrete Bridges*. Basel: Birkhäuser Verlag.
Schlaich J., Scheef H. (1982) *Concrete Box-girder Bridges*. Zurich: IABSE.
Schlaich J., Seidel J. (1988) *Die Fussgängerbrücke in Kelheim*. Bauingenieur 63.

15

Lateral response of multi-storey systems

15.1 Introduction

In this chapter the behaviour of a three-dimensional multi-storey building system under lateral forces is examined. These forces may result either through an imposed seismic ground motion, or as an incoming wind action, or even through a temperature variation which, as will be explained later, activates the lateral stiffness of the structural system. The reason for making the consideration of temperature change an essential design issue is not only because of the obvious environmental temperature influence on the structure but also because of the fact that the shrinkage that the concrete slabs at each storey of the load-bearing system are subjected to, after the concrete has hardened, may be expressed through a temperature fall, thus causing a response of the whole system.

15.2 Formation of the system

Multi-storey systems are considered here as horizontal diaphragms (slabs) with very high rigidity within their own plane, and arranged vertically at a regular distance of a storey height and connected with each other as well as with the foundation ground through vertical elements of rigidity, according to a particular layout plan. These vertical elements consist either of frames formed by the columns and the building girders, or of single columns or shear walls, or possibly of thin-walled elements having an open or closed section (Figure 15.1).

It should be noted that the diaphragm concept used below is not necessarily restricted to a solid concrete slab but can apply to any structural formation possessing a high stiffness in its own plane. The same naturally applies for all the vertical elements, which do not necessarily have to be made of concrete but can be formed of steel elements.

Each of the horizontal diaphragms under the action of horizontal forces moves as a rigid body within its own plane. This movement consists of a horizontal displacement as well as a rotation, and is obviously imposed unchanged on all vertical elements, causing a corresponding response (Figure 15.2).

In the present examination, which, as previously pointed out, facilitates, in principle, structural design rather than 'accurate' analysis, it may be assumed that all vertical elements, as the supporting skeleton of the diaphragms, constitute a set of elements of purely plane stiffness (Stavridis, 1986).

Structural systems: behaviour and design
978-0-7277-4105-9
Copyright Thomas Telford Limited © 2010
All rights reserved

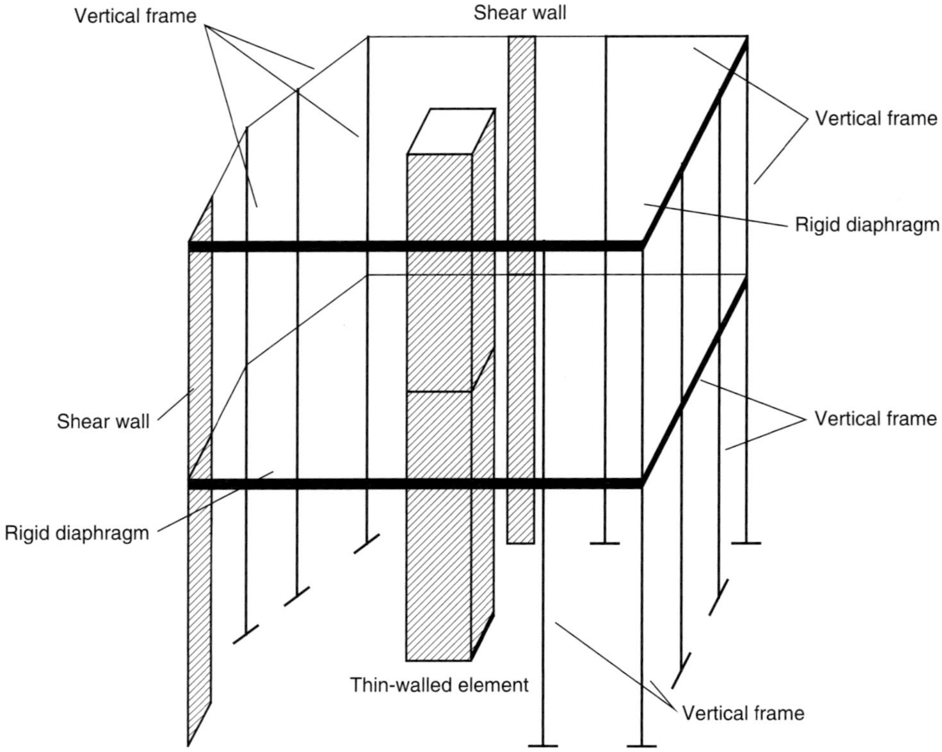

Figure 15.1 Layout of horizontal diaphragms and vertical stiffness elements

By the term *element of purely plane stiffness* or simply *plane element*, it is meant that the element is stressed only by that component of the diaphragm displacement which corresponds to its plane. Thus, for example, while the same displacement and rotation vector of the diaphragm is applied to each girder of all frames in Figure 15.3, each frame is

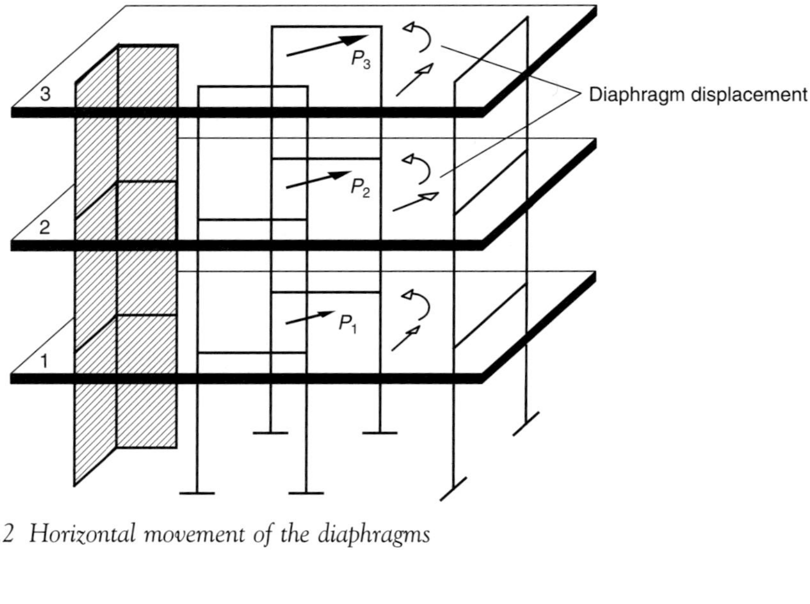

Figure 15.2 Horizontal movement of the diaphragms

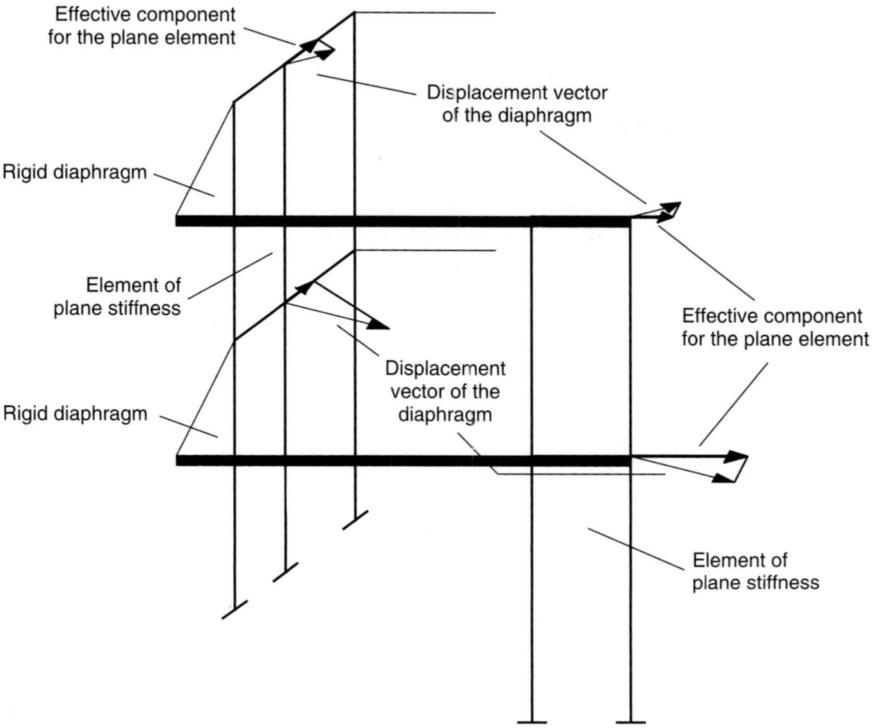

Figure 15.3 Activation of stiffness of the plane elements through the diaphragm

assumed to be stressed only by that component of the displacement vector corresponding to the specific frame plane.

Although the bending stiffness of a frame with respect both to its own plane and transversely to it can be described by two separate plane vertical elements, the clear preponderance of the stiffness in its own plane allows the omission of the corresponding transverse plane element. However, in a single column, two plane elements may be considered representative of its stiffness with respect to its two principal stiffness senses (Figure 15.4).

Moreover, it may be understood that a vertical thin-walled element of either an open or closed section may be represented by the plane elements constituting the whole section. These plane elements are also able to offer the torsional stiffness of the element, as any applied torsional load may be taken up by the plane state of stress of the constituent plane elements themselves (Figure 15.4).

15.3 Lateral response

15.3.1 Treatment of load-bearing action

The basic problem may be stated in the following way.

It is assumed that the diaphragm of the nth level is acted upon on its centroid by a horizontal force P (Figure 15.5). The diaphragm, with reference to the orthogonal

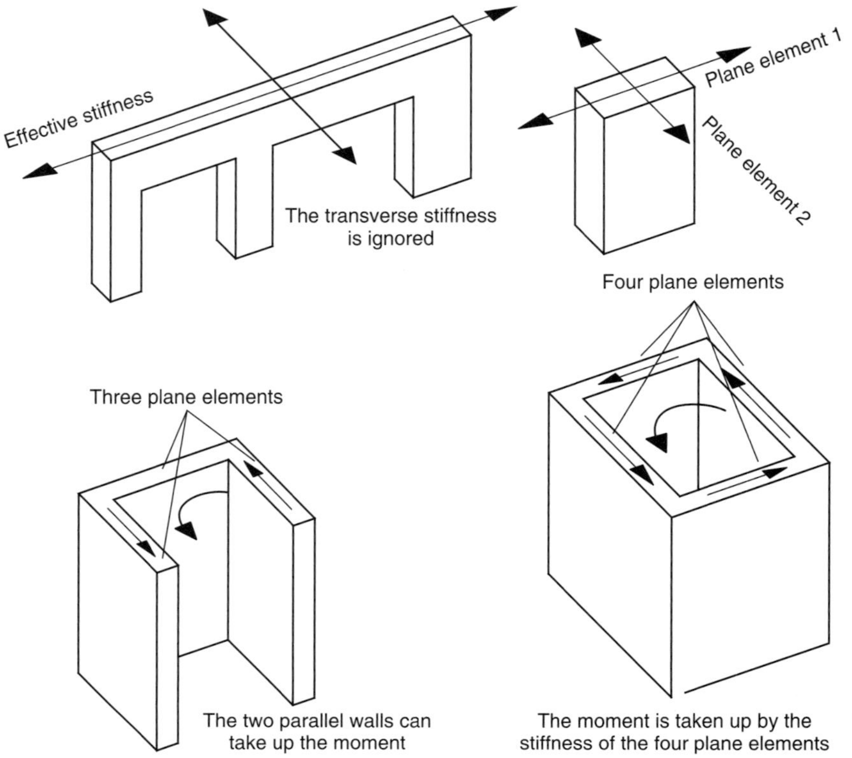

Figure 15.4 Expressing the stiffness of vertical members through the elements of the plane stress

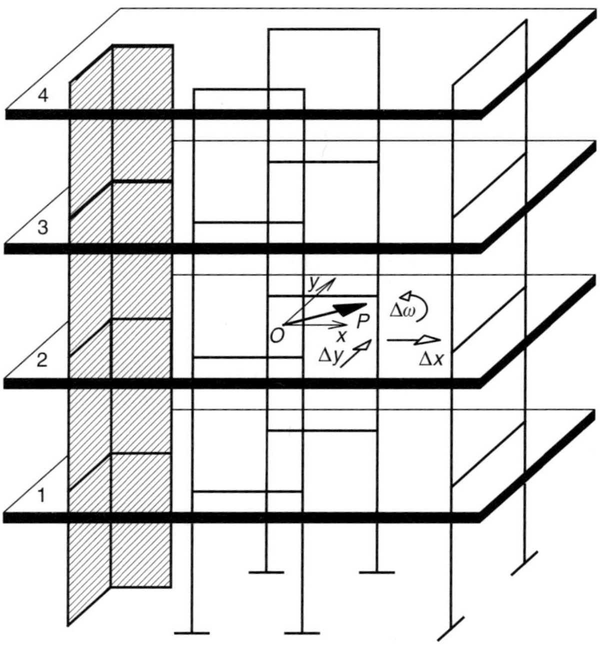

Figure 15.5 Horizontal loading of a single diaphragm

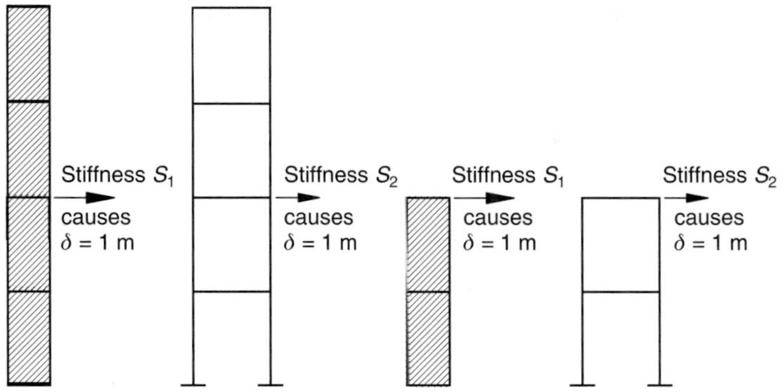

Figure 15.6 Lateral stiffness of a plane element

axes Oxy having their origin O at the centroid of the diaphragm (i.e. its centre of mass), undergoes, through the action of the force P, horizontal movement as a rigid body, represented by the components of displacement Δx and Δy as well as by the rotation $\Delta\omega$, taken as positive if anticlockwise. This displacement is obviously resisted by the lateral stiffness of plane elements, as explained above.

The lateral stiffness of each plane element is represented by the required horizontal force S acting on a specific level n, in order to induce a corresponding unit displacement (1 m) at that level. This force remains essentially the same, even if all the upper storey levels are omitted, as shown in Figure 15.6. This fact, apart from its simplifying effect, has also a further convenient consequence, as will become clear later.

The resistance of the plane elements to the diaphragm movement implies for each element the action of a certain force at the corresponding level, resulting from the imposed displacements on them, while the same forces are returned in an opposite sense to the diaphragm itself. This must be in equilibrium under the above forces as well as under the directly acting force P (Figure 15.7). Given that these 'resistance' forces may be expressed – as will be shown later – in terms of the displacement components Δx, Δy and $\Delta\omega$, these components can be determined through the three equations of equilibrium of the diaphragm. The above components thus allow the evaluation of the forces acting on each plane element at the specific level. However, it must be ensured that the diaphragm in question can safely take up the self-equilibrating coplanar forces that it is subjected to.

The simultaneous loading of some or all of the diaphragms subjected to corresponding forces P may be treated for each diaphragm separately, by neglecting all the above-lying levels of the plane elements involved, as has been previously indicated. Of course, the final response of each vertical element will be deduced from the total action of the corresponding forces determined at each level (Figure 15.8). These forces are used for the design of the plane elements and the assessment of their displacements at the various levels, as well as for the response of the corresponding diaphragm.

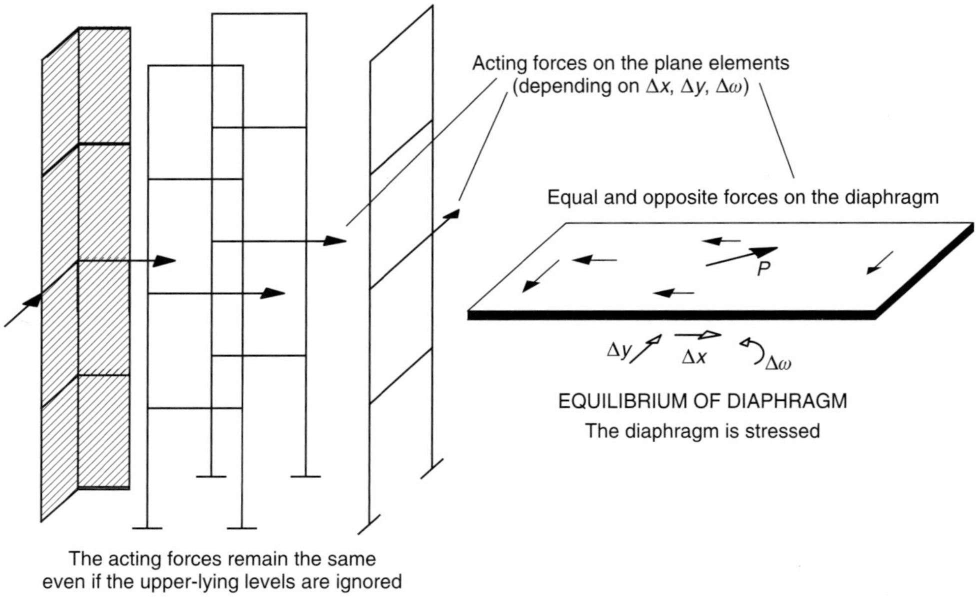

Figure 15.7 Equilibrium of the loaded horizontal diaphragm

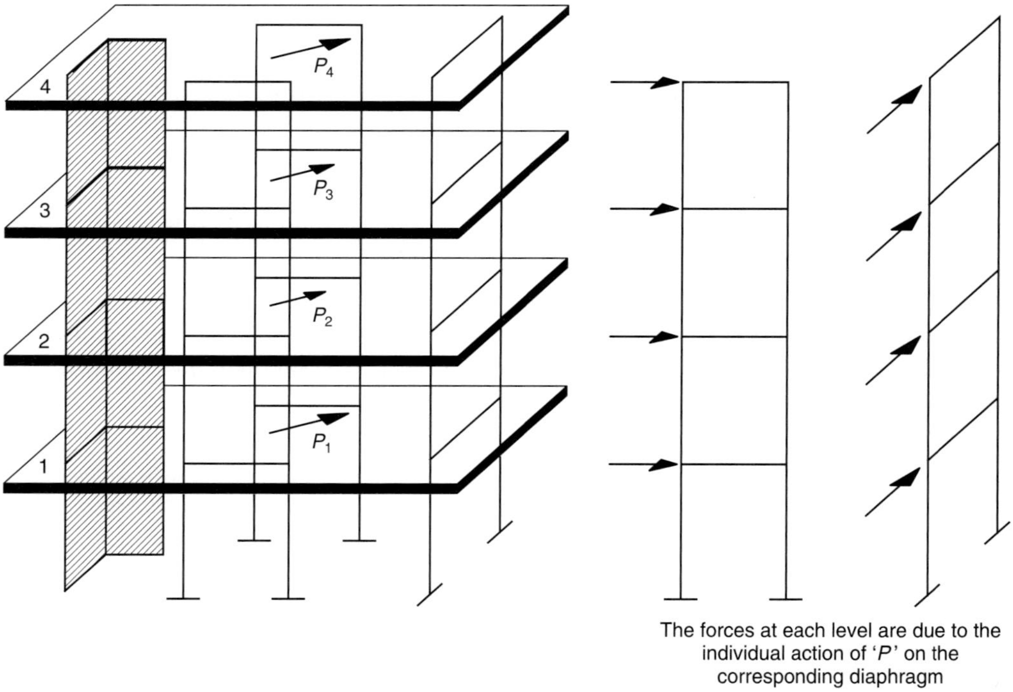

Figure 15.8 Treatment of the loading of all diaphragms

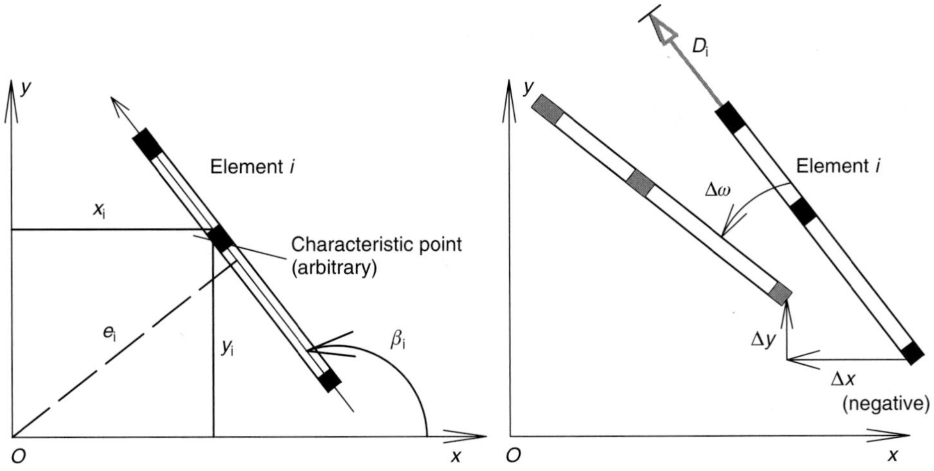

Figure 15.9 Geometric and kinematic characteristics of the plane element

15.3.2 Response of a plane element

Each plane element i is characterised by the angle β_i, described in an anticlockwise sense by the axis Ox with an arbitrarily determined positive direction on the trace of the element in the ground plan. On this trace a characteristic point with coordinates x_i and y_i is selected (Figure 15.9).

By considering a specific level i of an element where the displacement components Δx, Δy and $\Delta\omega$ of the corresponding diagram are imposed, the corresponding displacement D_i for the element may be determined, by assuming that the diaphragm, as already pointed out, remains undeformable during its movement as a rigid body:

$$D_i = \Delta x \cdot \cos\beta_i + \Delta y \cdot \sin\beta_i + \Delta\omega \cdot e_i$$

where e_i is the (signed) distance of the trace of the element from the origin of the coordinate axes O, being equal to

$$e_i = x_i \cdot \sin\beta_i - y_i \cdot \cos\beta_i$$

In the level under consideration, the element i may be structurally represented by the corresponding force S that must be applied in order to produce a unit displacement (1 m), known also as the *lateral stiffness*. This magnitude can be determined by calculating the displacement δ (m) due to a unit horizontal force (1 kN) applied to that level (Figure 15.10). Then,

$$S = 1/\delta \quad \text{(kN/m)}$$

As has already been pointed out in Section 15.3.1, neglect of the overlying levels does not affect the above displacement in practice, or, in consequence, the corresponding stiffness (see Figure 15.10).

Thus, in order for the plane element i to contribute to the imposed diaphragm displacement at the level under consideration, a horizontal force $F = S_i \cdot D_i$ must be

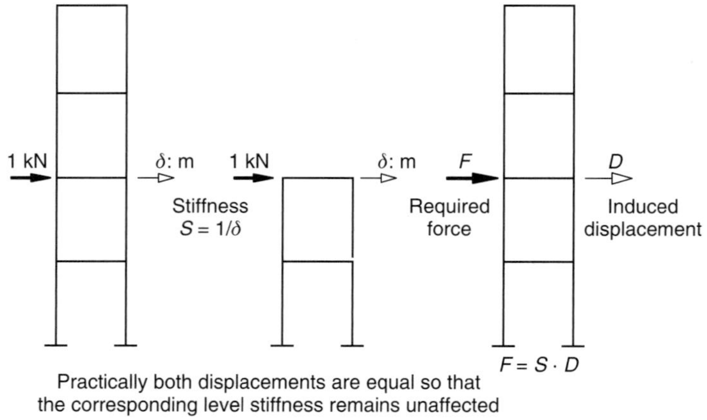

Figure 15.10 Determination and evaluation of the translational stiffness of a plane element

imposed on it, i.e.

$$F = S_i \cdot (\Delta x \cdot \cos\beta_i + \Delta y \cdot \sin\beta_i + \Delta\omega \cdot e_i)$$

15.3.3 General layout

If the diaphragm of a specific level is subjected to a horizontal force P applied to its centroid and represented by the components P_x and P_y, then it may be found from its equilibrium, as previously considered, that the components Δx, Δy and $\Delta\omega$ of its displacement must satisfy the following three-equation system:

$$K_1 \cdot \Delta x + K_2 \cdot \Delta y + K_3 \cdot \Delta\omega = P_x$$

$$K_2 \cdot \Delta x + K_4 \cdot \Delta y + K_5 \cdot \Delta\omega = P_y$$

$$K_3 \cdot \Delta x + K_5 \cdot \Delta y + K_6 \cdot \Delta\omega = 0$$

where

$$K_1 = \sum S_i \cdot \cos^2\beta_i, \quad K_2 = \sum S_i \cdot \sin\beta_i \cdot \cos\beta_i, \quad K_3 = \sum S_i \cdot e_i \cdot \cos\beta_i$$

$$K_4 = \sum S_i \cdot \sin^2\beta_i, \quad K_5 = \sum S_i \cdot e_i \cdot \sin\beta_i, \quad K_6 = \sum S_i \cdot e_i^2$$

Once the components Δx, Δy and $\Delta\omega$ are known, the horizontal forces acting at the corresponding levels for all plane elements can be readily determined, as described in Section 15.3.2 (Stavridis, 1986).

The stiffness S_i at any level may be thought of as a horizontal spring in the direction of the corresponding element, which may be connected to the diaphragm at the position of the corresponding characteristic point of the element (Figure 15.11). In this way, the diaphragm acted upon by the force P maintains its equilibrium because of the developed spring forces acting on it, while the same forces are also acting on the corresponding plane elements.

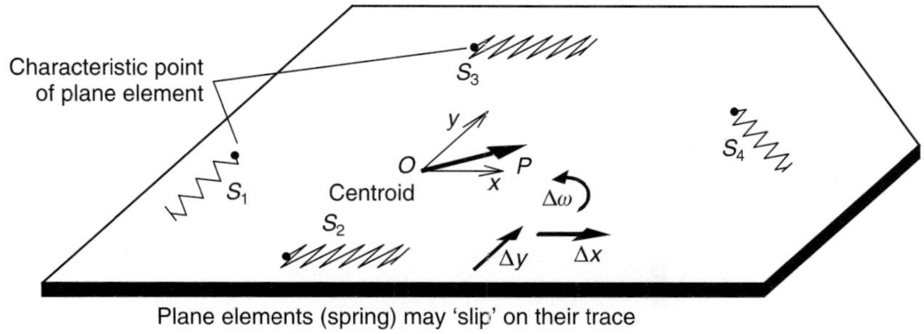

Figure 15.11 Visualisation of the support of the loaded diaphragm through the corresponding springs

Alongside the above considerations, it is also useful to clarify the structural significance of the coefficients K_1, K_2, K_3, K_4, K_5 and K_6 in the above system of equations. This group of coefficients characterises the resistance offered by the system of plane elements against the horizontal movement of the diaphragm. More specifically, the terms $K_1 \cdot \Delta x + K_2 \cdot \Delta y + K_3 \cdot \Delta\omega$, $K_2 \cdot \Delta x + K_4 \cdot \Delta y + K_5 \cdot \Delta\omega$ and $K_3 \cdot \Delta x + K_5 \cdot \Delta y + K_6 \cdot \Delta\omega$ represent, respectively, the forces P_x and P_y, and the torsional moment M_ω that must be applied at the origin O in order to cause a diaphragm displacement having the components Δx, Δy and $\Delta\omega$. In other words, the above terms represent the resistance offered by the diaphragm at point O if this is 'obliged' to be displaced by Δx, Δy and $\Delta\omega$. It is, of course, understood that specifying a 'point of application' for the torsional moment M_ω does not make much sense, as the moment may be applied anywhere in the diaphragm. However, the null value that is assigned to this moment in the above equation system is due to the fact that the displacement components Δx, Δy and $\Delta\omega$ are realised simply through the application of a single force P at point O, i.e. without imposing any moment M_ω. If for any reason the diaphragm is acted upon by an external moment M_ω, then this value – positive for an anticlockwise direction – must be used instead of zero.

From the above equations it can be deduced that the plane elements may be 'shifted' anywhere along their own trace, as only their corresponding angle β_i and their distance e_i from the origin of the coordinate axes matter. This may also be understood through the concept of spring supports introduced above, which, owing to the absolute rigidity of the diaphragm, may be connected at any point along their line of action.

15.3.4 Orthogonal layout

The vertical elements in building structures are usually arranged in mutually orthogonal directions. In this case, the axes Ox and Oy are placed parallel to these two directions (Figure 15.12). Then, all the elements in the x sense have $\beta_i = 0$ and $e_i = -y_i$, whereas all the elements in the y sense have $\beta_i = 90°$ and $e_i = x_i$.

Thus (see Section 15.3.3),

$$K_1 = \sum S_x, \quad K_2 = 0, \quad K_3 = -\sum (S_x \cdot y_i)$$

$$K_4 = \sum S_y, \quad K_5 = \sum (S_y \cdot x_i), \quad K_6 = \sum (S_x \cdot y_i^2) + \sum (S_y \cdot x_i^2)$$

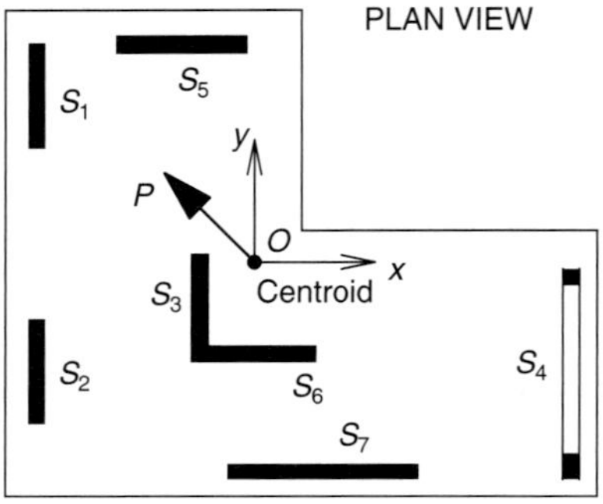

Figure 15.12 Orthogonal layout of the plane elements

while Δx, Δy and $\Delta\omega$ are determined from the following expressions:

$$\Delta x = \frac{P_x}{K_1} - \frac{K_3}{K_1}\cdot\frac{\dfrac{K_3}{K_1}+\dfrac{K_5}{K_4}}{\dfrac{K_3^2}{K_1}+\dfrac{K_5^2}{K_4}-K_6}, \qquad \Delta y = \frac{P_y}{K_4} - \frac{K_5}{K_4}\cdot\frac{\dfrac{K_3}{K_1}+\dfrac{K_5}{K_4}}{\dfrac{K_3^2}{K_1}+\dfrac{K_5^2}{K_4}-K_6}$$

$$\Delta\omega = \frac{\dfrac{K_3}{K_1}\cdot P_x + \dfrac{K_5}{K_4}\cdot P_y}{\dfrac{K_3^2}{K_1}+\dfrac{K_5^2}{K_4}-K_6}$$

For the forces acting on the elements in the *x* direction at the level under consideration (see Section 15.3.2),

$$F_x = S_x\cdot(\Delta x - \Delta\omega\cdot y_i)$$

whereas for the elements in the *y* direction the forces acting are

$$F_y = S_y\cdot(\Delta y + \Delta\omega\cdot x_i)$$

The magnitudes of K_3 and K_5, i.e. $\sum(S_x\cdot y_i)$ and $\sum(S_y\cdot x_i)$, respectively, play an important role. If both are zero, the diaphragm does not rotate at all ($\Delta\omega=0$), and Δx and Δy are simply

$$\Delta x = P_x/\sum S_x, \quad \Delta y = P_y/\sum S_y$$

This result may also be interpreted and evaluated according to the spring model of the rigid diaphragm (Figure 15.13).

A force P_x acting on the diaphragm at point O in the *x* direction, such that it causes no rotation but only a displacement Δ, must lie on the line of action of the resultant of the spring forces $(S_x\cdot\Delta)$, which means that $\sum(S_x\cdot\Delta)\cdot y=0$, i.e. $\sum(S_x\cdot y)=0$.

Moreover, for equilibrium of the diaphragm, $P_x=\sum(S_x\cdot\Delta)=\Delta\cdot(\sum S_x)$, as has been exactly obtained before. For this examination of the behaviour in the *x* sense, a rigid beam may substitute for the diaphragm, as shown in Figure 15.13. Analogous results are also obtained for the *y* sense.

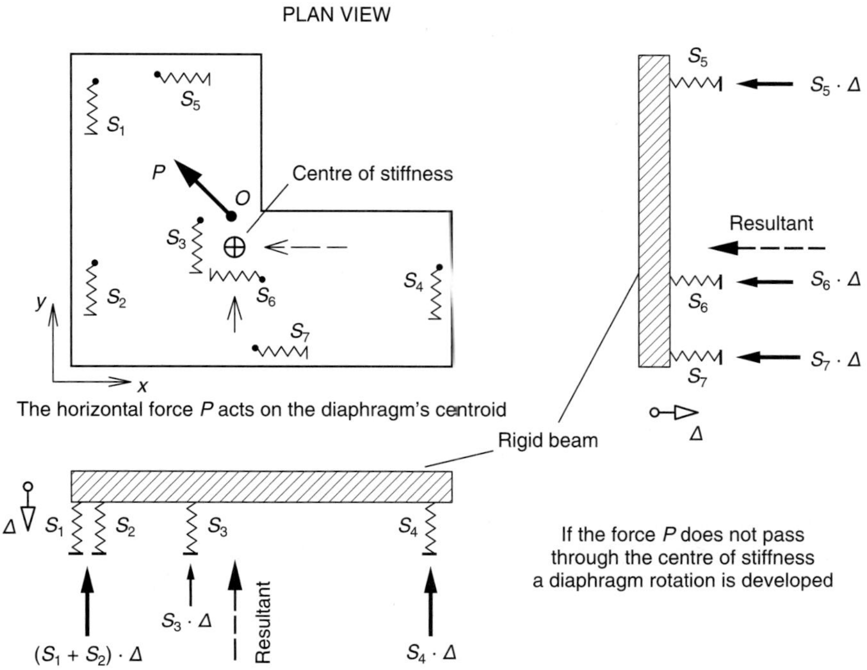

Figure 15.13 Simulating the diaphragm behaviour through a spring-supported rigid beam

In this way, for an arbitrary displacement Δ imposed in the x direction, the specific line of action of the resultant of the spring forces $(S_x \cdot \Delta)$ may be readily determined, while for another imposed arbitrary displacement Δ in the y direction, the spring forces $(S_y \cdot \Delta)$ allow the determination of the line of action of their own resultant (see Figure 15.13). It is obvious that the position of the above resultants is independent of the value Δ, and hence for any force passing through the intersection of the above action lines, the diaphragm will not rotate but only shift in the direction of the acting force. This point, which is known as the *centre of stiffness* or the *centre of resistance*, may be paralleled to the shear centre of thin-walled sections (see Section 13.3).

It is now clear that a zero value for the magnitudes K_3 and K_5, i.e. of $\sum (S_x \cdot y_i)$ and $\sum (S_y \cdot x_i)$, means that the point O – representing in practice the centre of gravity of the vertical element compressive forces – coincides with the centre of stiffness.

It should be made clear that all the foregoing equations are valid, independently of whether the point of application of the load P on the diaphragm – which may be its centroid – coincides or not with the centre of stiffness. However, it is clear that a possible application of a load P at a point different from the centre of stiffness means the action of a certain moment on the diaphragm, which must be balanced by the spring forces, i.e. the resistance forces offered by the plane elements. Moreover, the spring (plane element) layout must ensure the uptake of an arbitrary load P acting in any direction.

The above remarks are illustrated in Figure 15.14.

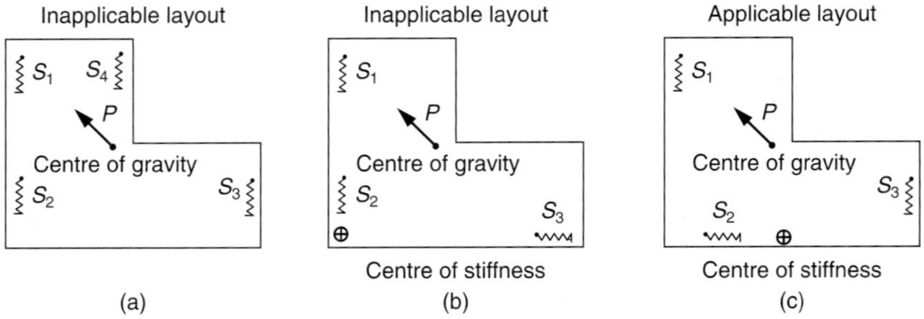

Figure 15.14 Criteria for the right or wrong layout of the plane stiffness elements

In case (a) the x component of the load P cannot be balanced by the spring forces, and consequently the layout is unsafe. In case (b), although springs, i.e. plane elements, are available to the diaphragm for the uptake of any component of the load P, these are unable to provide a couple of forces, in order to balance the moment of the load with respect to the centre of stiffness, and therefore the system may easily collapse. However, in case (c) the developed spring forces can balance any component of the acting load, as well as the moment of the load P with respect to the centre of stiffness through the offered forces S_1 and S_3.

It is now clear that, while the point of application of the load P is generally determined by the diaphragm itself, the centre of stiffness is dependent on the selected layout of the springs, i.e. the plane stiffness elements.

It must be pointed out that selecting a layout of vertical elements – frames, shear walls, etc. – that exhibits small or zero values of the magnitudes $\sum (S_x \cdot y)$ and $\sum (S_y \cdot x)$ has a definite advantage over a layout where the point of application of load P, i.e. the centre of gravity of the diaphragm, is some distance from the centre of stiffness, thus causing a rotation $\Delta\omega$ of the diaphragm. This rotation is obviously the same as that occurring in each of the rigid beam models in Figure 15.13. As can be seen from the above expressions, the rotation $\Delta\omega$ causes a non-uniform distribution in the response of the plane elements, depending on their position in the ground plan (for a generalised treatment of the problem see Section 16.3.1.4).

15.4 Temperature effect

15.4.1 Treatment of load-bearing action

Any temperature variation in a diaphragm always causes a certain stress state on the vertical elements. The reason for this is that a dilatation or a shrinkage of a diaphragm always imposes displacements at the corresponding level of the plane elements, which are consequently stressed (Figure 15.15).

More specifically, assuming that a diaphragm is subjected to a temperature increase of ΔT by retaining its centre of mass O absolutely fixed, it can be seen that some radial displacement will be imposed on all the vertical plane elements, which will lead to a certain shift along their trace (Figure 15.16).

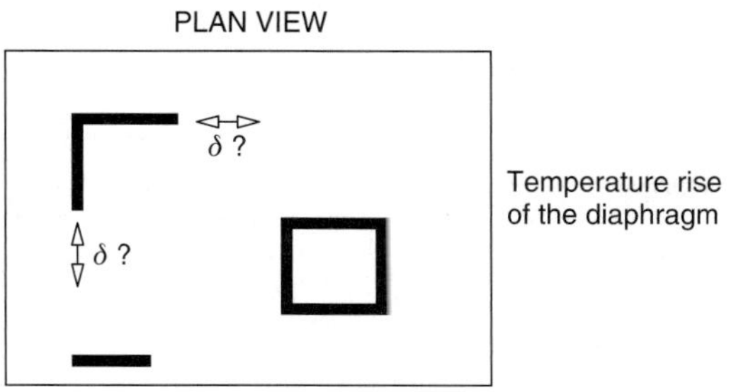

Figure 15.15 Displacements of the plane elements due to a temperature variation in the diaphragm

Due to this imposed displacement, the plane elements will offer resistance at their corresponding level that will be acted on by the diaphragm, being obviously equal and opposite to that acting on the plane element itself. It is clear that the diaphragm under the above forces will generally not be in equilibrium as a free body and, therefore, at the point of fixation O, a specific force and a moment have to be applied. The action of these forces ensures the fixation of point O under the temperature variation of the diaphragm (Figure 15.16). However, given that, in reality, no external actions are applied at point O, equal and opposite actions to those above must additionally be applied at this point, in order to restore the real situation. Of course, the last actions on the 'free' system will cause displacement of the diaphragm with the components Δx, Δy and $\Delta\omega$, which in turn will cause an additional stress state on the plane elements, through the corresponding horizontal forces acting on their corresponding levels (Figure 15.17).

The final response of the vertical elements will result as a superposition of the two above stress states (Stavridis, 1994), namely:

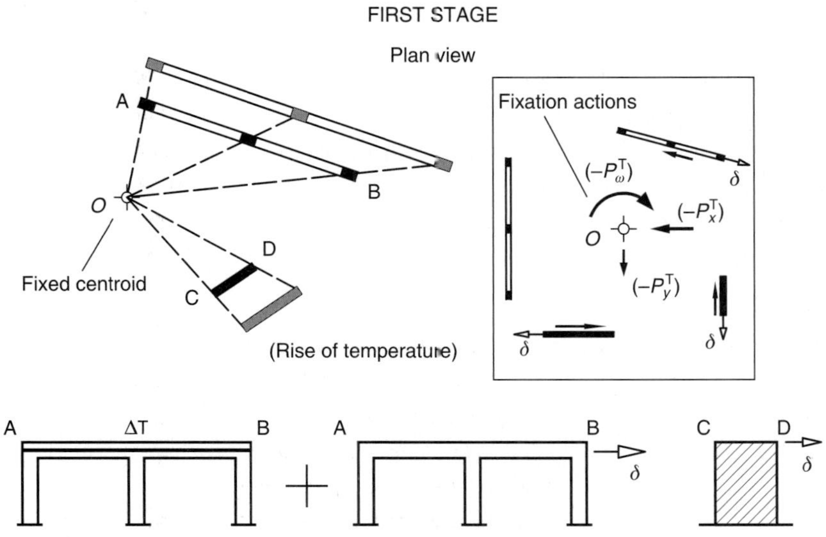

Figure 15.16 Temperature variation under fixation of the centroid of the diaphragm

SECOND STAGE

Imposition of effective actions on the free diaphragm

Developed: Δx, Δy, $\Delta\omega$

Plan view

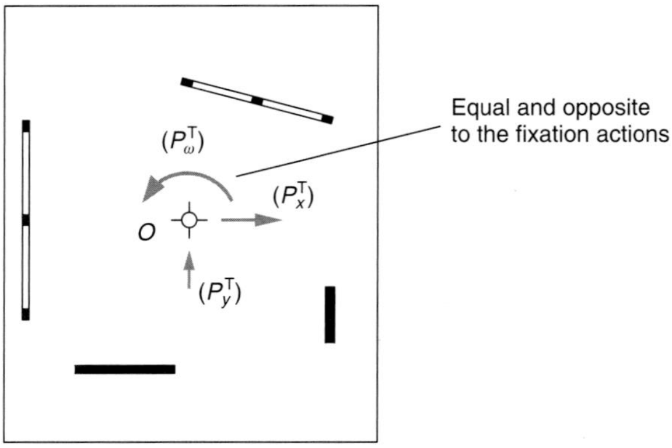

The induced element actions at the corresponding level are superposed on those of the first stage

Figure 15.17 Imposition of the opposite fixation actions to the free system

- *First stage*. The state of stress resulting from the free dilatation of the diaphragm with its centroid O fixed through externally applied actions at this point (see Figure 15.16). This stress state is due to the imposed displacements on the specific level of the plane elements because of the radial dilatation of the diaphragm on the one hand and the temperature variation itself imposed on the corresponding girder of the plane frame on the other hand.
- *Second stage*. The state of stress resulting in the – now – free system, through the imposition at the point O of the equal and opposite actions of the previous stage, which leads to the final displacement of the diaphragm, described by the components Δx, Δy and $\Delta\omega$ (see Figure 15.17).

15.4.2 General layout

According to the foregoing examination, and on the basis of Sections 15.2 and 15.3, it may be deduced that, for the specific level considered, the horizontal displacement D_i of the ith plane element referred to a characteristic point (x_i, y_i) selected at the middle of its trace may be expressed as follows, by conveying the influence of the two superposed states:

$$D_i = [\tau_i \cdot \alpha_T \cdot \Delta T] + [\Delta x \cdot \cos\beta_i + \Delta y \cdot \sin\beta_i + \Delta\omega \cdot e_i]$$

where $\tau_i = x_i \cdot \cos\beta_i + y_i \cdot \sin\beta_i$, and the temperature coefficient $\alpha_T = 10^{-5}/°C$.

The temperature variation ΔT is considered positive if the temperature of the diaphragm increases, and vice versa.

Thus, the force F acting on each plane element at the level examined is

$$F = S_i \cdot D_i = S_i \cdot \alpha_T \cdot \tau_i \cdot \Delta T + S_i \cdot (\Delta x \cdot \cos\beta_i + \Delta y \cdot \sin\beta_i + \Delta\omega \cdot e_i)$$

The determination of Δx, Δy and $\Delta\omega$ is by the solution of an analogously structured system of equations as in the previous case of lateral loading:

$$K_1 \cdot \Delta x + K_2 \cdot \Delta y + K_3 \cdot \Delta\omega = P_x^T$$

$$K_2 \cdot \Delta x + K_4 \cdot \Delta y + K_5 \cdot \Delta\omega = P_y^T$$

$$K_3 \cdot \Delta x + K_5 \cdot \Delta y + K_6 \cdot \Delta\omega = P_\omega^T$$

In the present case of temperature influence, however, the right-hand variables represent – with opposite signs – the required external actions at the point O, to comply with the aforementioned first stage:

$$P_x^T = -\alpha_T \cdot \Delta T \cdot \sum (\cos\beta_i \cdot S_i \cdot \tau_i)$$

$$P_y^T = -\alpha_T \cdot \Delta T \cdot \sum (\sin\beta_i \cdot S_i \cdot \tau_i)$$

$$P_\omega^T = -\alpha_T \cdot \Delta T \cdot \sum (S_i \cdot \tau_i \cdot e_i)$$

15.4.3 *Orthogonal layout*

In the usual case where the plane elements are arranged in an orthogonal layout, the above expressions are simplified as follows.

For the plane elements in the x sense, $\tau_i = x_i$ and $e_i = -y_i$, whereas for the elements in the y sense, $\tau_i = y_i$ and $e_i = x_i$.

Then,

$$P_x^T = -\alpha_T \cdot \Delta T \cdot \sum (S_x \cdot x_i)$$

$$P_y^T = -\alpha_T \cdot \Delta T \cdot \sum (S_y \cdot y_i)$$

and

$$P_\omega^T = -\alpha_T \cdot \Delta T \cdot \sum (S_y \cdot x_i \cdot y_i - S_x \cdot x_i \cdot y_i)$$

while Δx, Δy and $\Delta\omega$ are determined as

$$\Delta x = -\alpha_T \cdot \Delta T \cdot \frac{\sum (S_x \cdot x_i)}{K_1} - \frac{K_3}{K_1} \cdot \frac{\dfrac{K_3}{K_1} + \dfrac{K_5}{K_4}}{\dfrac{K_3^2}{K_1} + \dfrac{K_5^2}{K_4} - K_6}$$

$$\Delta y = -\alpha_T \cdot \Delta T \cdot \frac{\sum (S_y \cdot y_i)}{K_4} - \frac{K_5}{K_4} \cdot \frac{\dfrac{K_3}{K_1} + \dfrac{K_5}{K_4}}{\dfrac{K_3^2}{K_1} + \dfrac{K_5^2}{K_4} - K_6}$$

$$\Delta\omega = -\alpha_T \cdot \Delta T \cdot \frac{\dfrac{K_3}{K_1} \cdot \sum (S_x \cdot x_i) + \dfrac{K_5}{K_4} \cdot \sum (S_y \cdot y_i) + \sum (S_x \cdot x_i \cdot y_i - S_y \cdot x_i \cdot y_i)}{\dfrac{K_3^2}{K_1} + \dfrac{K_5^2}{K_4} - K_6}$$

According to the expression for D_i in Section 15.4.2, the elements in the x and y senses are subjected at the examined level to the following forces, respectively:

$$F_x = S_x \cdot D_i = S_x \cdot (x_i \cdot \alpha_T \cdot \Delta T + \Delta x - \Delta\omega \cdot y_i)$$

$$F_y = S_y \cdot D_i = S_y \cdot (y_i \cdot \alpha_T \cdot \Delta T + \Delta y + \Delta\omega \cdot x_i)$$

It is clear, however, that the consideration of an applied temperature difference at more than one diaphragm simultaneously may be treated by superposing the respective results for each separate level, exactly as in the case of lateral loading examined in Section 15.3.1. Thus, each plane element will be subjected to a corresponding force at each specific level where the temperature variation is applied.

From the above equations, it may be concluded that, contrary to what happens in the case of horizontal loading, the specific position of each plane element in the ground plan plays its own role in the development of the response, so that the element may not 'slip' along its axis.

The symmetry in the ground plan with respect to either the x or y axis simplifies the above expressions, as the coefficients K_3 and K_5 have null values.

The existence of a double symmetry for the axes Ox and Oy for the stiffnesses S_x and S_y, respectively, means that $K_3 = K_5 = 0$, and leads to an important simplification: $\Delta x = \Delta y = \Delta\omega = 0$. Hence, the plane elements in the x and y senses are acted upon by the forces $F_x = S_x \cdot x_i \cdot \alpha_T \cdot \Delta T$ and $F_y = S_y \cdot y_i \cdot \alpha_T \cdot \Delta T$, respectively.

The forces F_x and F_y must be taken into account for all cases during the design of the arrangement of the vertical elements – frames, shear walls – of a building structure. They meet precisely the criterion for the maximum building dimensions with regard to a monolithic construction, in order to withstand not only possible climatic variation associated with temperature change but also the ever-present concrete shrinkage of the diaphragms (slabs), being equivalent to a temperature fall of about 25°C. The typically considered 'maximum length' of 30 m, where the 'dilatation joints' are normally provided, must clearly always be adjusted for, depending in each case on the plane stiffness elements themselves, as well as on their specific layout in the ground plan.

It is thus clear that the design of the vertical members of a multi-storey building structure, apart from ensuring the safe uptake of any horizontal action either as seismic or wind loading, must also cover the 'temperature variation' of the diaphragms.

References

Stavridis L. (1986) Static and dynamic analysis of multistorey systems. *Technika Chronika Scientific Journal of the Technical Chamber of Greece* **6**(2), 187–219.

Stavridis L. (1994) Beanspruchung von mehrstöckigen Bauten infolge Temperaturänderung. *Bauingenieur* **69**(3), 117–122.

16

Dynamic behaviour of discrete mass systems

16.1 Introduction

The static loads, which all load-bearing structures have to take up, may be considered as completely independent of the structure itself, as far as its deformation is concerned. This uptake of static loads is realised through the stiffness of the structure, depending on its geometric and material characteristics. The closed process (load action) $\longleftrightarrow$ (uptake through stiffness) is accomplished through the process of the realisation of equilibrium, i.e. of the immobility of the structure. In this static process the *mass* (m) of the structure obviously remains inactive. It is activated, however, as a factor of generating loads according to Newton's second law of motion ($F = m \cdot \gamma$), from the moment that the structure is somehow set in motion. There are many factors that can set the structure – which is fixed to the ground – in motion, such as wind forces, earthquake ground movements, 'sudden' load applications, and moving loads (e.g. vehicles or pedestrians on roads or footbridges, respectively), etc.

The problem created through the motion of the mass of the structure is that it implies accelerations (γ) leading, according to d'Alembert's principle, to inertial forces equal to $-m \cdot \gamma$. The structure is thus stressed under these loads, which may clearly be treated as static ones in the usual manner. However, the problem lies in the fact that the above accelerations, which vary in time, are unknown from their onset. In fact, they depend on the masses and the stiffness of the system, and to determine their variation in every localised mass of the structure and at every instant of time constitutes a dynamic problem that is complicated from the onset of the motion.

Thus, the whole dynamic problem may be described as follows. The load-bearing system consists of specific masses that are connected to each other as well as to the ground through a massless *structural web*, which, apart from its own geometric layout, exhibits specific stiffness characteristics (Figure 16.1). When the structure is set in motion, the inertia forces of the masses ($-m \cdot \gamma$) on the one hand and the elastic forces of the structural web on the other are activated, depending on its stiffness characteristics and acting on the same masses. The solution to the problem of the determination of the accelerations 'γ' of the masses results from the requirement of equilibrium for each mass under the action of the inertia as well as the elastic forces. However, the difficulty in determining the motion of the masses arises from the fact that, while the elastic forces are generally proportional to the displacements of the

Structural systems: behaviour and design
978-0-7277-4105-9
Copyright Thomas Telford Limited © 2010
All rights reserved

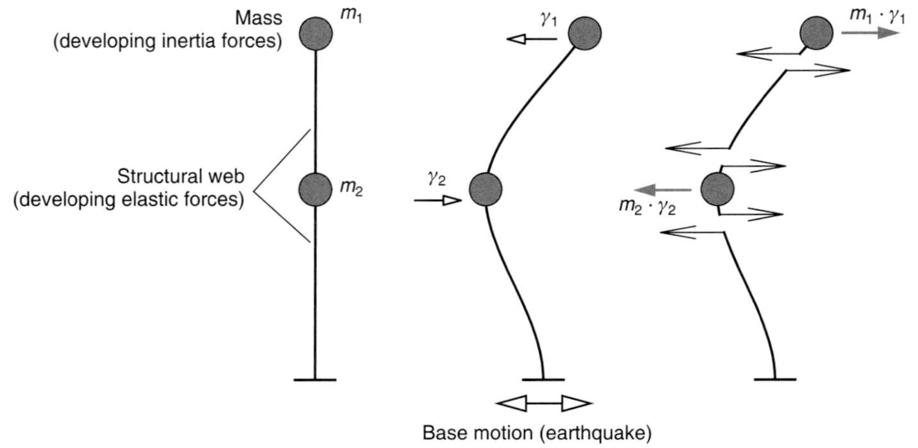

Figure 16.1 Masses in motion and the structural web

masses, the inertia forces are proportional to the second derivative of these displacements with respect to time.

The structures constitute *a priori* continuous systems in the sense that their mass is continuously distributed and not concentrated at discrete locations. Nevertheless, lumping the mass on the basis of behavioural criteria is always attempted for selected locations, as the analytical treatment of the continuous mass distribution is mathematically much more cumbersome. ■

The possible motions of the discrete (lumped) masses generally determine the inertia forces they are subjected to, and from this point of view the latter are distinguished between *translational* and *rotational inertia forces*. In the present examination, which is, however, oriented to preliminary design needs, the inertia forces due the rotation of the masses will be neglected, except for the horizontal diaphragms of multi-storey buildings undergoing a rotation $\Delta\omega$, as explained in Section 15.2.

Thus, in the plane frame shown in Figure 16.2, in order for a lateral excitation to apply, its masses are assumed – approximately – to be concentrated at each storey, only their horizontal displacement is recognised, and their rotations about an axis perpendicular to the plane of the frame, as well as their vertical displacements, are neglected as a source of inertia forces.

Similarly, in the continuous beam shown in Figure 16.2, where a transverse excitation is encountered, its mass is considered concentrated at specific points, and only the vertical displacement of these points is recognised, their rotations about their own axis being neglected as a source of inertia forces.

Finally, in the two-storey spatial system shown in Figure 16.2, where, a horizontal seismic excitation occurs, the mass is considered equally distributed in each horizontal diaphragm, and only the horizontal displacement components (Δx, Δy), producing inertia forces in the x and y directions, respectively, as well as the rotation $\Delta\omega$, producing an inertia moment at each storey, are recognised.

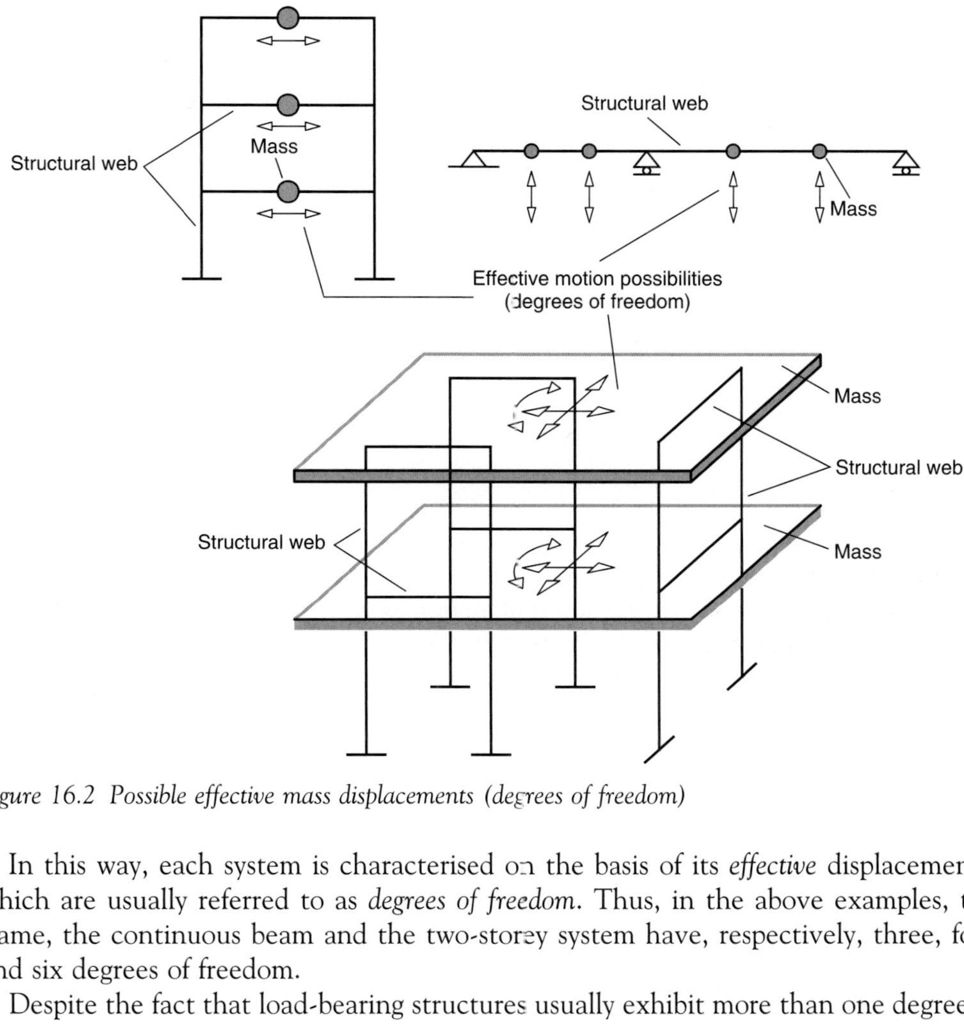

Figure 16.2 Possible effective mass displacements (degrees of freedom)

In this way, each system is characterised on the basis of its *effective* displacements, which are usually referred to as *degrees of freedom*. Thus, in the above examples, the frame, the continuous beam and the two-storey system have, respectively, three, four and six degrees of freedom.

Despite the fact that load-bearing structures usually exhibit more than one degree of freedom, the study of the single-degree-of-freedom system is absolutely crucial in order to understand all the essential characteristics and dynamic behaviour of multi-degree systems. It is necessary, then, to begin an elementary study of the dynamic behaviour of a single mass with only one 'active' displacement, given also that this model applies directly to the practical treatment of composite systems, particularly those with a continuous mass distribution, as shown in the last two sections of this chapter.

16.2 Single-degree-of-freedom systems

16.2.1 Dynamic equilibrium

A mass m connected to the free end of a cantilevered bar of length L will be used as an example of a single-degree-of-freedom system. The bending stiffness of the bar is much less than its axial stiffness, and therefore the only possibility for displacement of the mass is transversely to the axis of the bar. Consequently, the stiffness k (kN/m) of the bar is

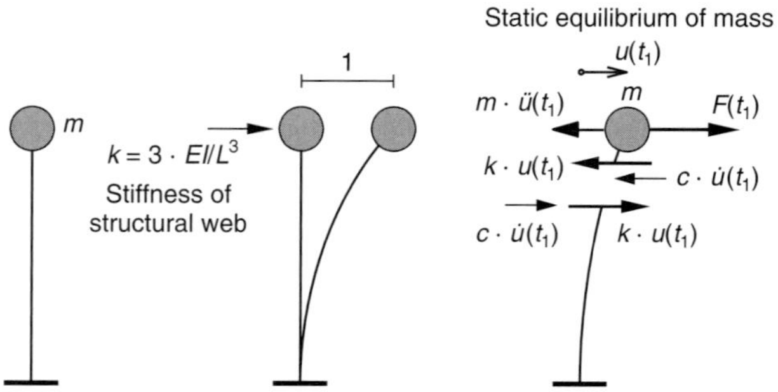

Figure 16.3 Dynamic equilibrium of a single mass in a cantilevered structural web

considered with respect to the displacement of its free end (Figure 16.3), and is equal to $k = 3 \cdot EI/L^3$.

It is assumed that the mass is acted upon by a horizontal force function $F(t)$ and that at a specific moment t_1 it undergoes a displacement $u(t_1)$ in the same direction.

Moreover, it is assumed that the mass receives an additional resistance to its motion due to what can be considered the 'internal friction' of the material of the bar, which is at first taken as proportional to the velocity of the mass $\dot{u}(t)$, and thus equal to $c \cdot \dot{u}(t)$. The coefficient c is called the *damping coefficient*, as by its nature it tends to dampen the motion of the mass.

Thus, at the time t_1 the mass m, according to d'Alembert's principle, may be considered in equilibrium under the following forces:

- the external force $F(t_1)$ in the same direction as $u(t_1)$
- the elastic force $k \cdot u(t_1)$ having a sense opposite to that of $u(t_1)$
- the d'Alembert's inertia force $m \cdot \ddot{u}(t_1)$ with a sense opposite to that of $u(t_1)$
- the damping force $c \cdot \dot{u}(t_1)$ with a sense opposite to that of $u(t_1)$.

The static equilibrium of the mass requires (see Figure 16.3)

$$m \cdot \ddot{u}(t_1) + k \cdot u(t_1) + c \cdot \dot{u}(t_1) = F(t_1)$$

It should be pointed out that the elastic force $k \cdot u(t_1)$ coincides with that needed to shift the mass m by $u(t_1)$, according to the stiffness of its structural web.

16.2.2 Free vibration

16.2.2.1 Undamped vibration

As the dynamic characteristics of the system are really independent of the external loading, their investigation is made easier if the force $F(t_1)$ is removed, and thus the free vibration of the mass m is considered after an initial arbitrary displacement is imposed at the rest position. By temporarily neglecting the 'damping' factor, it may be written for any time t_1 that

$$m \cdot \ddot{u}(t_1) + k \cdot u(t_1) = 0$$

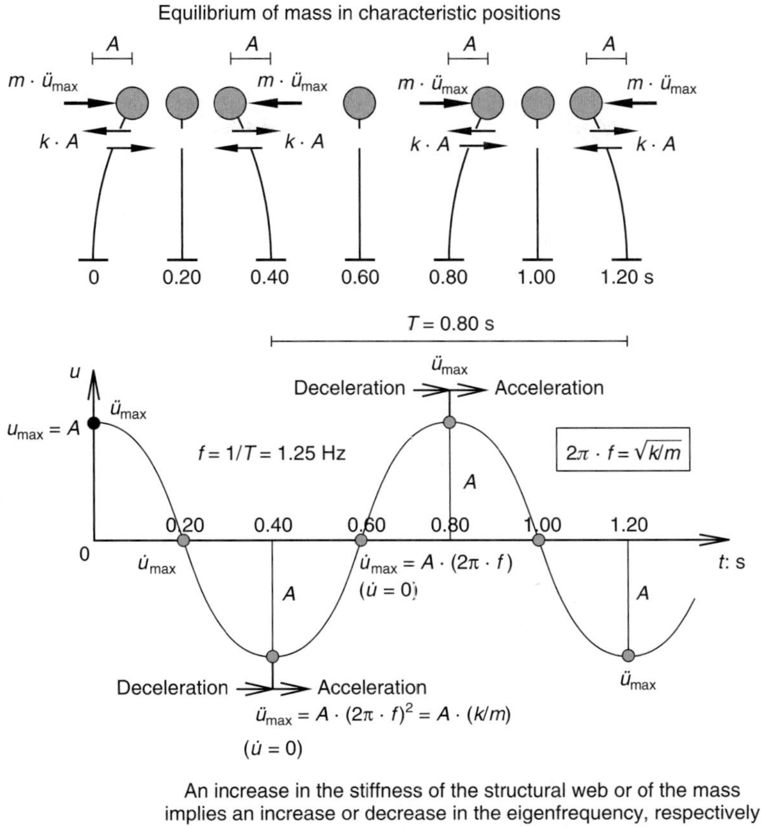

Figure 16.4 Free vibration of a single mass

For the determination of the displacement function u, the kinematic conditions of mass m at the start time are needed, i.e. its displacement $u_0 = u(t=0)$ and its velocity $\dot{u}_0 = \dot{u}(t=0)$. The solution of the above equation shows that the mass oscillates continuously in a periodic manner about the 'rest' position with a deviation $u(t)$ that varies sinusoidally with respect to time (Figure 16.4).

The time needed for a full oscillation of the mass, after which the motion is repeated in the exact same way, is called the *natural period* T (s). The inverse magnitude f of the period is called the *frequency*, and $f = 1/T$ (1/s). The frequency f represents the number of full oscillations that take place within a second, the unit of this quantity being the *Hertz*. It is common in vibration studies to also use the term *circular frequency* instead of 'frequency f', denoting the magnitude ω, on the basis of the relation:

$$f = \omega/2 \cdot \pi$$

The concept of frequency, or the *natural frequency*, of a system constitutes a basic characteristic of the system, directly influencing its dynamic behaviour towards the imposition of external forces. It is found that

$$\omega = \sqrt{\frac{k}{m}} \quad \text{(rad/s)}$$

and, consequently,

$$f = \frac{1}{2 \cdot \pi} \cdot \sqrt{\frac{k}{m}} \quad \text{(1/s = hertz)}$$

It should be noted that the mass unit used for structures is generally the tonne (t), which is equal to 1000 kg: 1 t = 1 kN s^2/m, and, given a weight W in kN, its mass is

$$m = W/9.81 \text{ t}$$

As seen from the last equation, the larger the mass, the fewer the vibrations that occur in 1 s, whereas the stiffer the connection of the mass with the ground – i.e. the structural web – the higher the frequency of the system. However, the presence of the square root attenuates the effect of the above parameters.

Of course, the frequency f of the system is independent of both the initial deviation u_0 and the initial velocity $\dot{u}_0$ of mass m.

The mass m executes a steadily varying, oscillating motion (see Figure 16.4). Its displacement $u(t)$ is expressed as

$$u(t) = \frac{\dot{u}_0}{\omega} \cdot \sin \omega \cdot t + u_0 \cdot \cos \omega \cdot t$$

Its maximum deviation A on both sides, also called the *amplitude*, is found to be

$$A = \sqrt{u_0^2 + \left(\frac{\dot{u}_0}{\omega}\right)^2}$$

At these positions, $u = \pm A$, the mass has zero velocity and maximum acceleration (or deceleration), $\ddot{u}_{max} = A \cdot \omega^2$, whereas at the 'rest position' ($u = 0$) it has the maximum velocity, $\dot{u}_{max} = A \cdot \omega$.

Moreover, it should be pointed out that the elastic forces $k \cdot u(t_1)$ have the opposite sense to the corresponding d'Alembert forces $m \cdot \ddot{u}(t_1)$, as is implied by the basic equation itself. More specifically, as the inertia force $m \cdot \ddot{u}(t_1)$ is opposite to the acceleration vector $\ddot{u}(t_1)$, just before the maximum deviation A, the inertia force is opposite to the maximum deceleration developed $\ddot{u}_{max}$, whereas just after this position the inertia force is opposite to the equal acceleration $\ddot{u}_{max}$. This means that in both cases the inertia force is directed to the right, i.e. in the opposite direction to the acting elastic force $(k \cdot A)$ (see Figure 16.4) and, finally, from the equilibrium equation $m \cdot \ddot{u}_{max} = k \cdot u_{max} = k \cdot A$, it can be checked that $\ddot{u}_{max} = A \cdot \omega^2$. ■

It should be made clear that the selection of the cantilever-type single-mass system is only indicative of a system with mass m and stiffness of structural web k, subjected to an external force $F(t)$, with respect to its active stiffness k. A system consisting of, for

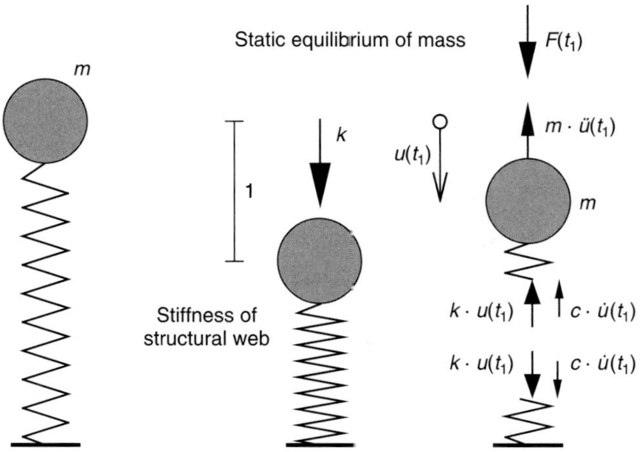

Figure 16.5 Dynamic equilibrium of a mass carried by a spring

example, a mass connected to a spring, as shown in Figure 16.5, would be equally representative, provided, of course, that the self-weight of the mass m is ignored.

However, it should be recognised that while the cantilevered mass is offered as a more illustrative model for the examination of systems consisting of a number of masses arranged in height, as in, for example, buildings or bridges, the mass–spring model is more representative of the examination of vertical dynamic actions on existing structures such as beams or plates, as in, for example, a machine permanently fixed in place. The behaviour of the mass–spring model should thus be recognised as analogous to the cantilever system, as shown in Figure 16.6.

16.2.2.2 Damped vibration

Consideration of the damping coefficient c in the equation

$$m \cdot \ddot{u}(t) + k \cdot u(t) + c \cdot \dot{u}(t) = 0$$

does not alter the periodicity of the oscillation of the mass; however, it does result in the continuous decrease of the amplitude in each vibration cycle. It should be noted that the system vibrates with a somewhat increased period T_{d}, which remains constant (Figure 16.7). It is found that the correspondingly decreased circular frequency ω_{d} is equal to

$$\omega_{\mathrm{d}} = \sqrt{\omega^2 - \left(\frac{c}{2 \cdot m}\right)^2}$$

At this point it is useful to clarify some facts concerning the damping, in order to better understand its treatment. Damping is the loss of energy in each vibration cycle. This energy is absorbed by the internal friction in the structure carrying the masses, i.e. the structural web, which gives stiffness to the system. Thus, the damping force $c \cdot \dot{u}(t)$ is acting on the mass through the structural web, as a restraining force in the

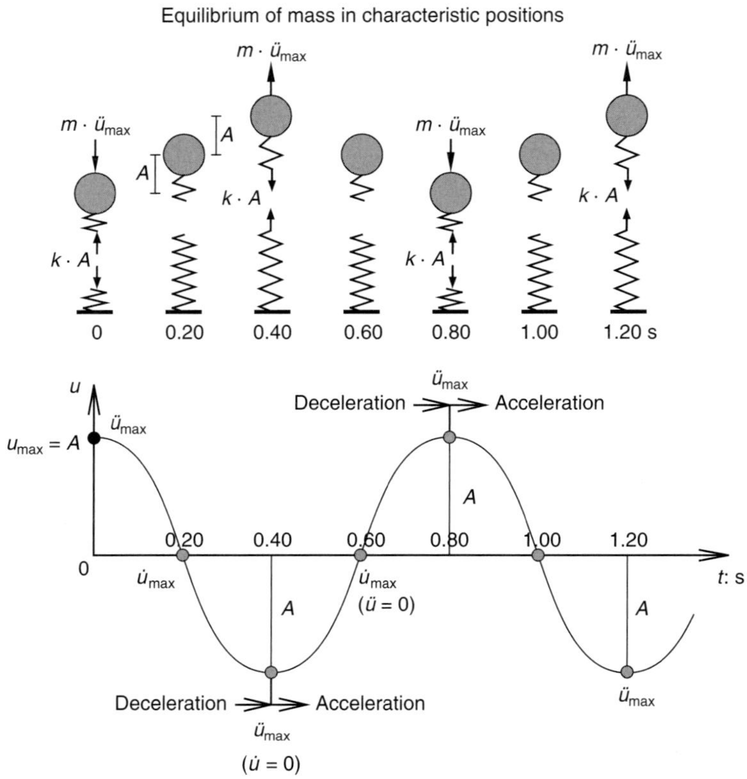

Figure 16.6 Free vibration of a mass–spring system

same direction as the acting elastic force. Of course, the direct result of damping is, as previously mentioned, a continuous decrease in the vibration's amplitude.

The description of damping as a force proportional to the velocity of the mass is not satisfactory from a physical point of view. Nevertheless, it permits – as will be shown later – a practical and effective quantitative description of the phenomenon, because of its convenient mathematical treatment.

Thus, it is found that if the damping coefficient c achieves the critical value $c_{cr} = 2 \cdot m \cdot \omega$, then no vibration can be developed, and the mass moves from its position of maximum deviation directly to its resting point. Although such a situation is impossible in a structure, its consideration is nevertheless useful because it represents a reference level for the applied damping coefficient in each case.

First, the so-called *damping ratio* ζ is defined, as the ratio of the damping coefficient to its critical value: $\zeta = c/c_{cr}$. On the basis of this equation,

$$\zeta = c/(2 \cdot m \cdot \omega)$$

and, consequently,

$$\omega_d = \omega \cdot \sqrt{(1 - \zeta^2)}$$

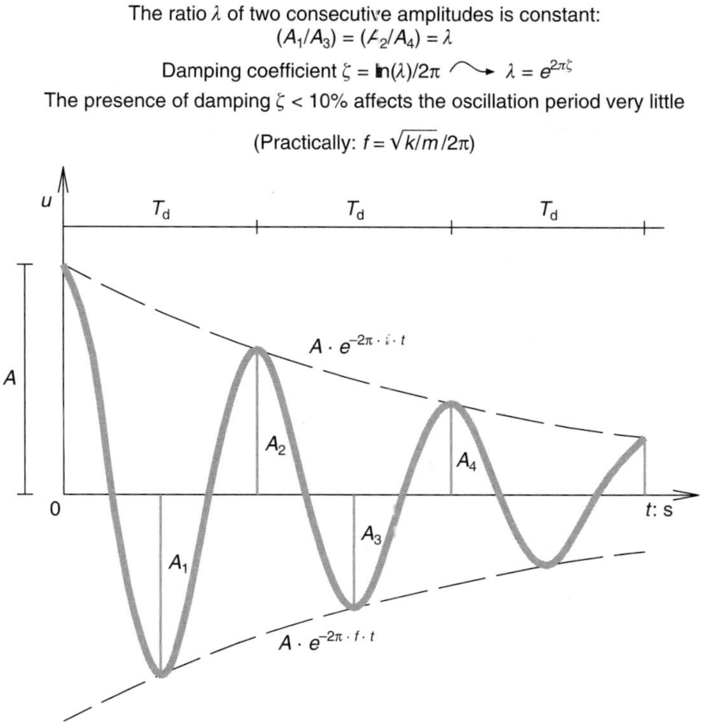

Figure 16.7 Characteristics of damped vibration

The ratio ζ, despite the fact that it refers to the unrealistic concept of the coefficient c, has an essential physical meaning, something that may not only be deduced from the last equation but also from the fact that between two consecutive amplitudes of the decaying vibration, with a 'time distance' T_d,

$$\delta = \ln\frac{u(t)}{u \cdot (t + T_d)} = \frac{c}{2 \cdot m} \cdot \frac{2 \cdot \pi}{\omega} = \frac{2 \cdot \pi \cdot \zeta}{\sqrt{1 - \zeta^2}}$$

The above equation is of particular practical importance, because, from the measurement of the consecutive amplitudes of a damped vibration of a system exhibiting a constant ratio, the value of ζ may be directly deduced. As this value, for structural systems encountered in practice, does not usually exceed 10%, it may be considered with a good approximation to be

$$\omega_d = \omega \quad \text{(hence } T_d = T\text{)}$$

on one side and

$$\delta = 2 \cdot \pi \cdot \zeta$$

on the other.

In real situations, however, for practical reasons as well as for accuracy, the amplitudes of two vibration cycles are measured as having a time distance rather equal to $n \cdot T$, in

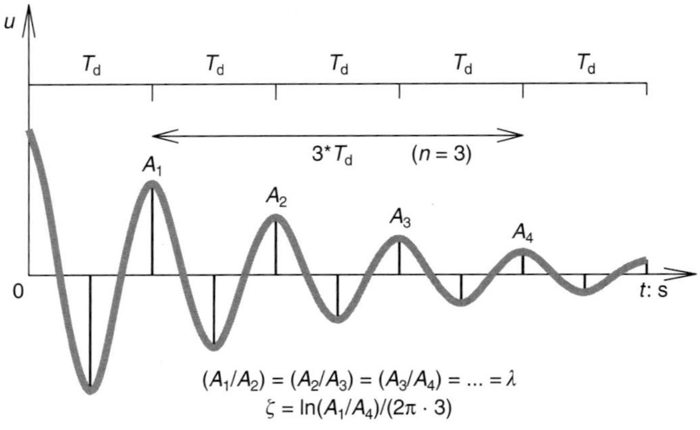

Figure 16.8 Practical determination of the damping ratio

which case (Figure 16.8)

$$\delta_n = \ln \frac{u(t)}{u(t + n \cdot T)} = n \cdot \delta = 2 \cdot \pi \cdot \zeta \cdot n$$

Hence,

$$\zeta = \frac{\delta_n}{2 \cdot \pi \cdot n}$$

Finally, the expression of the displacement $u(t)$ of a mass m, which, after being displaced by A, is left to vibrate freely with zero initial velocity, under a damping ratio ζ ($\omega_d \approx \omega$), is

$$u(t) = \frac{A}{e^{\zeta \cdot \omega \cdot t}} \cdot (\zeta \cdot \sin \omega \cdot t + \cos \omega \cdot t)$$

16.2.3 Forced vibration

The aim of the above analysis of free vibration was to depict the basic dynamic characteristics of a single-degree-of-freedom system. An understanding of these characteristics is necessary to understand the dynamic response of the system if the mass is acted on by a force varying in time, i.e. exactly as was initially considered, according to the equation of motion (Müller and Keintzel, 1984):

$$m \cdot \ddot{u}(t) + k \cdot u(t) + c \cdot \dot{u}(t) = F(t)$$

where, after the introduction of the damping ratio ζ,

$$c = 2 \cdot \zeta \cdot m \cdot \omega$$

It is useful to express the force $F(t)$ in the form $F(t) = F_1 \cdot f(t)$, where $f(t)$ describes the variation of the force during the whole period of the time examined, while F_1 represents the maximum value attained by this force. For the design of a structure, the response of the structural web connecting the mass with the ground is of primary interest. This response depends only on the displacement u. It is understandable that the maximum response corresponds to the maximum displacement u_{max}, which is clearly larger than that corresponding to the 'static' loading of the system with the force F_1, this being equal to

$$u_{stat} = F_1/k$$

Thus, if in a forced vibration the ratio $DMF = u/u_{stat}$ is defined as the *dynamic magnification factor*, then its maximum value $(DMF)_{max} = u_{max}/u_{stat}$ is of definite design importance.

Of course, the response of the system under the loading $F_1 \cdot f(t)$ is fully described by the exact solution of the above differential equation. This solution underlines the fact that the final response is the result of a superposition of the free vibration and the effect of the force imposed on the system, being expressed by the so called *Duhamel integral*, as shown below. Thus, the general solution of this equation is (Biggs, 1964)

$$u(t) = \mathrm{e}^{-\zeta \cdot \omega \cdot t}\left(u_0 \cdot \cos\omega_\mathrm{d} \cdot t + \frac{\dot{u}_0 + \zeta \cdot \omega \cdot u_0}{\omega_\mathrm{d}} \sin\omega_\mathrm{d} \cdot t\right)$$
$$+ \frac{F_1}{m \cdot \omega_\mathrm{d}} \cdot \int_0^t f(\tau) \cdot \mathrm{e}^{-\zeta \cdot \omega \cdot (t-\tau)} \cdot \sin\omega_\mathrm{d} \cdot (t-\tau)\,\mathrm{d}\tau$$

Obviously, direct use of this solution is not practical. However, the accessibility of special software dealing also with systems of multiple degrees of freedom allows the desired results to be obtained through a direct numerical approach rather than through the above analytic approach. Nevertheless, the above expression is very useful, because it permits direct analysis of some typical cases and allows particularly important conclusions to be obtained for the response of single-degree and, additionally, multi-degree systems. A few examples of such typical cases are listed below (see Figure 16.9):

- The 'abrupt' application of force F_1, followed either by its equally 'abrupt' removal after some period t_d, or by an immediate linearly varied decrease during the same period t_d.
- The progressive application of the force up to the value F_1 and, consecutively, its decrease down to zero over a total time period t_d, first in a linear fashion and then in a parabolic manner.
- The progressive linearly increasing application of force F_1, over a time period t_d, followed by maintenance of this acquired value for an unlimited time.

In all the above cases, the damping effect is ignored for simplicity.

In the curves shown in Figure 16.9 the behaviour of the maximum value of the dynamic magnification factor $(DMF)_{max}$ is depicted for each of the above cases, as a function of the ratio of the characteristic time duration t_d to the natural period T of the vibrating system.

From curves (a), it can be seen that in the case of a constant applied force, when the time reaches the value $T/2$, the factor $(DMF)_{max}$ takes its maximum value of 2, which is not exceeded in any other case. This value is also valid for any t_d greater than $T/2$.

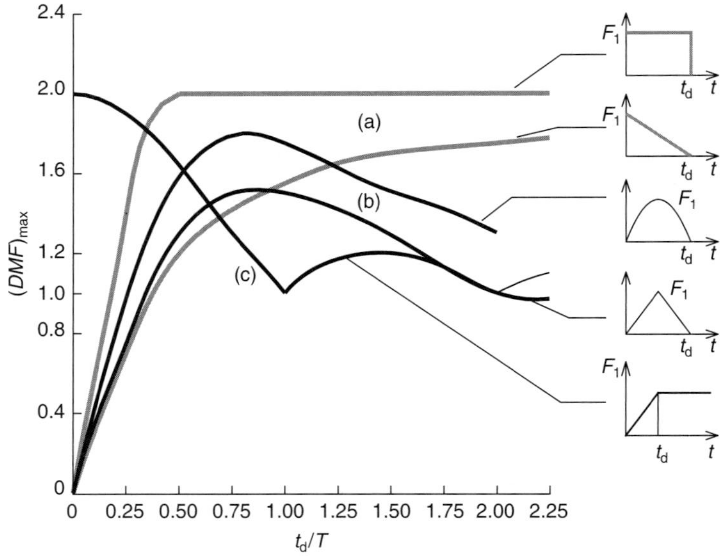

Figure 16.9 Behaviour of a single-mass system under a dynamically applied load

Otherwise, for a duration t_d less than about $T/5$, the factor $(DMF)_{max}$ is less than unity, i.e. the maximum deviation of the mass is even less than $u_{stat} = F_1/k$.

In contrast, the 'immediately occurring' linear decrease of the applied force has more moderate results. Only after an unloading duration t_d of about 10 times the natural period T does the factor $(DMF)_{max}$ begin to approach the value of 2.0, while for a duration less than $T/3$ the maximum deviation is less than the 'static' one u_{stat}.

From curves (b), it can be seen that the maximum value of $(DMF)_{max}$, i.e. the maximum mass deviation, occurs when the total loading–unloading duration approaches 90% of the natural period T. This value equals 1.8 or 1.5 for a parabolic or linear force variation, respectively. For lower or greater durations t_d, the maximum value of $(DMF)_{max}$ is always lower.

Curve (c) represents the typical force application scheme that is encountered in many practical cases. The more the duration t_d falls below T, the more the value of the factor increases, never exceeding, though, the absolute maximum value of 2.0, no matter how 'abrupt' the loading process is. However, for any duration t_d greater than T, the maximum value of factor DMF does not exceed the value 1.2. The above results confirm the definition of the static loading, as given in the introduction to Chapter 1.

16.2.4 Periodic sinusoidal acting force

Of particular importance for the design of structures is the case when the acting force $F(t)$ is acting on the system in a periodic manner, according to the expression $F(t) = F_0 \cdot \sin\Omega \cdot t$. The examination of the corresponding response permits one to draw useful conclusions about any kind of periodic excitation – machinery effects, human rhythmic activities, seismic action, etc. – as well as about measures required for the attenuation of their effects (dynamic isolation).

By applying the general solution of the previous section in the present case and ignoring the effect of damping, the following expression is obtained:

$$u(t) = U \cdot \sin(\omega \cdot t + \varphi) + \frac{F_0/k}{1-(\Omega/\omega)^2} \cdot \sin \Omega \cdot t$$

The first term on the right-hand side is essentially identical to the corresponding expression for the general solution, where

$$U = A \quad \text{and} \quad \tan\varphi = \frac{u_0}{\dot{u}_0/\omega}$$

This term represents, as already mentioned, that part of the response corresponding to the free vibration of the system, whereas the second term is exclusively due to the externally applied force.

Now, as consideration of even the least possible damping makes the contribution of the free vibration (i.e. the first term) soon vanish, the conclusion may directly be drawn that the maximum value of the dynamic magnification factor $(DMF)_{max}$ will be

$$(DMF)_{max} = \left(\frac{u}{u_{stat}}\right)_{max} = \frac{1}{\left|1-(\Omega/\omega)^2\right|}$$

It is clear that the ratio $r = \Omega/\omega$ plays a basic role. For values less than 0.20 and greater than 1.50, the dynamic influence is not important. More specifically, while for values of r tending to zero (static loading) the factor $(DMF)_{max}$ tends to 1, for a 'high-frequency' excitation ($r > 2.50$) this factor tends towards 0. These correlations are clearly shown in Figure 16.10. However, as can be deduced from the equation above, when $\Omega = \omega$, the factor $(DMF)_{max}$ is infinite. What is actually happening in this case is that in each cycle the factor DMF is steadily increasing in time, according to the equation

$$(DMF)_{\Omega=\omega} = (\sin\omega \cdot t - \omega \cdot t \cdot \cos\omega \cdot t)$$

On the basis of this equation:

$$(DMF)_{max,\Omega=\omega} = \frac{u_{max}}{F_0/k} = \frac{1}{2} \cdot n \cdot \pi$$

where n is the ever-increasing number of cycles.

The situation that occurs in the case when $\Omega = \omega$ is called *resonance*, and is something that should, of course, be avoided. However, as an infinite value of the factor DMF corresponds to a system without any damping effect, it should pointed out that this does not happen in reality, as even a small amount of damping action reduces the response significantly.

More precisely, for an existing damping factor equal to ζ (Biggs, 1964),

$$(DMF)_{max} = \frac{1}{\sqrt{[1-(\Omega/\omega)^2]^2 + 4 \cdot \zeta^2 \cdot (\Omega/\omega)^2}}$$

where the restrictive contribution of ζ to the response is clear (see Figure 16.10).

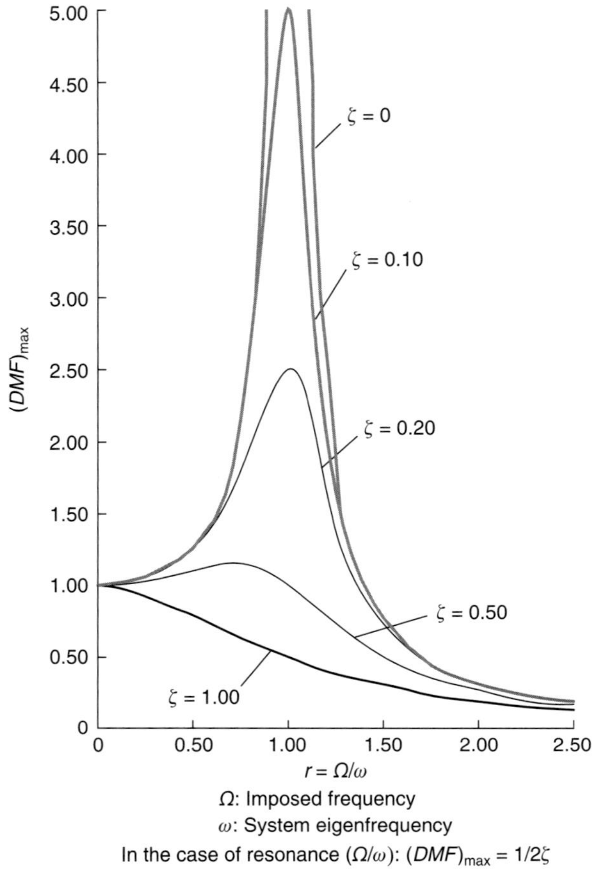

Figure 16.10 Behaviour of a system towards a periodic load action

Thus, in the case of resonance ($\Omega = \omega$), $(\mathrm{DMF})_{\mathrm{max}}$, $\Omega = \omega = 1/(2\zeta)$, and, consequently,

$$u_{\mathrm{max}} = \frac{1}{2 \cdot \zeta} \cdot \frac{F_0}{k}$$

However, until the system develops the above maximum deviation of mass, a large number of cycles is needed. ■

Of particular interest is the assessment of the transmitted forces to the base, i.e. to the support of the structural web.

From the examination of the dynamic equilibrium of a system – see Section 16.2.1 and also Figure 16.11 – the transmitted force F_g at the base of the structural web, consists of its elastic force $(k \cdot u)$ plus the damping force $(c \cdot \dot{u})$ acting in the same sense.

The maximum force transmitted to the base max F_g can be obtained according to the equation

$$\max F_g = F_0 \cdot \left(\frac{1 + 4 \cdot \zeta^2 \cdot r^2}{(1 - r^2)^2 + 4 \cdot \zeta^2 \cdot r^2} \right)^{1/2}$$

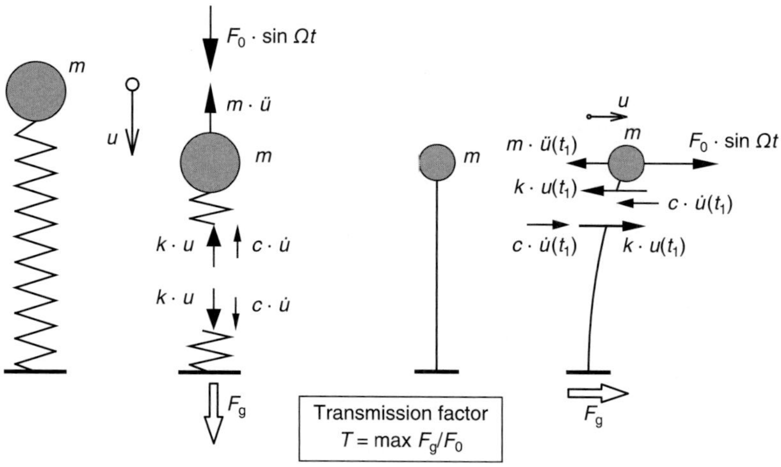

Figure 16.11 Transmission of dynamic actions at the base of a structural web

The ratio $T = \max F_g/F_0$ is called the *transmission factor*, and its dependence on the ratio $r = \Omega/\omega$, as well as on the damping factor, is depicted in Figure 16.12.

According to Figure 16.12, given the frequency Ω of an imposed load, due to, for example, an installed machine, by providing to the system – through appropriate means – a natural circular frequency ω, which is less than about $\Omega/2.50$, the transmitted force to the base is drastically limited.

16.2.5 Seismic excitation

Seismic excitation consists of an imposed oscillating motion of the support of a structure. The cause of this motion is the abrupt (brittle) rupture in some region of an already

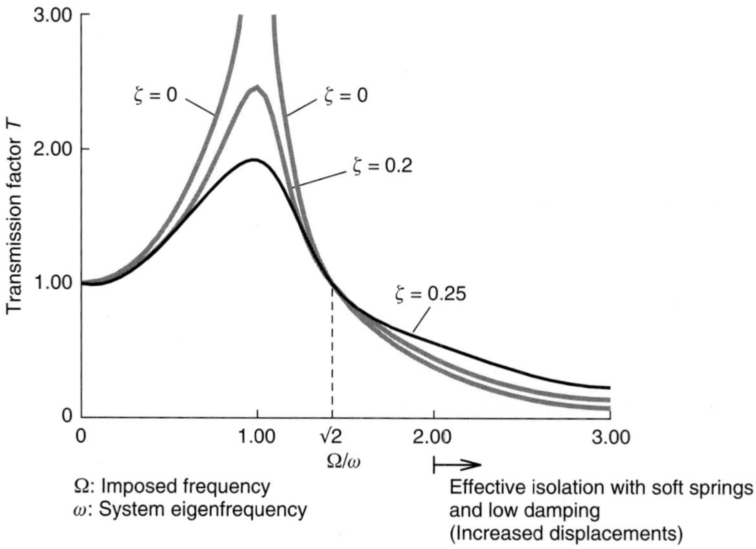

Figure 16.12 Transmissibility of dynamic actions at the base of a system

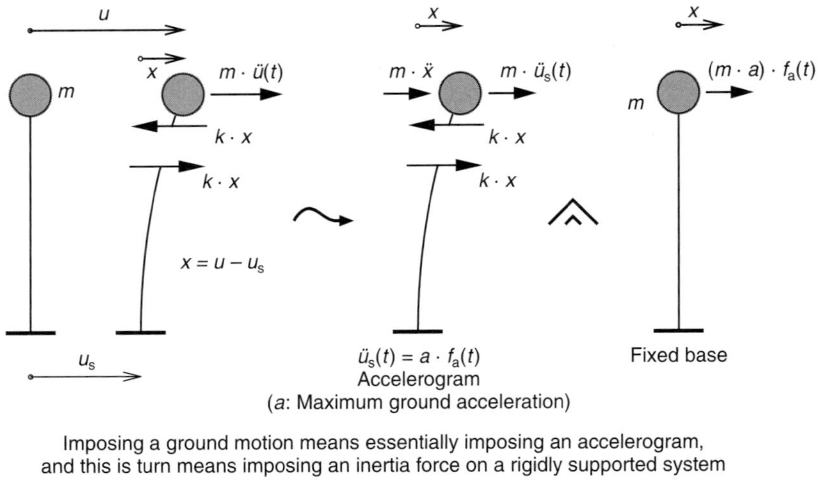

Figure 16.13 Imposition of seismic ground motion and dynamic equilibrium

existing fault in the earth's crust, as a result of accumulated stresses. This rupture is instantaneously spread out along the fault, causing propagation of a vibration wave within the earth's crust lasting for a period of about 15–30 s, and then coming to rest.

First, it is assumed that the imposed horizontal ground motion $u_s(t)$ on the support of the structure may be expressed as $u_s(t) = u_{s,0} \cdot f(t)$, where $u_{s,0}$ represents the maximum value of the ground displacement in the region of the support. In the present case of a single-mass system, the direction of u_s corresponds to the sense for that of the stiffness k of the system (Figure 16.13).

It is clear that the elastic force acting on the mass m will be due to the deformation of the whole system, i.e. to the relative displacement of the mass to the ground, expressed as $u - u_s$. Given that the mass m is not acted on by any external load, but is subjected to the inertia force only, obviously with reference to its absolute displacement u, as well as to the elastic force, it can be written, if the damping force is ignored, that

$$m \cdot \ddot{u}(t) + k \cdot [u(t) - u_s(t)] = 0$$

The absolute displacement $u(t)$ of the mass m cannot describe the deformation of the structure depending, as pointed out, exclusively on the relative displacement:

$$x(t) = u(t) - u_s(t)$$

Then, substituting the last equation,

$$m \cdot (\ddot{x} + \ddot{u}_s) + k \cdot x = 0$$

It is clear that the response $k \cdot x$ of the structure depends solely on the ground acceleration $\ddot{u}_s(t)$ (see Figure 16.13).

Thus, the characteristic of an earthquake, which is relevant for the design of a structure, is only the so-called *accelerogram*, which is recorded by special equipment each time. For a specific earthquake and within a major region, different accelerograms may be recorded for different locations, depending on the distance from the respective source of the event (hypocentre) as well as on the existing soil properties along the path followed by the seismic waves to the place recorded.

The accelerogram $\ddot{u}_s(t)$ may be expressed as

$$\ddot{u}_s(t) = a \cdot f_s(t)$$

where a represents the maximum occurring ground acceleration, while the function $f_a(t)$ describes its evolution in time. Now, the last equation may be written as

$$m \cdot \ddot{x} + k \cdot x = -(m \cdot a) \cdot f_a(t)$$

The form of this equation is exactly the same as that for the forced vibration of a single-degree-system on a fixed basis (see Figure 16.13), and, as the existing damping obviously affects the relative velocity $\dot{x}$ of the mass m with respect to the ground, the equation may be accordingly modified as follows:

$$m \cdot \ddot{x} + k \cdot x + c \cdot \dot{x} = -(m \cdot a) \cdot f_a(t)$$

It should be noted that the right-hand term refers to a 'fictitious force' – representing along with the force $m \cdot \ddot{x}$ the total inertia force that the mass m is subjected to.

By assuming initial conditions corresponding to 'rest', i.e. $x = 0$ and $\dot{x} = 0$, the solution of the last equation, analogously to Section 16.2.3, may be written as (Müller and Keintzel, 1984)

$$x(t) = \frac{a}{\omega_d} \cdot \left[\int_0^t f_a(\tau) \cdot e^{-\zeta \cdot \omega \cdot (t-\tau)} \cdot \sin \omega_d \cdot (t-\tau) \, d\tau\right]$$

As previously explained (see Section 16.2.2.2), for values of ζ up to 10%, the fundamental circular frequency ω_d does not differ substantially from the natural frequency of system ω without damping. In any case, however, for the design of a structure, it is the maximum value S_d of the relative deviation that is of paramount interest, as it obviously leads to the maximum response:

$$S_d = x_{max} = \frac{a}{\omega} \cdot \left[\int_0^t f_a(\tau) \cdot e^{-\zeta \cdot \omega \cdot (t-\tau)} \cdot \sin \omega \cdot (t-\tau) \, d\tau\right]_{max}$$

It can be seen that at the time that the maximum deviation x_{max} occurs, the acceleration (or deceleration) $\ddot{x}$ also takes its maximum value $\ddot{x}_{max}$, whereas the velocity $\dot{x}$ – and hence the damping force – vanishes. Thus, at this moment, the elastic force $k \cdot x_{max}$ acting on the mass m balances its inertia force, depending, of course, on its absolute acceleration, and is equal to $m \cdot (\ddot{x}_{max} + \ddot{u}_s)$.

Assuming that at this same moment the maximum ground acceleration is essentially attained, it can be written (Figure 16.14) that

$$m \cdot (\ddot{x} + \ddot{u}_s)_{max} = k \cdot x_{max} \quad \text{or} \quad (\ddot{x} + \ddot{u}_s)_{max} = \omega^2 \cdot x_{max}$$

Thus, the maximum value S_a of the absolute acceleration of the mass is

$$S_a = (\ddot{x} + \ddot{u}_s)_{max} = a \cdot \omega \cdot \left[\int_0^t f_a(\tau) \cdot e^{-\zeta \cdot \omega \cdot (t-\tau)} \cdot \sin \omega \cdot (t-\tau) \, d\tau\right]_{max}$$

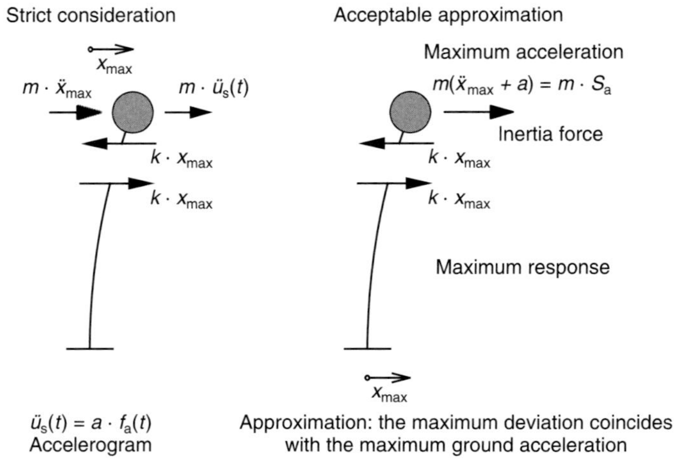

Figure 16.14 Position of maximum response in a single-degree-of-freedom system

This expression is strictly valid provided that the damping coefficient ζ does not exceed 10%. ■

It is thus clear that the response of a single-degree system to a specific seismic excitation – i.e. with known characteristics regarding the time development $f_a(t)$ of the ground acceleration, as well as its maximum value a – depends only on the natural frequency, or natural period, of the system and the corresponding damping factor.

On the basis of the last equation and for a given accelerogram, the maximum absolute acceleration S_a can therefore be plotted as a function of the natural period T, for different values of the damping coefficient ζ. Such a diagram represents the *acceleration spectrum*, and it permits, for any oscillator with a known natural period T, the direct determination of its maximum absolute acceleration. It may be easily seen that $S_a = S_d \cdot \omega^2$.

It is useful to point out here that, according to the above expression, S_a – as well as the *spectral displacement* S_d – is proportional to the maximum ground acceleration a accompanying the specific accelerogram, so that if divided by a, it represents a spectrum corresponding to an accelerogram exhibiting the same evolution in time but with respect to unit maximum acceleration.

Figure 16.15 shows an accelerogram recorded from the El Centro earthquake (California, USA, 1940), together with the acceleration spectrum (S_a/a) corresponding to unit ground acceleration, for two different damping coefficients.

It is can now be seen that for the practical evaluation of a specific earthquake being described by its accelerogram, the acceleration spectrum S_a – or the unit spectrum S_a/a together with the ground acceleration a – represents the examined excitation quite sufficiently. Thus, for a specific system with a known natural period T, on the basis of the corresponding spectral value S_a, the maximum inertia force can be directly obtained as $m \cdot S_a$. This force represents the maximum shear force V_a developed at

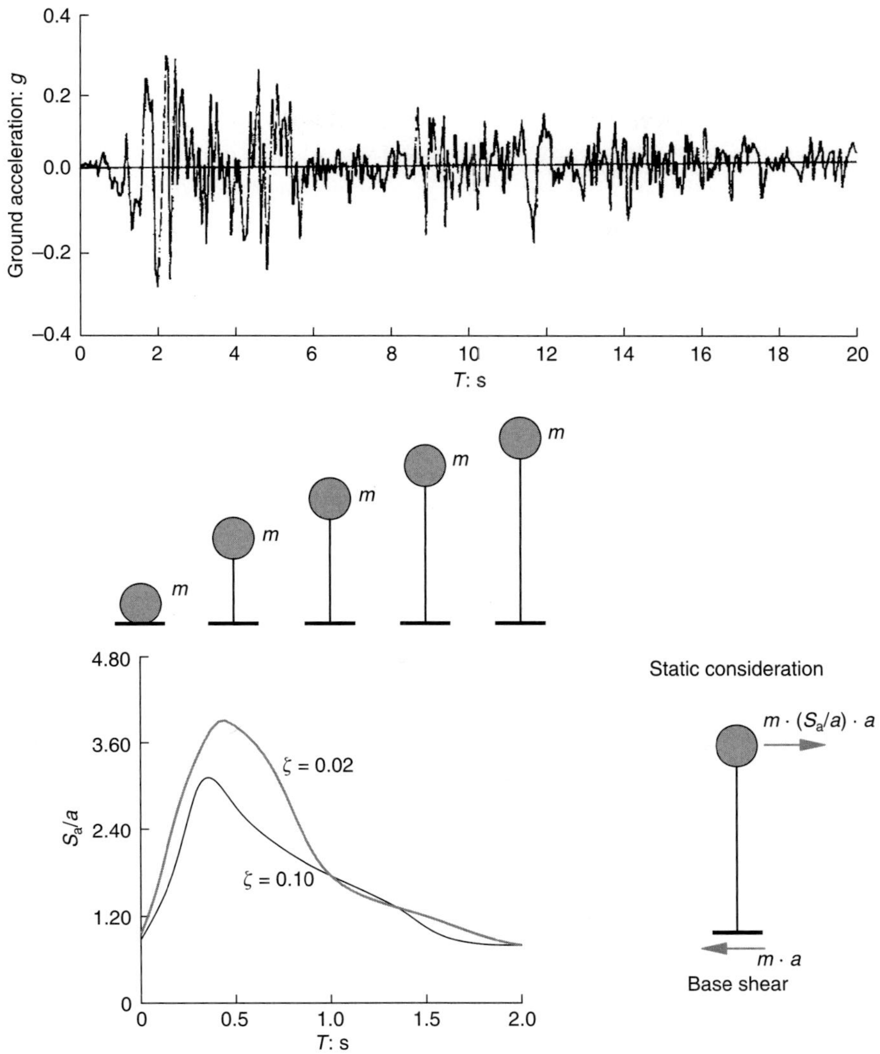

Figure 16.15 Accelerogram and acceleration spectrum (El Centro, California)

the base of the structure, called the *base shear*:

$$V_a = m \cdot S_a$$

It is clear that the above inertia force ($m \cdot S_a$) is identical to the maximum elastic force ($k \cdot S_d$) acting on the mass m (see Figure 16.15). Thus,

$$(m \cdot S_a) = k \cdot S_d = k \cdot S_a/\omega^2$$

The use of the acceleration spectrum therefore allows the determination of the maximum response in the structural web through a static loading, which is, of course, of particular practical importance.

What can be deduced from the spectrum of a specific earthquake is that there is always a range of periods which is more sensitive towards the development of the maximum inertia force, or – in other words – of the maximum shear base. As can be seen in Figure 16.15, structures with a lower stiffness being reflected in values of a natural period T greater than, say, 1.0 sec, develop a clearly limited response.

The dynamic magnification factor $(DMF)_{max}$, which as defined in Section 16.2.3 is the ratio of the maximum relative displacement S_d to the 'static' displacement $(m \cdot a)/k$, is now examined:

$$(DMF)_{max} = \frac{S_d}{(m \cdot a/k)} = \frac{\omega^2 \cdot S_d}{a} = \frac{S_a}{a}$$

$$= \omega \cdot \left[\int_0^t f_a(\tau) \cdot e^{-\zeta \cdot \omega \cdot (t-\tau)} \cdot \sin \omega \cdot (t-\tau) \, d\tau \right]_{max}$$

Written differently:

$$S_d = (a/\omega^2) \cdot (DMF)_{max}$$

$$S_a = S_d \cdot \omega^2 = a \cdot (DMF)_{max}$$

It should be noted that the dynamic factor $(DMF)_{max}$, as can be seen from the above expression, is identical to the acceleration spectral value (S_a/a). ■

Assuming now, in an approximate approach, that the variation of the acceleration a in time is sinusoidal, i.e. $f_a(t) = \sin \Omega \cdot t$, then, given that the two last equations are generally valid, the previously found expression for the dynamic factor (see Section 16.2.4) may be applied directly:

$$(DMF)_{\Omega,max} = \frac{1}{\sqrt{[1-(\Omega/\omega)^2]^2 + 4 \cdot \zeta^2 \cdot (\Omega/\omega)^2}}$$

Thus,

$$S_{d,\Omega} = (a/\omega^2) \cdot (DMF)_{\Omega,max}$$

and

$$S_{a,\Omega} = a \cdot (DMF)_{\Omega,max}$$

The diagram in Figure 16.10 may therefore also be interpreted as an acceleration spectrum – with respect to unit ground acceleration – for any value of the ratio of the frequency Ω (taken as constant) to the natural frequency ω of the single-degree system. The magnitude of the assumed frequency Ω of the maximum ground acceleration depends on the geotechnical characteristics of the foundation soil. For stiff soils and rocks, Ω ranges from 20 to 60 rad/s, for medium stiff soils, from 12 to 50 rad/s, and for soft or very soft soils, between 10 and 3 rad/s. ■

Considering the influence of the foundation soil, it can be seen from Figure 16.16 that its deformability influences, in addition to the maximum developed ground acceleration,

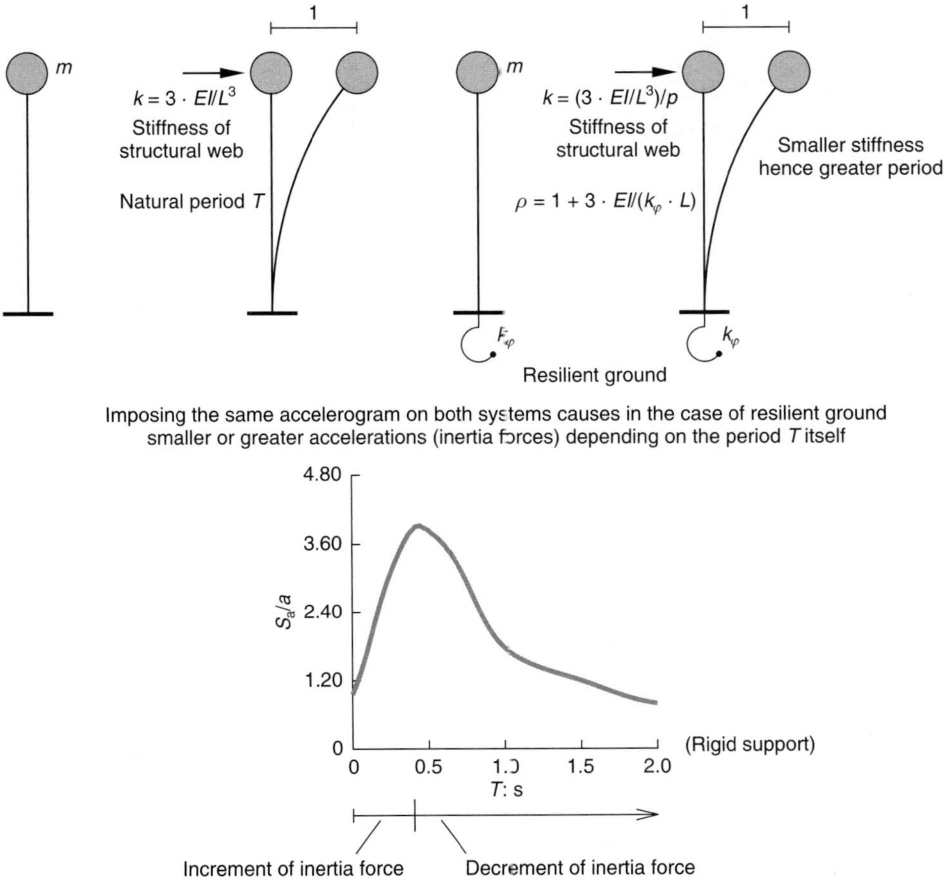

Figure 16.16 Influence of the deformability of the soil on the seismic response

the response of the structure itself, due to the variation of the effective inertia force of the mass. This force increases or decreases according to the region of the corresponding acceleration spectrum that the natural period of the structure belongs to, by assuming a fixed base (see Figure 16.16).

In actual design practice, the so-called *design spectrum*, which contains the effects of a large number of relevant accelerograms for the region examined, is used as an acceleration spectrum, also making a distinction between the various soil types, given the influence of the foundation soil deformability, as explained above.

Of course, the elaboration of the design spectral curves is a major issue for the regulations in a particular country. Examples of these curves are shown in Figure 16.17. The influence of the soil type on the maximum effective acceleration S_a is obvious.

16.2.6 Influence of plastic behaviour on the seismic response

According to the previous section, the basic equation of motion of a single-degree-of-freedom system under a seismic excitation, with respect to a fixed base, can be

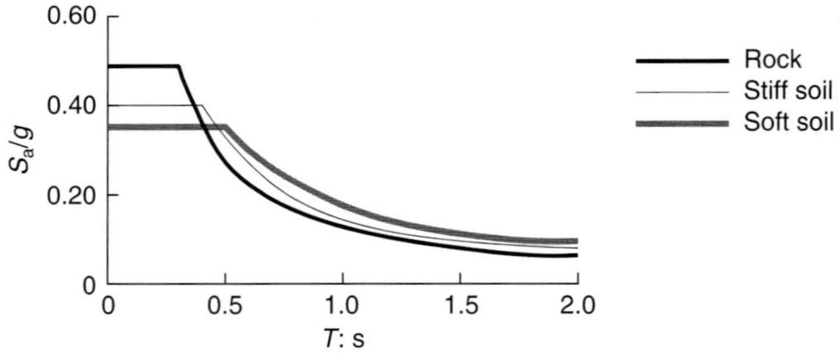

Figure 16.17 Design spectrum

written as

$$m \cdot \ddot{x} + k \cdot x + c \cdot \dot{x} = -(m \cdot a) \cdot f_a(t)$$

This equation expresses the static equilibrium between the externally acting inertia force $(m \cdot \ddot{x} + m \cdot a \cdot f_a(t))$ together with the damping force on one side and the elastic force developed by the structural web due to its deformation on the other. This last force $(k \cdot x)$ increases for as long as the displacement x of the mass increases, i.e. as for as long as the deformation of the structural web increases, according to the existing stiffness factor k.

However, it is clear that a possible plastic hinge at the base of the cantilever, where the maximum bending response is developed, places an upper limit on the increase of the above-mentioned force, which remains constant afterwards, despite further increase in the displacement x (Figure 16.18). More specifically, if L is the cantilever height and M_{pl} represents the plastic bending strength of the structural web, then the limit F up to which the force can be transmitted to the mass, i.e. the 'elastic force', may be increased, $F = M_{pl}/L$. This, of course, is under the condition that the relevant

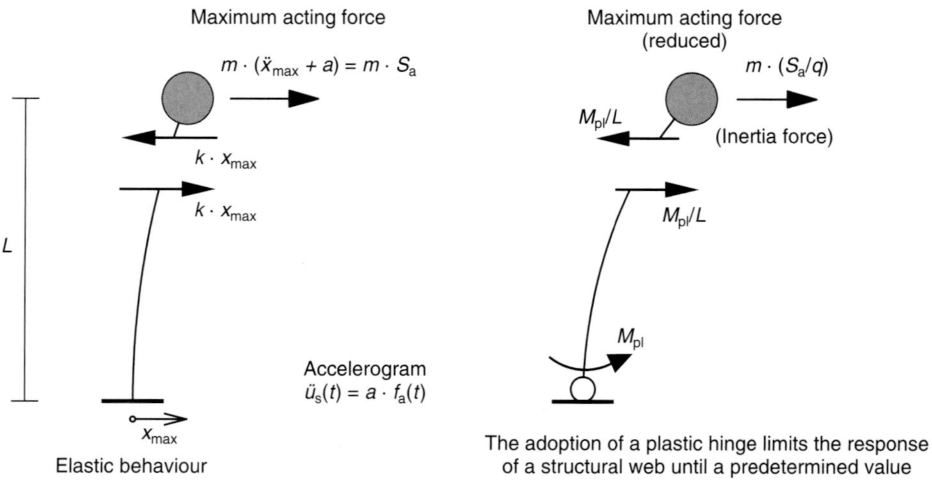

Figure 16.18 Limitation of the inertia force due to plastification

section of the structural web affords the required ductility (rotational deformability), so that the augmented displacement x can be realised.

It is clear that the limitation of the internal – elastic – force expressed by the term $k \cdot x$ in the above equation restricts, accordingly, the level of the effective inertia force too. This means that, although the mass displacement x may increase, the above equilibrium equation can be satisfied with less internal force and, correspondingly, less inertia force, than required by a purely elastic behaviour of the structural web.

This fact may be taken into account through a corresponding reduction in the spectral acceleration value S_a, through the factor q, known as the *behaviour factor*. Thus, in the corresponding equations in the previous section, the magnitude S_a/q is used instead of S_a. The selected value of the factor q is essentially the designer's decision, depending on the estimated – or even desired – ductility of the structural web. A value of q equal to 1.0 would indicate the acceptance of elastic behaviour throughout the web without any reduction in the inertia seismic forces. A steel structure may allow consideration of a behaviour factor up to 4.0. In reinforced concrete structures, with an appropriate layout of reinforcement in order to ascertain good deformability (ductility) of the 'critical' sections, a value of q of even 3.0 may be adopted, leading to a corresponding decrease in the acting seismic forces. However, it should be pointed out that in the design process the consideration (or not) of q greater than unity – i.e. the adoption (or not) of plastic hinges in the structural web to account for a given maximum ground acceleration – relies exclusively on the designer's engineering judgement.

16.3 Multi-degree systems

As *multi-degree systems* imply, they are systems that contain more than one discrete mass. These masses are 'carried' by a massless structural web exhibiting a clearly determinable behaviour regarding its deformability and stiffness. Within this frame of behaviour the masses show displacements or rotations involving either negligible or non-negligible inertia forces. These last displacements are considered as 'effective' for the corresponding masses, and their number represents the so-called *effective degrees of freedom* of the system. Examples of such multi-degree systems have been mentioned in the introductory section of this chapter (see Figure 16.2).

16.3.1 The stiffness matrix

16.3.1.1 The concept of the stiffness matrix

Of decisive importance for the description of the behaviour of a multi-degree system is the so-called *stiffness matrix*, i.e. a set of magnitudes, describing the stiffness of its structural web, which are determined as shown in the example of a three-storey plane model in Figure 16.19. Here, it is assumed that the masses m_1, m_2 and m_3 develop in a horizontal excitation (e.g. earthquake) inertia forces with respect only to their horizontal displacements. Thus, the effective displacements of the system are the lateral storey shifts Δ_1, Δ_2 and Δ_3.

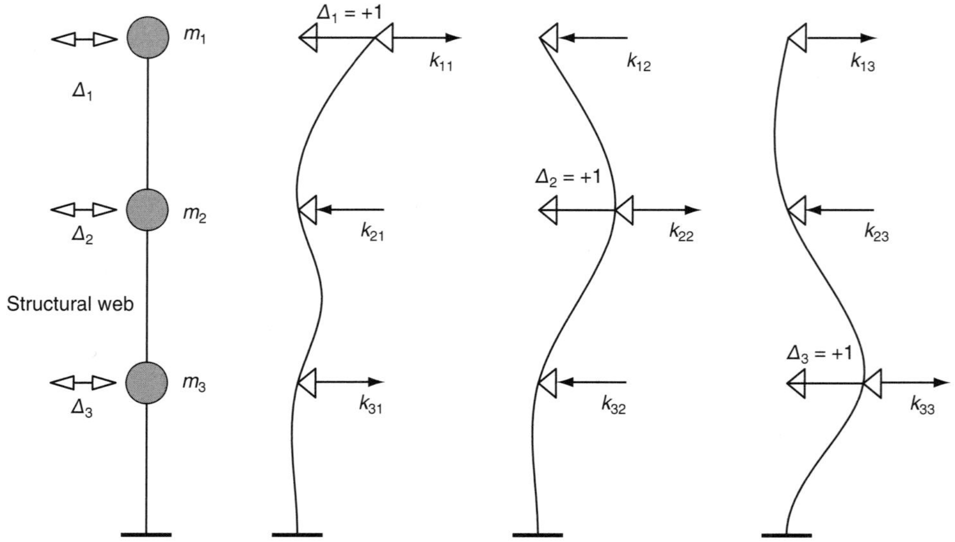

$$\mathbf{k} = \begin{bmatrix} k_{11} & k_{12} & k_{13} \\ k_{21} & k_{22} & k_{23} \\ k_{31} & k_{32} & k_{33} \end{bmatrix}$$

The stiffness matrix represents the stiffness of the structural web
with respect to the effective mass displacements

Figure 16.19 The elements of a stiffness matrix

The stiffness matrix consists in the horizontal juxtaposition of three groups of forces that are determined as follows (see Figure 16.19). The first group consists of those three – horizontal – forces (k_{11}, k_{21}, k_{31}) that are required to be applied correspondingly to the effective displacements in order to produce a unit shift in level 1 and zero shift in the two others (i.e. $\Delta_1 = 1$, $\Delta_2 = 0$, $\Delta_3 = 0$).

Analogously, the second group consists of those forces (k_{12}, k_{22}, k_{32}) needed to be acted on in the sense of the effective displacements, in order to produce a unit shift in level 2 and zero shift in the two others (i.e. $\Delta_1 = 0$, $\Delta_2 = 1$, $\Delta_3 = 0$). Finally, the third group contains those forces (k_{13}, k_{23}, k_{33}), which, acting along the effective displacements, produce $\Delta_1 = 0$, $\Delta_2 = 0$, $\Delta_3 = 1$.

Each one of the above groups comprises the reactions developed if, at first, fictitious simple supports are inserted into the system, in order to restrain all its effective displacements, and then the corresponding unit-support settlement (1 m) is imposed, as shown in Figure 16.19. In order to achieve this, the use of a computer is clearly necessary.

Defining the positive senses of these forces as coinciding with those of the displacements (Δ_1, Δ_2, Δ_3), i.e. to the right, it is clear that the forces of each group will exhibit alternating signs.

The structural web which carries the three masses may be represented by the *stiffness matrix* **k** by juxtaposing horizontally the above three force groups as previously mentioned, according to the following formulation:

$$\mathbf{k} = \begin{bmatrix} k_{11} & k_{12} & k_{13} \\ k_{21} & k_{22} & k_{23} \\ k_{31} & k_{32} & k_{33} \end{bmatrix}$$

It is useful to become familiar with matrix operations. However, it must be pointed out that the use of matrices contributes very little, if at all, to the understanding of the static or dynamic behaviour of any system – at least at this introductory level. Nevertheless, matrices are a necessary tool for the compact writing of equilibrium equations and the relevant equations, thus facilitating their methodical numerical treatment through computer programs.

16.3.1.2 Matrix operations

In the present context, of particular interest are either the so-called *square matrices* – having an equal number of rows and columns – or the so-called *column matrices* – consisting of the vertical juxtaposition of numerical elements.

A square matrix **k** of the order $N \times N$, has N rows and N columns (see Figure 16.19). The elements of the square matrix along the diagonal starting from the upper left element constitute the so-called *diagonal elements*. A column matrix $\boldsymbol{a}$ of the order $N \times 1$ has N rows and 1 column.

Interchange of the rows and columns of a matrix – or vice versa – leads to the *transpose matrix*, denoted by the letter T. Thus, while the matrix $\mathbf{k}^{\mathrm{T}}$ still has N rows and N columns (i.e. $N \times N$), matrix $\boldsymbol{a}^{\mathrm{T}}$ has 1 row and N columns (i.e. $1 \times N$).

The basic physical concept of a matrix is directly related to the column matrix. This is the recording of quantities in correspondence to specific magnitudes, e.g. the storey shifts of the structure described previously, or the forces acting at each storey level. The appropriate horizontal juxtaposition of such column matrices leads to square matrices, like the stiffness matrix which has been previously defined.

The sum of two column matrices is a matrix of the same order, and comprises the sums of their corresponding terms. Although the addition also applies to two square matrices of the same order, the physical meaning of this operation is clearer and better understood in this context with respect to the column matrices.

The multiplication of two matrices is of basic importance, and it is accomplished as shown in Figure 16.20. The multiplication of two matrices is feasible only when the number of columns in the left matrix is equal to the number of rows in the right matrix. This may also be understood from the rule that the final order of the product matrix is found by eliminating the 'interior' dimensions of the matrices – as shown below.

The product $\mathbf{k} \cdot \boldsymbol{a}$ is written as $(N \times N) \cdot (N \times 1)$, and is a matrix of order $N \times 1$, whereas product $\boldsymbol{a} \cdot \mathbf{k}$, which is written as $(N \times 1) \cdot (N \times N)$, does not make sense. Of

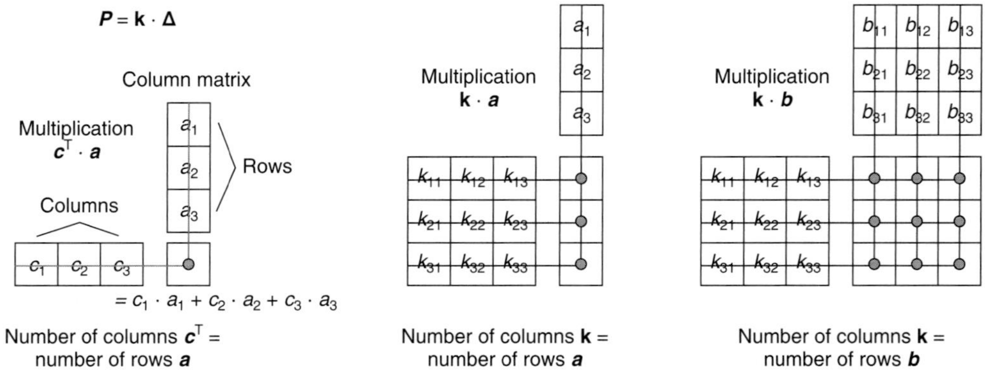

Figure 16.20 Layout for the multiplication of matrices

course, product $\boldsymbol{a}^{\mathrm{T}} \cdot \mathbf{k}$ does have a meaning, because it is written as $(1 \times N) \cdot (N \times N)$, which leads to a matrix of order $1 \times N$, whereas it can be seen that product $\boldsymbol{a}^{\mathrm{T}} \cdot \boldsymbol{a}$, written as $(1 \times N) \cdot (N \times 1)$, leads to a matrix of order 1×1, i.e. a single number.

It is thus clear, also taking into account the schematic layout of Figure 16.20, that the multiplication of two matrices consists of the consecutive formation of matrix products of the last type, $(1 \times N) \cdot (N \times 1)$, which, of course, represent a single number.

The square matrix, the diagonal elements of which consist of unity while all other elements are null, is called the *identity matrix*, and is denoted by **I**. Consideration of the identity matrix allows reference to the basic concept of the so-called *inverse matrix* of a square matrix **A** of the order $N \times N$, which is denoted by $\mathbf{A}^{-1}$. The inverse matrix is one that satisfies the equation

$$\mathbf{A} \cdot \mathbf{A}^{-1} = \mathbf{I}$$

or (identically)

$$\mathbf{A}^{-1} \cdot \mathbf{A} = \mathbf{I}$$

Obviously, both matrix $\mathbf{A}^{-1}$ and matrix **I** are of the order $N \times N$.

The concept of the inverse matrix is directly connected to the solution of a linear system of N equations with N unknowns. Such a system may be represented through the following matrix notation:

$$\mathbf{A} \cdot \boldsymbol{X} = \mathbf{C}$$

If matrix **A** consists of the coefficients of the unknowns, then column matrices $\boldsymbol{X}$ and **C** contain the unknowns and the 'known terms' of the right-hand side of the equations, respectively. Multiplying both sides by the inverse matrix $\mathbf{A}^{-1}$ yields

$$\mathbf{A}^{-1} \cdot \mathbf{A} \cdot \boldsymbol{X} = \mathbf{A}^{-1} \cdot \mathbf{C}$$

that is,

$$\mathbf{I} \cdot \boldsymbol{X} = \mathbf{A}^{-1} \cdot \mathbf{C}$$

and,

$$X = A^{-1} \cdot C$$

Of course, the inversion of matrix **A** requires a numerical procedure that is equally cumbersome as the solution of the linear system, so that the advantage in the use of the matrix notation lies only in the compact formulation and not in the computing effort required, which in any case will be undertaken using a computer.

16.3.1.3 Correlation of loading and displacements

The usefulness of the matrix notation in describing in a compact manner the behaviour of a system is illustrated using the stiffness matrix **k** (3×3) of the structural model shown in Figure 16.21. From the physical definition of **k** in Section 16.3.1.1, it is easily seen that the three forces (P_1, P_2, P_3) required to be applied to the three storey levels in order produce the corresponding displacements (Δ_1, Δ_2, Δ_3) can be obtained from the equation

$$\mathbf{P} = \mathbf{k} \cdot \boldsymbol{\Delta} \quad \text{(a)}$$

where matrix **P** (3×1) represents the forces (P_1, P_2, P_3), and matrix $\boldsymbol{\Delta}$ (3×1) represents the displacements (Δ_1, Δ_2, Δ_3).

According to the introduced 'matrix language', it can be seen that if the loads **P** are known and the displacements $\boldsymbol{\Delta}$ caused by them are sought, these can be obtained by multiplying both sides of the above equation by the inverse $\mathbf{k}^{-1}$ of the stiffness matrix **k**, i.e.

$$\mathbf{k}^{-1} \cdot \mathbf{P} = \mathbf{k}^{-1} \cdot \mathbf{k} \cdot \boldsymbol{\Delta}$$

so that

$$\boldsymbol{\Delta} = \mathbf{k}^{-1} \cdot \mathbf{P}$$

Clearly, this result represents nothing more than the solution of the linear system (a), without this compact matrix formulation helping the solution itself.

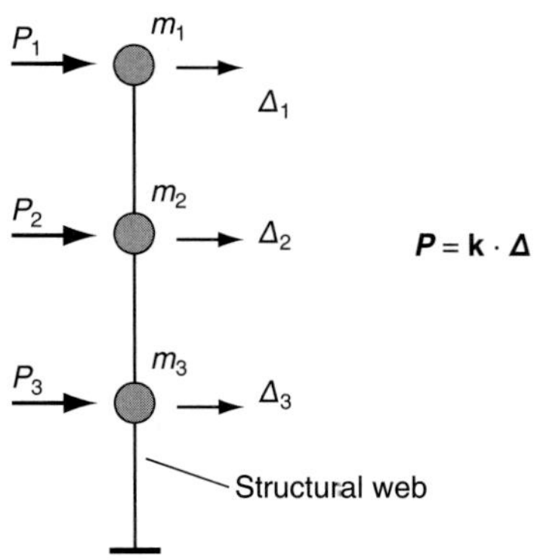

Figure 16.21 Correlation of loading and deformation in a multi-degree system

16.3.1.4 Multi-storey systems

At this point it is useful – for later use too – to describe through the compact matrix methodology the overall response of a multi-storey spatial system under horizontal loads, according to the previous examination in Chapter 15 (Figure 16.22).

From Chapter 15, we know that the structural web of a multi-storey spatial system consists of elements of plane stiffness. Each of the plane elements is represented by its own stiffness matrix $\mathbf{S}_i$ – as previously defined in Section 16.3.1.1 and shown in Figure 16.23.

The overall stiffness matrix $\mathbf{K}$ of the order 9×9, referring to the whole system, relates the loads $\mathbf{P}_x$ and $\mathbf{P}_y$ (order 3×1) to the developed displacements $\boldsymbol{\Delta}_x$, $\boldsymbol{\Delta}_y$ and $\boldsymbol{\Delta}_\omega$ at each diaphragm. Matrix $\mathbf{K}$ is constructed as follows:

$$\mathbf{K} = \begin{bmatrix} \mathbf{K}_1 & \mathbf{K}_2 & \mathbf{K}_3 \\ \mathbf{K}_2 & \mathbf{K}_4 & \mathbf{K}_5 \\ \mathbf{K}_3 & \mathbf{K}_5 & \mathbf{K}_6 \end{bmatrix}$$

where

$$\mathbf{K}_1 = \sum \mathbf{S}_i \cdot \cos^2 \beta_i$$

$$\mathbf{K}_2 = \sum \mathbf{S}_i \cdot \sin \beta_i \cdot \cos \beta_i$$

$$\mathbf{K}_3 = \sum \mathbf{S}_i \cdot e_i \cdot \cos \beta_i$$

$$\mathbf{K}_4 = \sum \mathbf{S}_i \cdot \sin^2 \beta_i$$

$$\mathbf{K}_5 = \sum \mathbf{S}_i \cdot e_i \cdot \sin \beta_i$$

$$\mathbf{K}_6 = \sum \mathbf{S}_i \cdot e_i^2$$

all matrices being of the order 3×3 (Stavridis, 1986).

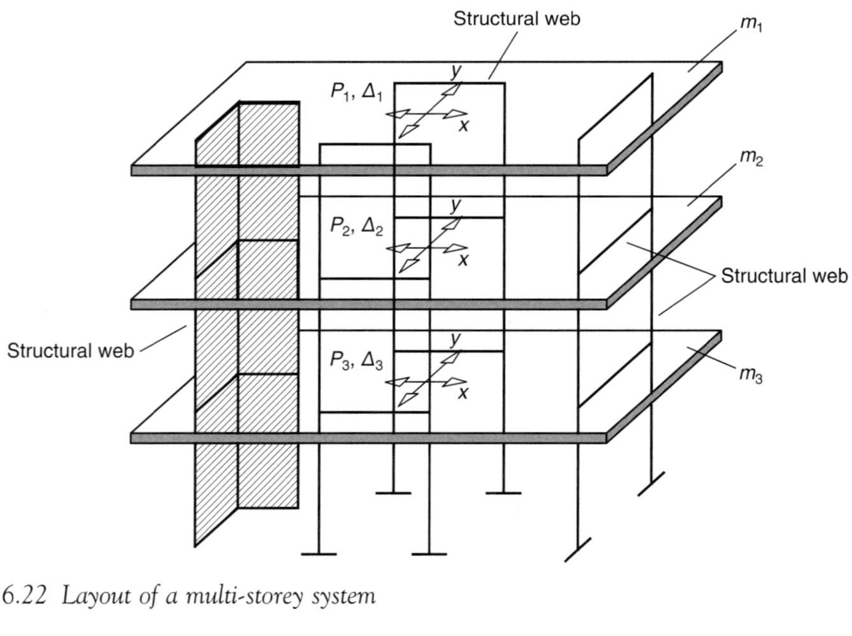

Figure 16.22 Layout of a multi-storey system

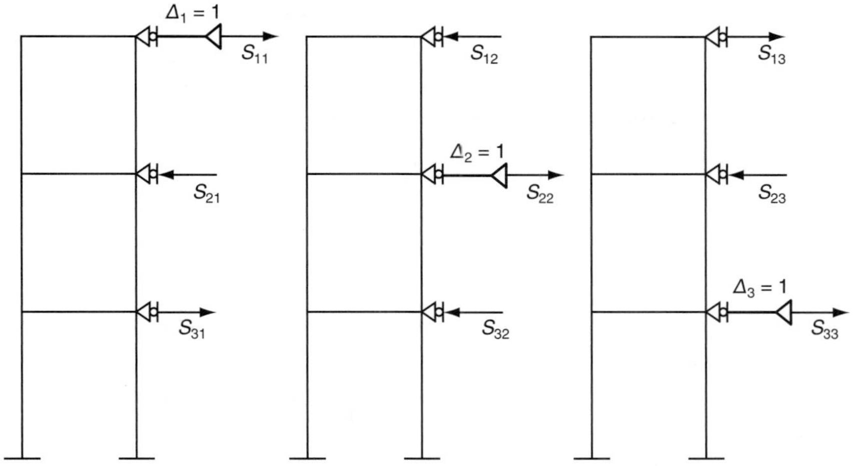

Figure 16.23 Determination of the matrix of lateral stiffness for a plane element

The correspondence with the relevant expressions in Section 15.3.3 – at least with reference to a single storey – is absolute.

Thus, the equation between loading and deformation of the system is written as follows:

$$\begin{Bmatrix} \boldsymbol{P}_x \\ \boldsymbol{P}_y \\ 0 \end{Bmatrix} = \begin{bmatrix} \mathbf{K}_1 & \mathbf{K}_2 & \mathbf{K}_3 \\ \mathbf{K}_2 & \mathbf{K}_4 & \mathbf{K}_5 \\ \mathbf{K}_3 & \mathbf{K}_5 & \mathbf{K}_6 \end{bmatrix} \cdot \begin{Bmatrix} \boldsymbol{\Delta}_x \\ \boldsymbol{\Delta}_y \\ \boldsymbol{\Delta}_\omega \end{Bmatrix}$$

It should be noted that the matrices $\mathbf{P}_x$ and $\boldsymbol{\Delta}_x$ consist of the components in the x sense of the magnitudes (P_1, P_2, P_3) and (Δ_1, Δ_2, Δ_3), respectively (see Figure 16.22). This is analogous for matrices $\mathbf{P}_y$ and $\boldsymbol{\Delta}_y$ too. The matrix $\boldsymbol{\Delta}_\omega$ consists of the diaphragm rotations (see Section 15.3.1).

It is, of course, clear that, given a multi-storey system with N diaphragms, the above equation allows the determination of the storey displacements, through inversion of matrix $\mathbf{K}$ of the order ($3N \times 3N$), as examined above, through use of suitable computer software. However, in the context of the examination of dynamic behaviour from a preliminary design point of view, of prime interest is the orthogonal layout of plane elements about the axes x and y. Then, the constituent matrices of the global stiffness matrix $\mathbf{K}$ are expressed as follows (see Section 15.3.4):

$$\mathbf{K}_1 = \sum \mathbf{S}_x$$

$$\mathbf{K}_2 = 0$$

$$\mathbf{K}_3 = -\sum (\mathbf{S}_x \cdot y)$$

$$\mathbf{K}_4 = \sum \mathbf{S}_y$$

$$K_5 = \sum (S_y \cdot x_i)$$

$$K_6 = \sum (S_x \cdot y_i^2) + \sum (S_y \cdot x_i^2)$$

Thus, the above condensed equation may be split up as follows:

$$P_x = K_1 \cdot \boldsymbol{\Delta}_x + K_3 \cdot \boldsymbol{\Delta}_\omega$$

$$P_y = K_4 \cdot \boldsymbol{\Delta}_y + K_5 \cdot \boldsymbol{\Delta}_\omega$$

$$0 = K_3 \cdot \boldsymbol{\Delta}_x + K_5 \cdot \boldsymbol{\Delta}_y + K_6 \cdot \boldsymbol{\Delta}_\omega$$

If the plane elements layout leads to nullification – even approximately – of the matrices $\mathbf{K}_3$ and $\mathbf{K}_5$, which should be the basic intention according to Section 15.3.4, then, from the last equation, $\boldsymbol{\Delta}_\omega = 0$, i.e. the diaphragms do not rotate, thus not subjecting plane elements to an additional response. Recall that the horizontal loads are acting on the centroid of each diaphragm, to which the coordinates x and y are also referred. By taking into account the above,

$$P_x = K_1 \cdot \boldsymbol{\Delta}_x = (\sum S_x) \cdot \boldsymbol{\Delta}_x$$

and

$$P_y = K_4 \cdot \boldsymbol{\Delta}_y = (\sum S_y) \cdot \boldsymbol{\Delta}_y$$

The above equations mean that, in the examined case, the displacements in both the x and y senses are independent of each other. They are obtained by applying the loads $\mathbf{P}_x$ and $\mathbf{P}_y$ on the two separate systems, respectively, which are formed through coupling of all the respective plane elements in each sense, carrying at each level the corresponding mass (Figure 16.24). This coupling is accomplished at each level through hinged undeformable bars, as shown in the figure. The above conclusion is of particular practical importance for both the assessment of the dynamic characteristics of the multi-storey

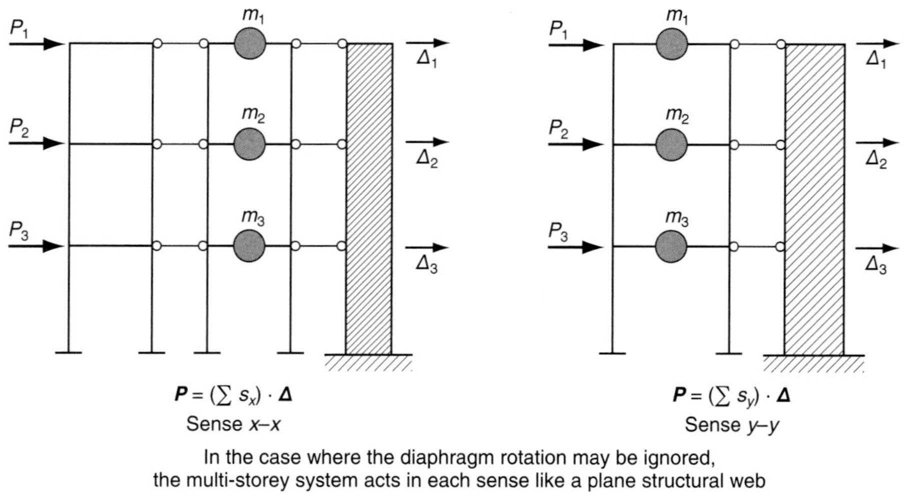

Figure 16.24 Equivalent plane system in both directions

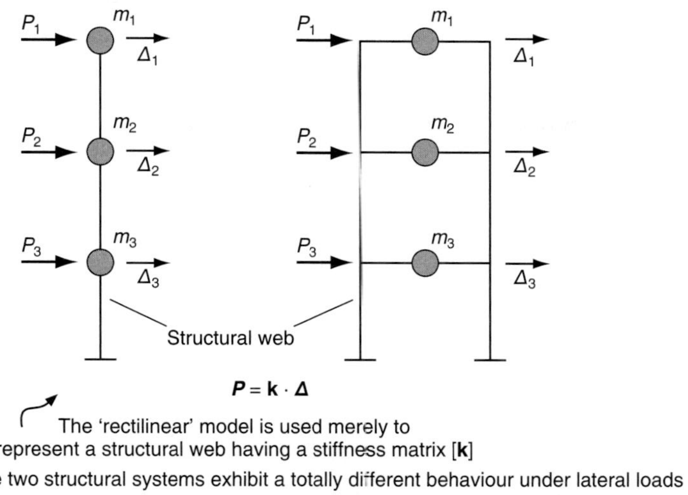

Figure 16.25 Simulation of a multi-degree frame with a rectilinear model

mass system and the determination of the equivalent static loading because of seismic excitation, as will be examined later in this chapter.

16.3.2 Free vibration

16.3.2.1 Plane systems

The study of multi-degree systems will be continued on an illustrative basis for the three-degree model, without thereby limiting the general validity of any conclusions for systems with more degrees of freedom.

At this point, it is noted that the 'rectilinear' structural web in Figure 16.25 should not necessarily be considered merely as a straight member carrying at some points the masses m_1, m_2 and m_3 but as a structural system described by its stiffness matrix **k**. Consequently, the model may also be considered to represent a plane frame, such as that shown in the figure, and having the same stiffness matrix **k** (see Figure 16.23). This clarification is necessary because the lateral shift behaviour of a vertical cantilever is quite different from that of the frame, as has already been pointed out in Section 6.4.3 (see Figure 6.16).

First, it is assumed that the mass m_i, acted on by an external force $F_i(t)$ in the sense of its own degree of freedom, i.e. horizontally, undergoes a displacement $x_i(t)$. The positive sense for all the forces and displacements is considered common (Figure 16.26).

Both the set of forces $F_i(t)$ and the set of displacements $x_i(t)$ may be represented by the column matrices $\mathbf{F(t)}$ and $\mathbf{x}$, respectively. For ease of matrix manipulation only, the system of masses m_i is represented by the square diagonal matrix $\mathbf{m}$ of order $N \times N$. In this matrix, the masses m_i constitute the diagonal elements, with all the other elements being zero. It can be written as

$$\mathbf{m} = \mathbf{I} \cdot m$$

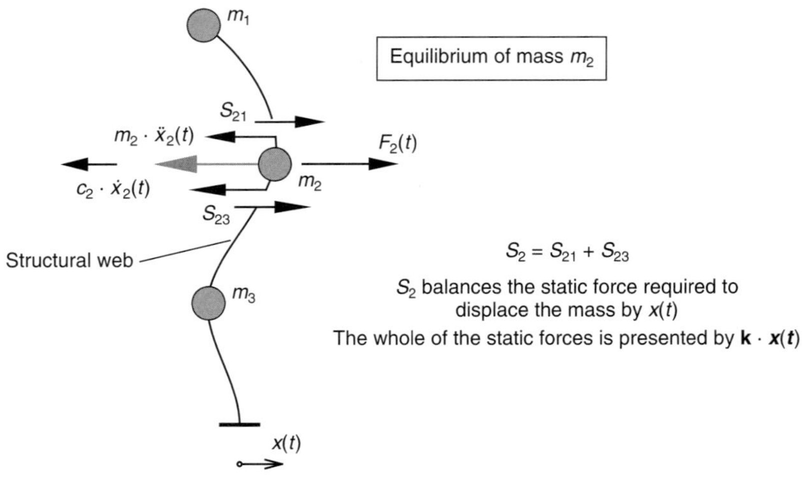

Figure 16.26 Dynamic equilibrium in a multi-degree system under external loading

Just as in a single-degree system, a specific mass m_i may be considered as being always in equilibrium under the following forces (see Figure 16.26):

(1) the external force $F_i(t)$
(2) the total of the elastic forces S_i transmitted from the structural web to the mass, as a consequence of its deformational configuration $\boldsymbol{x}(\boldsymbol{t})$
(3) the inertia force $m_i \cdot \ddot{x}_i(t)$
(4) the damping force $c_i \cdot \dot{x}_i(t)$.

As forces (2), (3) and (4) have the opposite sense to the force $F_i(t)$, for the equilibrium of each mass m_i (see Figure 16.26),

$$m_i \cdot \ddot{x}_i(t) + c_i \cdot \dot{x}_i(t) + S_i = F_i(t)$$

It is clear that the deformational configuration $\boldsymbol{x}(\boldsymbol{t})$ for the specific time instant t may be produced by a unique set of external forces depending on the stiffness of the structural web. According to the foregoing, this set of forces can be given by the expression $\mathbf{k} \cdot \boldsymbol{x}(\boldsymbol{t})$ (column matrix $N \times 1$), and it can be seen that each of these forces equals the respective sum S_i of the elastic forces acting directly on the mass m_i.

The simultaneous consideration of the equilibrium of all masses can now be expressed by the following matrix formulation:

$$\mathbf{m} \cdot \ddot{\boldsymbol{x}} + \mathbf{c} \cdot \dot{\boldsymbol{x}} + \mathbf{k} \cdot \boldsymbol{x} = \boldsymbol{F}(\boldsymbol{t})$$

The damping coefficients c_i are represented by the square diagonal matrix $\mathbf{c}$. ■

As in the single-degree systems, it is expedient here to also seek the basic dynamic characteristics of the multi-degree system, through the examination of its undamped free vibration.

Free vibration is evoked by imposing an arbitrary displacement at a certain mass, according to its corresponding degree of freedom (e.g. a horizontal displacement $\Delta_3 = \delta$) and then leaving the system free to vibrate.

In this case, the global equilibrium condition is written as

$$\mathbf{m} \cdot \ddot{\boldsymbol{x}} + \mathbf{k} \cdot \boldsymbol{x} = 0$$

By trying out the *harmonic* solution $\boldsymbol{x} = \boldsymbol{\varphi} \cdot \sin \omega \cdot t$, where the magnitude ω is, of course, unknown and, moreover, the column matrix $\boldsymbol{\varphi}$ also consists of unknown quantities, and given that $\ddot{\boldsymbol{x}} = -\omega^2 \cdot \boldsymbol{\varphi} \cdot \sin \omega \cdot t$, the above equation can be written as (Tedesco *et al.*, 1999)

$$[\mathbf{m}^{-1} \cdot \mathbf{k} - \omega^2 \cdot \mathbf{I}] \cdot \boldsymbol{\varphi} = 0$$

This equation is known as the *eigenvalue equation* or *frequency equation* of the system, and represents a linear system of N equations with respect to the N unknowns $\boldsymbol{\varphi}$. According to the laws of linear algebra, in order for a non-zero solution to exist in a linear system having all of its right-hand side members equal to zero, the determinant of the coefficients of the unknowns must be equal to zero. This determinant, as can be concluded from the last equation, consists of a polynomial of Nth degree with respect to ω^2, for which a number of N values (roots) can be determined. It is, of course, obvious that for each *eigenvalue* ω^2 the corresponding solution of the above linear system does not simply consist of a well-defined set of values $\boldsymbol{\varphi}$ but rather of an infinite group of values $\lambda \cdot \boldsymbol{\varphi}$, where λ is an arbitrary number.

It is clear that because of the computational effort required for the determination of the above magnitudes, these can only be obtained through the use of appropriate computer software.

From a physical point of view, the above signifies that the N-degree system has N different *eigenfrequencies* ω, and each one of them corresponds to a certain displacement configuration $\boldsymbol{\varphi}$ for each degree of freedom, which is multiplied by an arbitrary factor, either positive or negative. This displacement configuration is called the *eigenform* or *natural mode* (Figure 16.27). The physical meaning of any of the eigenforms $\boldsymbol{\varphi}$ is that

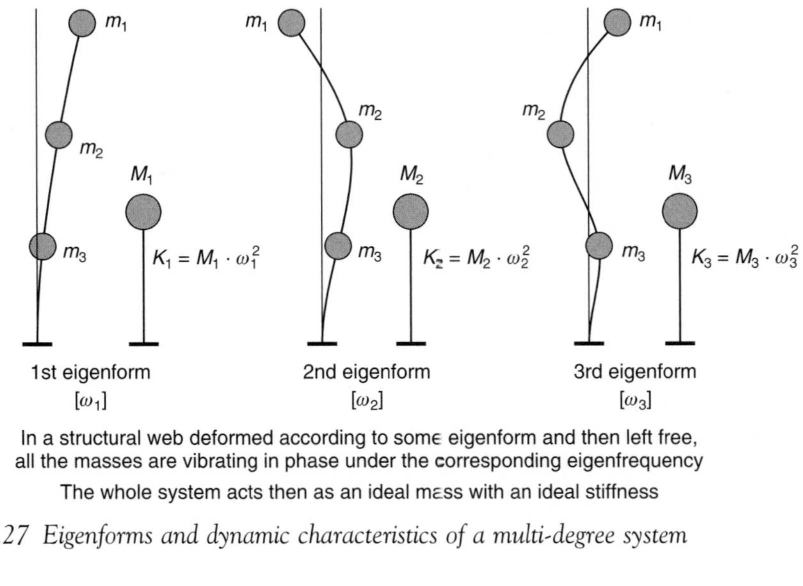

Figure 16.27 Eigenforms and dynamic characteristics of a multi-degree system

if it is somehow imposed on the system (i.e. through an appropriate application of horizontal forces) and the system is left free, then every mass of the system will vibrate like a single-mass oscillator under the same natural frequency ω, which corresponds to the selected eigenform.

An additional analytical property of the eigenforms φ should be noted. For any pair of different eigenforms φ_r and φ_s the following is valid:

$$\varphi_r^{\mathrm{T}} \cdot \mathbf{m} \cdot \varphi_s = 0$$

$$\varphi_r^{\mathrm{T}} \cdot \mathbf{k} \cdot \varphi_s = 0$$

which is known as the *orthogonality condition* of the eigenforms φ.

The fact that the whole system may vibrate at any specific frequency of the N realisable eigenfrequencies ω_r exhibited for each mass amplitude (represented by the eigenform φ_r) means that all masses are found at the same time either at the rest position or at their maximum deviation, and suggests the idea of considering an ideal (fictitious) mass M_r connected to an ideal (fictitious) stiffness K_r (see Figure 16.27). It is indeed found that for the corresponding eigenform φ_r the above magnitudes may be determined as

$$M_r = \varphi_r^{\mathrm{T}} \cdot \mathbf{m} \cdot \varphi_r = \sum_{1}^{N} m_r \cdot \varphi_r^2$$

$$K_r = \varphi_r^{\mathrm{T}} \cdot \mathbf{k} \cdot \varphi_r$$

According, of course, to the characteristics of a typical single-degree system (Biggs, 1964):

$$\omega_r^2 = K_r/M_r$$

In this way, for a given multi-degree system, N single-mass oscillators may be considered, each with specific characteristics of mass, stiffness and frequency. The lowest frequency ω is called the *fundamental frequency*, and has more physical importance than the others, as will be shown later.

More specifically, if an arbitrary horizontal displacement is imposed on any mass in any of the two systems shown in Figure 16.28, and the structure is then left free, each mass in the system will vibrate at a frequency that essentially coincides with the fundamental frequency. This fact, which may apply to any multi-degree system, leads to an important result that allows the determination of the fundamental frequency of the system, as will now be shown.

If to every mass in a multi-degree system an arbitrary static force F_r of the same sense is applied, then the masses will be displaced by Δ_r (Figure 16.29), and, if the system is then left free, a harmonic motion for every mass with a natural frequency ω will be immediately established, whereby all masses take both their rest position and their maximum deviation Δ_r at the same time instant.

According to the characteristics of the single-degree systems, the maximum velocity of every mass at its rest position will be $\omega \cdot \Delta_r$, and, consequently, the kinetic energy of the masses as a whole will be $\sum \frac{1}{2} \cdot m_r \cdot (\omega \cdot \Delta_r)^2$.

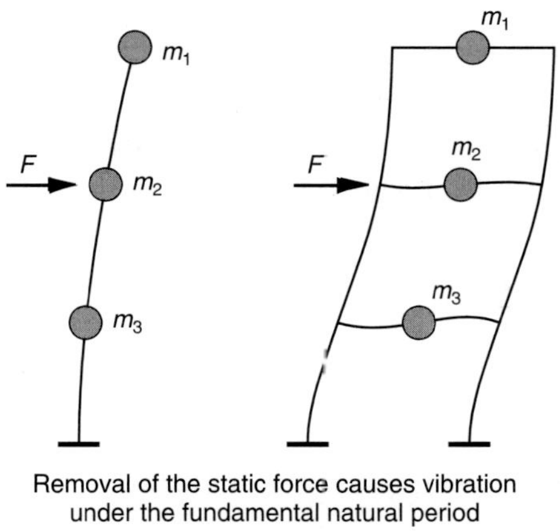

Figure 16.28 The fundamental frequency

Clearly, this kinetic energy should be equal to the total work done by the forces F_r, which is $\sum \frac{1}{2} \cdot F_r \cdot \Delta_r$. Thus, the value of the fundamental frequency ω may be readily obtained as (Biggs, 1964)

$$\omega^2 = \frac{\sum F_r \cdot \Delta_r}{\sum m_r \cdot \Delta_r^2}$$

The computation is facilitated if the forces F_r are selected as equal (see Figure 16.29). ■

The damping action is logically expressed through the damping coefficient ζ with regard to each eigenform. Usually, the value of ζ applied to the first eigenform corresponding to

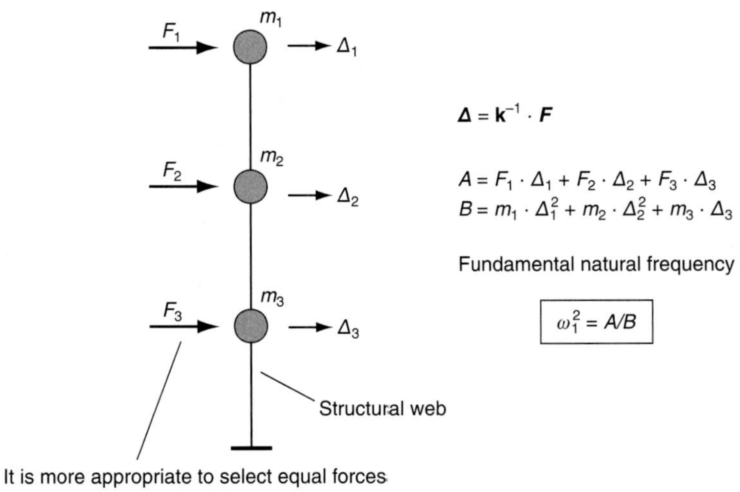

Figure 16.29 Practical evaluation of the fundamental eigenfrequency

the fundamental frequency ω is that directly relating to the nature of the structural web material, in accordance with Section 16.2.2.2. For each additional eigenform, the value of ζ may be considered to increase by 1%.

For the experimental determination of the damping coefficient, as well as of the eigenfrequency of each eigenform, the same procedure may be followed as in the case of a single-degree system, i.e. by creating through appropriate forces the corresponding displacement configuration φ_r and following the corresponding free vibration of the system (see Section 16.2.2.2).

The measured eigenfrequencies for each eigenform should not differ essentially from those obtained theoretically, as explained above, provided that the damping coefficients do not exceed 10%.

16.3.2.2 Multi-storey spatial systems

In the case of a multi-storey building with horizontal diaphragms arranged vertically over each other, and which are carried by the structural web consisting of vertical plane elements, as examined in Chapter 15, it becomes clear that the dynamic equilibrium of the diaphragm masses is related to the displacements Δx and Δy of each diaphragm, as well as to its rotation $\Delta\omega$. While displacements Δx and Δy activate the inertia forces $m \cdot \Delta\ddot{x}$ and $m \cdot \Delta\ddot{y}$ of the mass of each diaphragm, respectively, the rotation $\Delta\omega$ activates the rotational inertia moment $(\Theta \cdot \Delta\ddot{\omega})$ on the basis of the rotational moment of inertia Θ of the mass of the diaphragm with respect to its centroid.

For the case of an equally distributed mass in the diaphragm,

$$\Theta = m \cdot (I_x + I_y)$$

where I_x and I_y are the moments of inertia of the diaphragm area with respect to the axes x and y, respectively.

By taking into account the 'expanded' mass matrix ($3N \times 3N$),

$$\bar{\mathbf{m}} = \begin{bmatrix} \mathbf{m} & 0 & 0 \\ 0 & \mathbf{m} & 0 \\ 0 & 0 & \boldsymbol{\Theta} \end{bmatrix}$$

the eigenfrequency equation is analogous to that obtained previously (Stavridis, 1986):

$$[\bar{\mathbf{m}}^{-1} \cdot \mathbf{K} - \omega^2 \cdot \mathbf{I}] \cdot \boldsymbol{\varphi} = 0$$

The eigenfrequencies ω can be determined only through the use of a computer.

For the case where the diaphragm rotations are negligible, even approximately, according to Section 15.3.4 the fundamental eigenfrequency of the whole system may be assessed with satisfactory accuracy on the basis of the expression for ω^2 given in Section 16.3.2.1, in both the x and y directions, according to the following procedure (Figure 16.30):

- A plane formation is created separately for the x and y senses, consisting of the corresponding plane elements in each direction, as shown in Section 16.3.1.4.

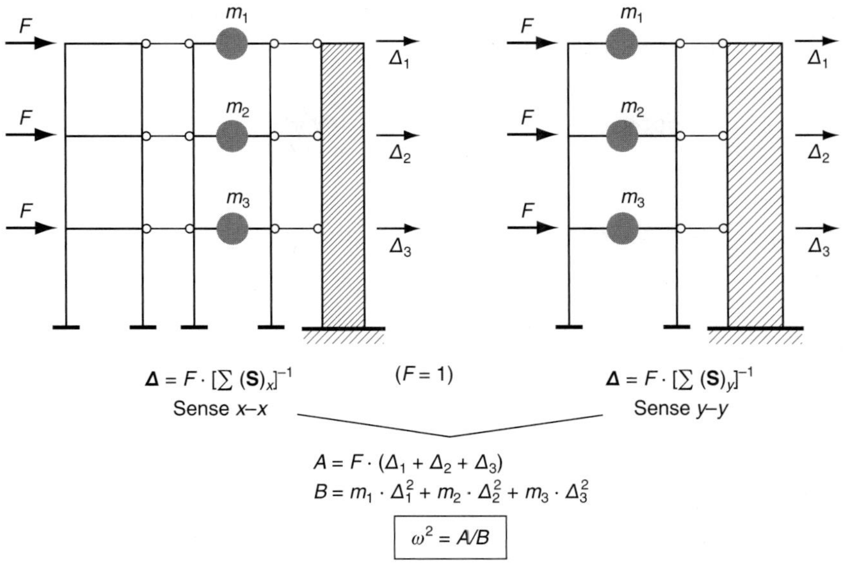

Figure 16.30 Practical determination of the fundamental eigenfrequency of a multi-storey system

- Unit loads are applied to the different levels of the structural web, for which the respective displacements are calculated. The fundamental eigenfrequency for the considered sense (x or y) is obtained by direct application of the last expression for ω^2 in Section 16.3.2.1.

Numerical example

The ground plan shown in Figure 16.31 applies for three consecutive storeys, each having a height of 3.50 m and a slab thickness equal to 25 cm. All of the frame girders have a width and depth equal to 30 cm and 80 cm, respectively, while all the columns have a section of 50/30 cm. The wall elements have a thickness of 20 cm.

Considering as the effective mass m only that corresponding to the self-weight of the slab:

$$\text{slab self-weight} = 15.40 \cdot 8.0 \cdot 0.25 \cdot 25.0 = 770\,\text{kN}$$

$$m = 770/9.81 = 78.5\,\text{t}$$

The distinction between the two plane systems is made as shown in Figure 16.30.

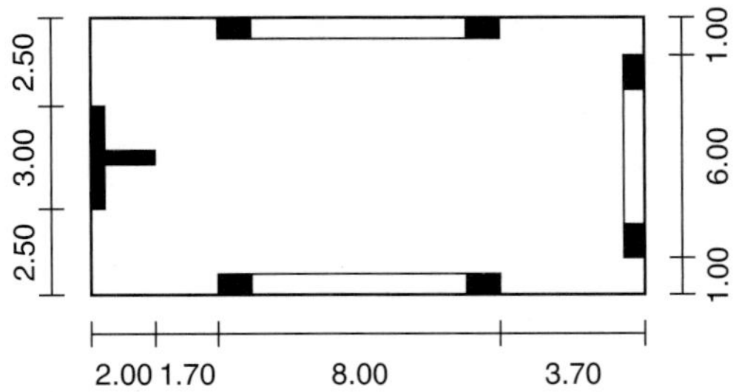

Figure 16.31 Plan view layout of vertical elements of stiffness

By applying horizontal unit loads at the corresponding levels of each plane system ($F = 1$ kN), the following displacements result:

- Plane system in the x direction:

  $\Delta_1 = 0.0477$ mm, $\Delta_2 = 0.0298$ mm, $\Delta_3 = 0.0117$ mm

- Plane system in the y direction:

  $\Delta_1 = 0.0773$ mm, $\Delta_2 = 0.0455$ mm, $\Delta_3 = 0.0161$ mm

According to the expression for ω^2 (see Section 16.3.2.1)

- For the x–x sense:

$$\omega^2 = \frac{1 \cdot 0.0477 + 1 \cdot 0.0298 + 1 \cdot 0.0117}{78.5 \cdot (0.0477^2 + 0.0298^2 + 0.0117^2)} = 344.3\ \mathrm{s}^{-2}$$

$$\omega = 18.50/\mathrm{s}$$

$$T = 2 \cdot \pi/\omega = 0.465\ \mathrm{s}$$

- For the y–y sense:

$$\omega^2 = \frac{1 \cdot 0.0773 + 1 \cdot 0.0455 + 1 \cdot 0.0161}{78.5 \cdot (0.0773^2 + 0.0455^2 + 0.0161^2)} = 213.1\ \mathrm{s}^{-2}$$

$$\omega = 14.60/\mathrm{s}$$

$$T = 2 \cdot \pi/\omega = 0.43\ \mathrm{s}$$

16.3.3 Forced vibration

The response of a multi-degree system subjected to external forces is governed, as analysed in Section 16.3.2.1, by means of a system of differential equations, which has been expressed in a condensed matrix form as (Müller and Keintzel, 1984)

$$\mathbf{m} \cdot \ddot{\boldsymbol{x}} + \mathbf{c} \cdot \dot{\boldsymbol{x}} + \mathbf{k} \cdot \boldsymbol{x} = \boldsymbol{F}(\boldsymbol{t})$$

As in the case of free vibration, it is also found here that the multi-degree system behaves as a system of N discrete and independent single-mass systems, having a fictitious mass M, as well as a fictitious stiffness K, according to the expressions in Section 16.3.2.1 (Figure 16.32). Each such system is governed by the equation

$$\mathrm{M_r} \cdot \ddot{q}_\mathrm{r} + 2 \cdot \zeta_\mathrm{r} \cdot \mathrm{M_r} \cdot \omega_\mathrm{r} \cdot \dot{q}_\mathrm{r} + K_\mathrm{r} \cdot q_\mathrm{r} = \varphi_r^\mathrm{T} \cdot \boldsymbol{F}(\boldsymbol{t})$$

which, for the usually encountered case where $\boldsymbol{F}(\mathbf{t}) = \boldsymbol{P} \cdot f(t)$, i.e. when all loads exhibit the same variation in time, may be rewritten as

$$\mathrm{M}_r \cdot \ddot{q}_r + 2 \cdot \zeta_r \cdot \mathrm{M}_r \cdot \omega_r \cdot \dot{q}_r + K_r \cdot q_r = \varphi_r^\mathrm{T} \cdot \boldsymbol{P} \cdot f(t) = \left(\sum_1^N \varphi_i^{(r)} \cdot P_i\right) \cdot f(t)$$

Although this equation can be treated analytically, as shown in Section 16.2.3, the numeric solution obtained using suitable computer software is more practical, provided

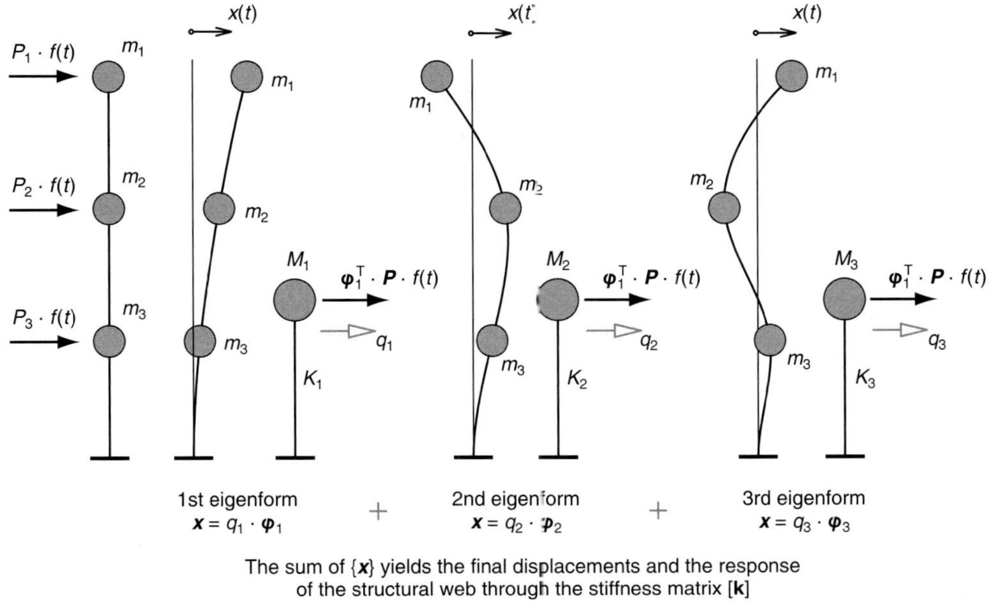

Figure 16.32 Dynamic response of a multi-degree system under external loads

that the time steps used in the solution are smaller than 1/10th of the corresponding eigenperiod $T_r = 2\pi/\omega_r$.

It should be noted that the determination of the displacements $q_r(t)$ of the N fictitious single-degree systems allows the calculation of the displacements $\mathbf{x}(\mathbf{t})$ of the masses in the multi-degree system, according to the equation $\mathbf{x}(\mathbf{t}) = \mathbf{\Phi} \cdot \boldsymbol{q}(\mathbf{t})$, where $\mathbf{\Phi}$ is a square matrix $(N \times N)$, obtained from the juxtaposition of the N column matrices φ_r (see Figure 16.32). Moreover, the fact that each eigenform may be considered to be multiplied by an arbitrary factor λ does not affect the resulting displacements $\mathbf{x}(\mathbf{t})$ at all, as this factor, according to the above equations, is eliminated.

16.3.4 Seismic excitation

16.3.4.1 Dynamic analysis

The governing factor for the response of a system to seismic excitation is, just as in the single-degree case (see Section 16.2.5), the accelerogram $\ddot{u}_s(t)$, expressed as

$$\ddot{u}_s(t) = a \cdot f_a(t)$$

where a represents the maximum-occurring ground acceleration (Figure 16.33).

It is obvious that the inertia mass forces are caused by their absolute acceleration, whereas the elastic as well as the damping forces acting on the masses depend only on their relative displacements to the ground. Given that no external forces are acting on the system, the condensed equation for dynamic equilibrium according to

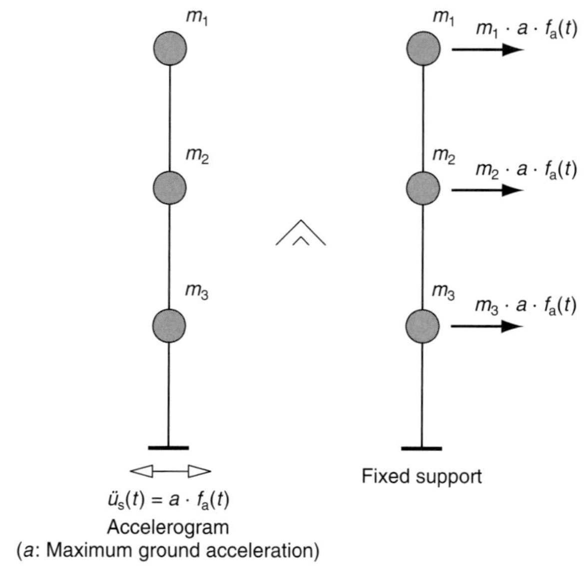

Figure 16.33 Seismic loading of a multi-degree system

the previous section is written as

$$\mathbf{m} \cdot \ddot{\boldsymbol{x}} + \boldsymbol{I} \cdot a \cdot f_a(t) + \mathbf{c} \cdot \dot{\boldsymbol{x}} + \mathbf{k} \cdot \boldsymbol{x} = 0$$

This equation can also be written as

$$\mathbf{m} \cdot \ddot{\boldsymbol{x}} + \mathbf{c} \cdot \dot{\boldsymbol{x}} + \mathbf{k} \cdot \boldsymbol{x} = -\mathbf{m} \cdot \boldsymbol{I} \cdot a \cdot f_a(t)$$

where $\boldsymbol{I}$ represents a column matrix containing N unit elements.

Thus, the problem of seismic excitation is posed as a problem of forced vibration according to the previous section, the multi-degree system being supported on fixed ground and subjected to the external loads expressed by the right-hand side of the last equation (see Figure 16.33).

According to the foregoing, the multi-degree system is 'split' into N single-mass oscillators, each one corresponding to a respective eigenfrequency ω with the accompanying eigenform φ. On this basis, the corresponding fictitious mass M_r may be determined as well as the fictitious stiffness K_r. The equation of motion for the rth eigenfrequency is written as (Müller and Keintzel, 1984)

$$M_r \cdot \ddot{q}_r + 2 \cdot \zeta_r \cdot M_r \cdot \omega_r \cdot \dot{q}_r + K_r \cdot q_r = -\varphi_r^{\mathrm{T}} \cdot \mathbf{m} \cdot \boldsymbol{I} \cdot a \cdot f_a(t)$$

and, dividing by M_r,

$$\ddot{q}_r + 2 \cdot \zeta \cdot \omega_r \cdot \dot{q}_r + \omega_r^2 \cdot q_r = -\varphi_r^{\mathrm{T}} \cdot \mathbf{m} \cdot \boldsymbol{I} \cdot \frac{1}{M_r} \cdot a \cdot f_a(t)$$

This equation is identical to the equation of motion of the single-mass system in Section 16.2.5, which is repeated here:

$$\ddot{x} + (2 \cdot \zeta \cdot \omega) \cdot \dot{x} + \omega^2 \cdot x = -a \cdot f_a(t)$$

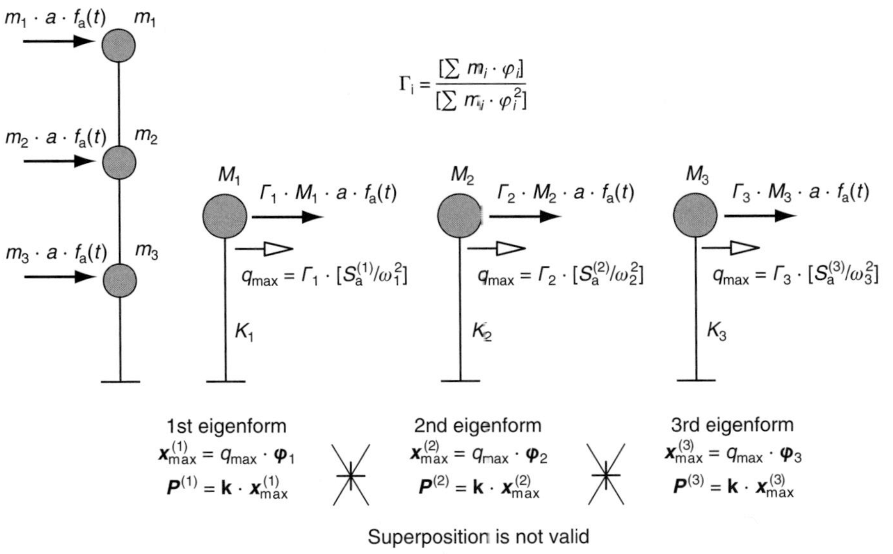

Figure 16.34 Seismic analysis of a multi-degree system

It is observed that the maximum ground acceleration a is now multiplied by the factor

$$\Gamma_r = \boldsymbol{\varphi}_r^{\mathrm{T}} \cdot \mathbf{m} \cdot \boldsymbol{I} \cdot \frac{1}{\mathrm{M}_r} = \frac{\sum m_i \cdot \varphi_i^{(r)}}{\sum m_i \cdot [\varphi_i^{(r)}]^2}$$

which is a pure number and expresses the degree to which the rth eigenform is participating in causing the response of the system (Figure 16.34), and is thus called the *participation factor*. Its value is less than unity, while the sum of all the N participation factors Γ_r is logically equal to 1. Usually, the first (fundamental) eigenform exhibits the greatest participation factor, which gradually decreases for the higher eigenfrequencies.

According to the results for the single-mass oscillator, it is possible to determine the corresponding $(q_r)_{\max}$ through the acceleration spectrum for an earthquake, on the basis of the corresponding value $S_a^{(r)}$ for the rth eigenfrequency, considering as maximum ground acceleration the value $(\Gamma_r \cdot a)$ instead of a. Hence, according to Section 16.2.5,

$$(q_r)_{\max} = \frac{1}{\omega^2} \cdot (\ddot{q}_r + \ddot{u}_s)_{\max}$$

$$= \frac{1}{\omega^2} \cdot \Gamma_r \cdot a \cdot \omega \cdot \left[\int_0^t f_a(\tau) \cdot e^{-\zeta \cdot \omega \cdot (t-\tau)} \cdot \sin \omega \cdot (t-\tau)\, d\tau\right]_{\max}$$

that is,

$$(q_r)_{\max} = \frac{S_a^{(r)}}{\omega_r^2} \cdot \Gamma_r$$

Now, for each oscillator r, the mass displacements $(\boldsymbol{x}_{\max})_r$ of the multi-degree system may be determined on the basis of the computed $(q_r)_{\max}$ (see Figure 16.34):

$$(\boldsymbol{x}_{\max})_r = \boldsymbol{\varphi}_r \cdot (q_r)_{\max} = \frac{\Gamma_r \cdot S_{\mathrm{a}}^{(r)}}{\omega_r^2} \cdot \boldsymbol{\varphi}_r$$

Thus, a maximum deviation is assigned to each mass, corresponding to each of the N fictitious oscillators. Of course, the actual displacement of a specific mass cannot be deduced as the sum of the above 'component' maximum deviations $(\boldsymbol{x}_{\max})_r$ on the basis of the N oscillators, because the corresponding $(q_r)_{\max}$ do not refer to the same time instant. Nevertheless, according to an acceptable probabilistic consideration, this mass deviation may be obtained as the square root of the sum of the squares of the above 'components'.

Considering now the determination of the maximum response in the structural web, it can be seen that for the realisation of the above displacements $(\boldsymbol{x}_{\max})_r$ of the N masses, the action of those forces $\boldsymbol{P}_r$ is required, which are suggested by the stiffness of the structural web itself according to the equation

$$\boldsymbol{P}_r = \mathbf{k} \cdot (\boldsymbol{x}_{\max})_r = \frac{\Gamma_r \cdot S_{\mathrm{a}}^{(r)}}{\omega_r^2} \cdot \mathbf{k} \cdot \boldsymbol{\varphi}_r$$

The determination of the stiffness matrix $\mathbf{k}$ follows Section 16.3.1.1

It is thus possible to determine the N systems of external forces $\mathbf{P}$ for the maximum (real) deviations $\boldsymbol{x}_{\max}$ corresponding to the N (fictitious) oscillators. However, as these forces do not correspond to simultaneously developed displacements, their superposition is not meaningful, and one has to use the previous treatment (the square root of the sum of their squares) in order to obtain an acceptable approximation for loading of the structural web.

16.3.4.2 Equivalent static loads

The bulk of the calculations in Section 16.3.4.1 that take into account the contribution of all the eigenforms of the system in its seismic excitation are not especially manageable, particularly for preliminary design purposes. This inconvenience led to the search for a static loading acting on the masses of the system such that the resulting response approximates the 'exact' dynamic analysis. A practical suggestion to this end is the fact that, for a regular distribution of mass and stiffness in a structural web, it is the first eigenform that plays the governing role in the dynamic response of the web (Müller and Keintzel, 1984).

Thus, in seeking these equivalent static loads, it is assumed, as an approximation, that these may be represented by the inertia forces $\boldsymbol{H}$ to which the masses in the system are subjected, if only the first eigenform ($r = 1$) is taken into account. According to Section 16.3.4.1 and Figure 16.35 (see also Section 16.2.5),

$$\boldsymbol{H} = \mathbf{m} \cdot [(\ddot{\boldsymbol{x}})_{\max} + \boldsymbol{I} \cdot a] = \omega^2 \cdot \boldsymbol{m} \cdot (\boldsymbol{x})_{\max}^{(1)} = \omega^2 \cdot (q_1)_{\max} \cdot \mathbf{m} \cdot \boldsymbol{\varphi}$$

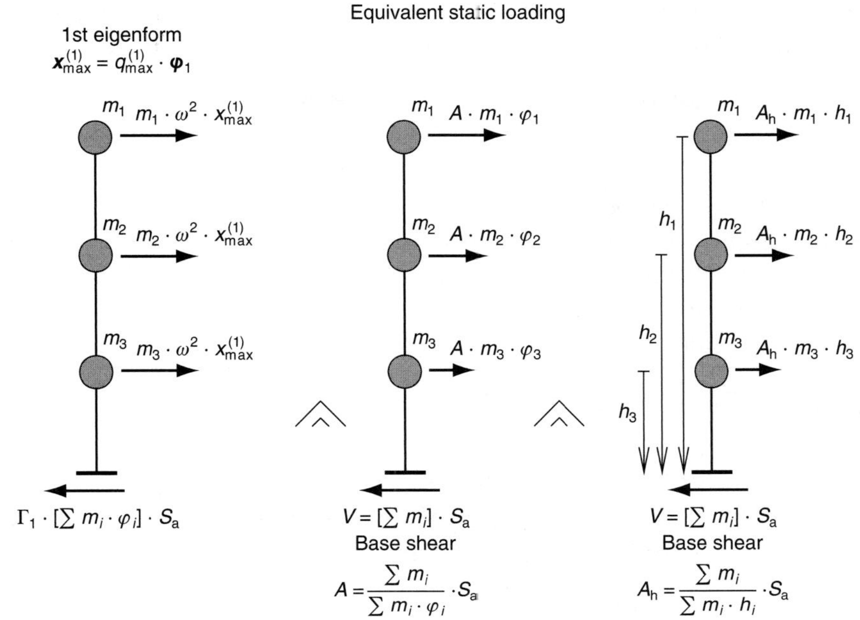

Figure 16.35 Determination of equivalent static loading in a multi-storey system

On the basis of the expression for $(q_1)_{max}$ in Section 16.3.4.1,

$$\boldsymbol{H} = \mathbf{m} \cdot \boldsymbol{\varphi} \cdot \frac{\sum m_i \cdot \varphi_i}{\sum m_i \cdot \varphi_i^2} \cdot S_a$$

Looking now for a more convenient expression for these forces, their sum ($\sum H$) over the whole height of the structural web is considered. It is clear that this force constitutes the sum of the actions to which the masses are subjected, transmitted directly to the ground. It is thus found that

$$\sum H = \frac{(\sum m_i \cdot \varphi_i)^2}{\sum m_i \cdot \varphi_i^2} \cdot S_a$$

This expression reveals that the factor

$$\frac{(\sum m_i \cdot \varphi_i)^2}{\sum m_i \cdot \varphi_i^2} = \Gamma \cdot \sum (m_i \cdot \varphi_i)$$

according to Section 16.2.5, represents a mass $\bar{m}$ which, given that the value of Γ for the first eigenform is near unity, may be considered as approximately equal to the sum M of the masses in the system, i.e.

$$\bar{m} = \Gamma \cdot \sum (m_i \cdot \varphi_i) \cong \sum m_i = M$$

Thus,

$$\sum H = V = M \cdot S_a$$

This V force is called the *base shear,* and is used in the exact same sense as in Section 16.2.5 (see Figure 16.35).

Now, the expression of the equivalent static loads H is written on the basis of the above as

$$\boldsymbol{H} = \mathbf{m} \cdot \boldsymbol{\varphi} \cdot \frac{1}{\sum m_i \cdot \varphi_i} \cdot \frac{(\sum m_i \cdot \varphi_i)^2}{\sum m_i \cdot \varphi_i^2} \cdot S_a = \mathbf{m} \cdot \boldsymbol{\varphi} \cdot \frac{V}{\sum m_i \cdot \varphi_i}$$

and, for each mass separately,

$$H_i = V \cdot \frac{m_i \cdot \varphi_i}{\sum m_i \cdot \varphi_i}$$

This simpler expression, though less accurate than the previous one, has nevertheless been adopted in most of the regulations (see Figure 6.35). However, in the search for an even simpler expression for the equivalent static loads for preliminary design purposes, a further simplification is undertaken, namely the consideration of the first eigenform as a straight line. Thus, by using the level heights h_i instead of the eigenform coefficients φ, the following final expression is obtained:

$$H_i = V \cdot \frac{m_i \cdot h_i}{\sum m_i \cdot h_i} \quad \text{(see Figure 16.35)}$$

More specifically, for a multi-storey building – as considered in Section 16.3.1.4 and from a dynamic viewpoint in Section 16.3.2.2 – which is subjected to a given ground acceleration, in order to determine the equivalent horizontal static load for each level, the following procedure should be followed:

(1) On the basis of the fundamental eigenfrequency for both the x and y senses, calculated according to Section 16.3.2.2, as well as of the maximum adopted ground acceleration, first the value $S_a^{(1)}$ and then the base shear V are determined from the acceleration spectrum used separately for both senses in the two discrete plane systems.
(2) The corresponding equivalent static loads are obtained through direct application of one of the two last-obtained expressions. In the case where the eigenform values φ_i are used, these may be considered to be the already determined displacements for the equivalent plane system under horizontal unit loads, applied in order to assess its fundamental frequency, as described in Section 16.3.2.2. It is clear that for a uniform distribution of masses, these static loads increase continuously for higher storeys.
(3) On the basis of the acting forces on each diaphragm of a storey, the response of the vertical – as well as the horizontal – elements of the multi-storey system may be readily assessed, according to Sections 15.3.1 and 15.3.2.

Numerical example

Continuing the example in Section 16.3.2.2, on the basis of the design spectrum in Eurocode EC8 and for soil class B, both for the sense x–x with a fundamental period $T = 0.34$ s and for the sense y–y with a fundamental period $T = 0.43$ s, a maximum spectral acceleration $S_a = 2.50$ is determined. Assuming a maximum ground acceleration equal to 0.16g, the following total seismic force (base shear) is obtained:

$$V = 3 \cdot 78.5 \cdot 0.16 \cdot 9.81 \cdot 2.50 = 924\,\text{kN}$$

Considering an earthquake in the x direction, the equivalent static loads acting on the centroid of each of the three diaphragms (slabs) are estimated on the basis of deviations φ. According to Section 16.3.2.2:

$$\sum m_i \cdot \varphi_i = 78.5 \cdot (0.0000477 + 0.0000298 + 0.0000117) = 0.007\,\text{t}$$

Thus, the loads acting in the x direction are (see Figure 16.35)

$$P_1 = 924 \cdot 78.5 \cdot 0.0000477/0.007 = 494\,\text{kN}$$

$$P_2 = 924 \cdot 78.5 \cdot 0.0000298/0.007 = 309\,\text{kN}$$

$$P_3 = 924 \cdot 78.5 \cdot 0.0000117/0.007 = 121\,\text{kN}$$

Considering an earthquake in the y direction, the equivalent static loads are analogously estimated. According to Section 16.3.2.2:

$$\sum m_i \cdot \varphi_i = 78.5 \cdot (0.0000773 + 0.0000455 + 0.0000161) = 0.011\,\text{t}$$

The corresponding forces acting in the y direction are (see Figure 16.35)

$$P_1 = 924 \cdot 78.5 \cdot 0.0000773/0.011 = 512\,\text{kN}$$

$$P_2 = 924 \cdot 78.5 \cdot 0.0000455/0.011 = 302\,\text{kN}$$

$$P_3 = 924 \cdot 78.5 \cdot 0.0000161/0.011 = 110\,\text{kN}$$

i.e. essentially with the same distribution as the x sense.

The assessment of the response of a multi-storey system under the above loads, for both the x and y senses, may be deduced by following the procedure described in Section 15.3.1.

16.4 Approximate treatment of continuous systems

As pointed out at the beginning of this chapter, the load-bearing structures have *de facto* a continuous mass distribution. However, the corresponding inertia forces that have to be considered in order to express the equilibrium of each element leads to unmanageable differential equations, which excludes their use in structural design.

However, the discrete systems examined so far, although not corresponding absolutely to physical reality, are nevertheless able to express the dynamic behaviour of any structure, by considering its mass lumped in as many places of the structural web as desired and by following the previously described procedures.

The total (continuous) mass m of a beam, for example, may be partitioned into five lumped equidistant masses ($m/5$) distributed over the structural web, thus creating a system with five degrees of freedom, if only those inertia forces, which correspond to transverse displacements, are counted (Figure 16.36).

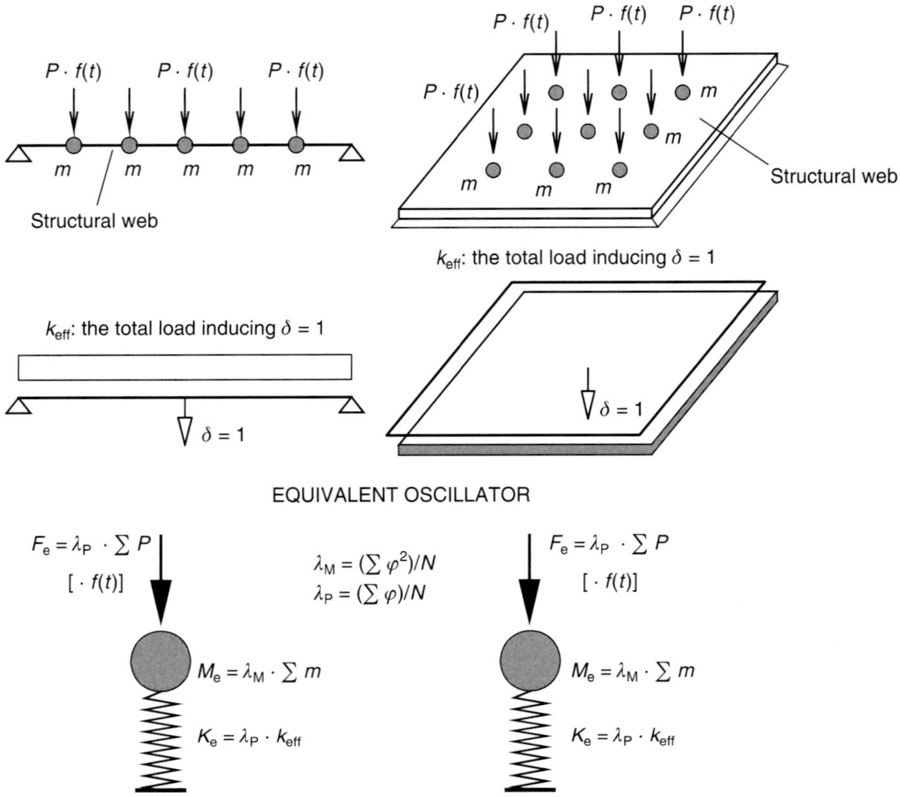

Figure 16.36 Equivalent substitution of a continuous structure through a single-mass system

In the same sense, the mass m of a plate may be divided in 25 equidistant masses ($m/25$), corresponding to a system with 25 degrees of freedom based only on the transverse deflections (Figure 16.36).

However, the analysis of the above discrete systems requires the use of appropriate computer programs. This does not directly meet the needs of a preliminary design with respect to basic design parameters, such as, for instance, the fundamental frequency of the system, or its maximum deformation, or even its maximum acceleration under specific dynamic loads. Thus, it is appropriate – wherever possible and at least for the usual cases – to pursue the equivalent dynamic behaviour of the continuous systems through single-mass systems exhibiting the same frequency as the fundamental frequency of the examined system. This idea is based on the fact that, in most practical cases, the fundamental eigenfrequency of a system plays the governing role in its dynamic behaviour, as pointed out in Section 16.3.4.2.

Figure 16.36 shows the 'substitution' of a beam and plate by a corresponding 'equivalent' single-mass system, for which the mass, stiffness and equivalent dynamic loading are to be determined (Biggs, 1964).

The determination of the above parameters is based on the fact that, as has been discussed in Sections 16.3.2 and 16.3.3, the dynamic analysis of an N-degree of freedom system goes back to the analysis of N independent single-mass oscillators, and, as explained above, it is the oscillator with the lowest frequency that can most successfully represent the whole system.

As described in Section 16.2.2, the mass M_e of this oscillator can be expressed as

$$M_e = \varphi_1^T \cdot \mathbf{m} \cdot \varphi_1 = \sum_{s=1}^{N} m_s \cdot \varphi_s^2 = \lambda_M \cdot \sum_{s=1}^{N} m_s$$

where φ_s are the deviation values of the first eigenform of the N-degree system and, in the case of a uniformly distributed total mass M in the discrete model, the *mass factor* λ_M is

$$\lambda_M = \frac{\sum_{s=1}^{N} \varphi_s^2}{N} \quad \text{(dimensionless number)}$$

Thus, the mass M_e of the equivalent oscillator is $M_e = \lambda_M \cdot M$, expressed as a part ($\lambda_M < 1$) of the total mass M of the system (see Figure 16.36).

Assume, now, that the discrete system is subjected to the dynamic loading

$$\mathbf{F(t)} = \mathbf{P} \cdot f(t)$$

where matrix $\mathbf{P}$ consists of the specific loads acting on the N masses of the system.

According to Section 16.3.3, the loading term of the single-mass 'component' oscillator with the lowest frequency (i.e. the 'fundamental oscillator') of the N-degree system is

$$F_1 = \varphi_1^T \cdot \boldsymbol{P} \cdot f(t) = \left(\sum_{s=1}^{N} P_s \cdot \varphi_s \right) \cdot f(t)$$

Considering that load $F_e \cdot f(t)$ will act on the equivalent oscillator, then

$$F_e = \sum_{s=1}^{N} P_s \cdot \varphi_s = \lambda_P \cdot \sum_{s=1}^{N} P_s \quad \text{(see Figure 16.36)}$$

and, in the case of an equally distributed total load to the N masses of the discrete system, the load factor λ_P will be

$$\lambda_P = \frac{\sum_{s=1}^{N} \varphi_s}{N} \quad \text{(dimensionless number)}$$

It is clear that in the case of a uniform distribution of load p (kN/m) on a beam of length L, $\sum P_s = (p \cdot L)$, whereas for a uniformly distributed load q (kN/m^2) on an orthogonal slab with dimensions (a, b), $\sum P_s = (q \cdot a \cdot b)$.

It should be noted that the basic aim of seeking an equivalent single-mass oscillator is to determine the kinematic behaviour (displacement, velocity, acceleration) of a specific critical point of the initial system, and this, for preliminary design purposes in the case of a single span beam or a plate, is obviously the central point of the structure. To this end, and extending the initially adopted notion of stiffness in Section 2.3.8, the concept of the effective stiffness k_{eff} is introduced as the total load which, under a given and prescribed distribution, causes unit displacement at the reference point of the structure

– i.e. at its midpoint (see Figure 16.36). It is, then, clear that the deflection δ at that point, due to the total load $\sum P_s$, is, for the considered distribution (see Figure 16.37),

$$\delta = \frac{\sum P_s}{k_{eff}}$$

For a uniform load distribution, in the case of a beam with length L, $k_{eff} = 76.8 \cdot EI/L^3$, whereas in the case of a simply supported square plate with side L, $k_{eff} = 271 \cdot EI/L^2$.

Thus, the sought-after stiffness K_e of the equivalent oscillator results from the requirement that the transverse displacement δ at the considered reference point of the continuous system, due to the static loads P_s, should be identical to the displacement of the equivalent oscillator under the load $F_e = \lambda_P \cdot \sum P_s$:

$$\delta = \frac{\sum P_s}{k_{eff}} = \frac{\lambda_P \cdot \sum P_s}{K_e}$$

and, hence,

$$K_e = \lambda_P \cdot k_{eff} \quad \text{(see Figure 16.36)}$$

The fundamental frequency f_1 of the continuous system, being equal to that of the equivalent oscillator, is thus obtained as

$$f_1 = \frac{1}{2\pi}\sqrt{\frac{K_e}{M_e}} = \frac{1}{2\pi} \cdot \sqrt{\frac{\lambda_P}{\lambda_M}} \cdot \sqrt{\frac{k_{eff}}{M}}$$

where the factors λ_P and λ_M refer to a uniform distribution of load and mass, respectively.

The two tables in Figure 16.37 give the values of the mass and load factors λ_P and λ_M for various cases of beams with distributed or concentrated loads, as well as for perimetrically supported plates (Biggs, 1964).

16.5 Design for avoiding annoying vibrations

16.5.1 Human activities

Apart from the seismic disturbance of multi-storey systems, which has been examined in Section 16.3.4, checking for transverse vibrations of horizontal load-carrying elements – such as beams or plates in normal use by humans – constitutes a major design issue in several cases (Bachmann and Ammann, 1987). A pedestrian bridge, for example, or an office floor, or even a floor designed for sport or dance activities, should in no case undergo vibrations which are felt by the people in a disturbing way. Such vibrations occur when the frequency with which the foot strikes the floor lies near the fundamental frequency of the load-carrying structure.

Within the framework of the preliminary design goals set out in this book, and without the need in the present context to dynamically analyse the human activity in itself, footbridges should be designed with a fundamental frequency outside the range 1.6–2.4 Hz and, if possible, not lower than 4.50 Hz. For floors intended for sport activities, this frequency should not be less than 6.8 Hz, whereas for dance floors the

Type of loading and beam Length L	Mass factor λ_M	Load factor λ_P	Effective stiffness k_{eff} $(\cdot EI/L^3)$
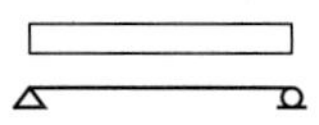	0.5	0.637	76.80
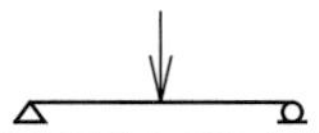	0.5	1.0	48
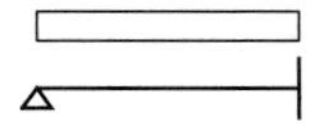	0.479	0.595	185
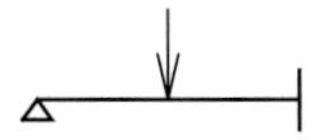	0.479	1.0	107
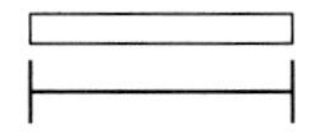	0.396	0.523	384
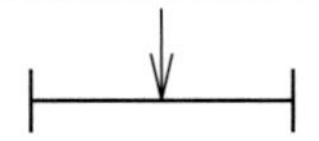	0.396	1.0	192

(a)

Uniform load on plate	a/b	Mass factor λ_M	Load factor λ_P	Effective stiffness k_{eff} $(\cdot Eh^3/12 \cdot a^2)$
Plate perimetrically simply supported	1.0	0.31	0.45	271
	0.9	0.33	0.47	248
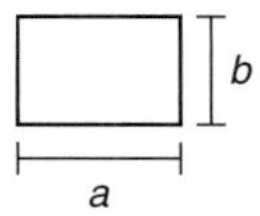	0.8	0.35	0.49	228
	0.7	0.37	0.51	216
	0.6	0.39	0.53	212
Plate thickness: h	0.5	0.41	0.55	216
Plate perimetrically fixed	1.0	0.21	0.33	870
	0.9	0.23	0.34	798
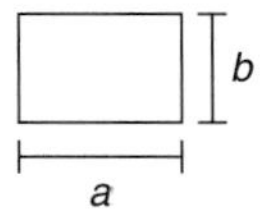	0.8	0.25	0.36	757
	0.7	0.27	0.38	744
	0.6	0.29	0.41	778
Plate thickness: h	0.5	0.31	0.43	866

(b)

Figure 16.37 Equivalent substitute parameters of single-mass systems for beams and plates

minimum value is 6.0 Hz. It is worth mentioning that, for the above uses, the maximum tolerable acceleration of the human body is about 5% of acceleration due to gravity g. Finally, in the case of office floors, where lower acceleration limits are required, a fundamental frequency of at least 7.50 Hz should be targeted (Bachmann and Ammann, 1995).

16.5.2 Vibrations induced by machines

The running of machinery generally causes vibrations in the part of the structure in which it is installed, which may be felt by and annoy people standing near it. The installed machines produce continuous periodical sinusoidal loading, characterised by the maximum value of force F_0 under a frequency f_0, according to the data supplied by the vendor.

As a measure of the acceptability level of accelerations felt by a person, the so-called *sensitivity factor* K can be considered, which is calculated using the empirical formula (Bachmann and Ammann, 1987)

$$K = d\,\frac{0.80 \cdot f^2}{\sqrt{1 + 0.032 \cdot f^2}}$$

where d is the maximum amplitude of vibration (mm) and f is the fundamental eigenfrequency (Hz) of the structural part supporting the machine.

Machines are supported on a floor structure either directly or through appropriate springs. The design goal is usually to limit the deflection δ_{max} and/or the velocity v_{max} of the supporting structural part at values tolerable by the machine, as well as to limit

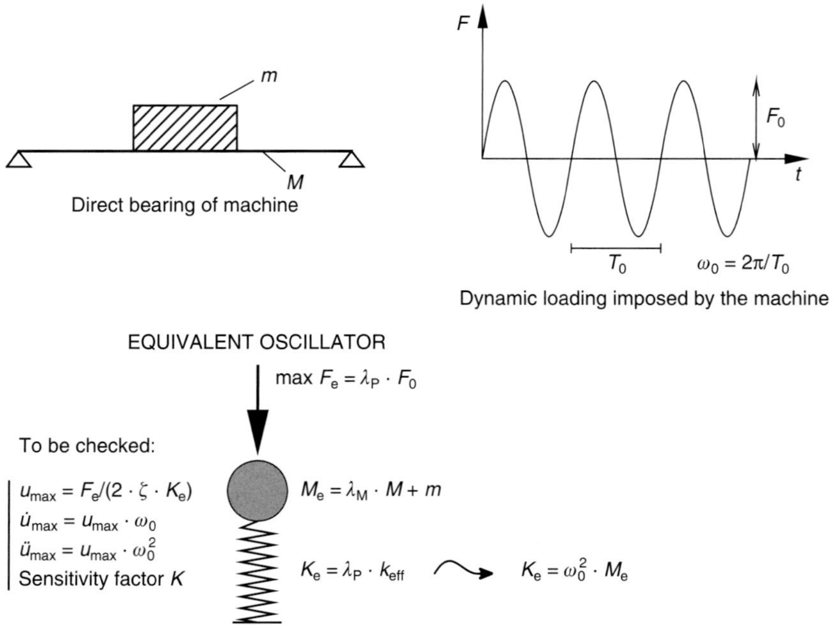

Figure 16.38 Direct support of a machine in a continuous system

the sensitivity factor K to under 2.0, so that the floor vibration is tolerable by people standing nearby. ■

In the case where the machine is directly supported on the floor, the whole system may be described by a fictitious oscillator with a determinable mass M_e and spring stiffness K_e, under an equivalent dynamic loading F_e (Bachmann and Ammann, 1987). This oscillator exhibits a damping coefficient identical to that of the supporting structure of the floor (Figure 16.38).

The mass M_e of the equivalent oscillator is determined as the sum of the equivalent vibrating mass $\lambda_M \cdot M$ of the supporting structural part (usually a slab) having the mass M and the mass m of the machine itself. So,

$$M_e = \lambda_M \cdot M + m$$

The spring stiffness K_e is expressed according to Section 16.4 through the effective stiffness k_{eff} of the supporting part, multiplied by the factor λ_P (see Figure 16.38).

If the fundamental eigenfrequency f_1 of the supporting structure being calculated, according to Section 16.4,

$$f_1 = \frac{1}{2\pi} \cdot \sqrt{\frac{\lambda_P}{\lambda_M}} \cdot \sqrt{\frac{k_{eff}}{M}}$$

is lower than the operation frequency f_0 of the machine, instead of calculating the stiffness K_e of the oscillator as $\lambda_P \cdot k_{eff}$, it is more useful to consider it equal to

$$K_e = \omega_0^2 \cdot M_e = (2 \cdot \pi \cdot f_0)^2 \cdot M_e$$

in order to take account of higher frequency values being applied to the supporting structural part, which can lead to resonance.

Then, on the basis of the adopted damping coefficient ζ for the structure, the determination of its maximum deflection u_{max} – and consequently of the machine itself – is possible, according to Section 16.2.4, as

$$u_{max} = \frac{1}{2 \cdot \zeta} \cdot \frac{F_e}{K_e} = \frac{1}{2 \cdot \zeta} \cdot \frac{\lambda_P \cdot F_0}{K_e}$$

which may be directly compared to the maximum δ_{max} supplied by the vendor. Moreover, the maximum velocity and the maximum acceleration developed can be checked, according to Section 16.2.2.1, as

$$\dot{u}_{max} = u_{max} \cdot \omega = (2 \cdot \pi \cdot f_0) \cdot u_{max}$$

and

$$\ddot{u}_{max} = u_{max} \cdot \omega^2 = (2 \cdot \pi \cdot f_0)^2 \cdot u_{max}$$

respectively (see Figure 16.38).

However, of decisive importance for the acceptance (or not) of the dynamic behaviour of the system is the value of the sensitivity factor K, according to the foregoing formula:

$$K = u_{max} \frac{0.80 \cdot f_0^2}{\sqrt{1 + 0.032 \cdot f_0^2}}$$

This factor should not exceed a value of 2.0. ■

If the above solution does not lead to acceptable results, then, in order to limit the response of the supporting structure, the machine should be placed on springs with appropriately selected stiffness. The two masses, of the machine and of the structure, are separated as shown in Figure 16.39, thus constituting a two-degree system consisting of an 'upper' mass m_M supported through a spring having stiffness k_M on the 'lower' mass m_S, being in turn supported through a spring of stiffness k_S on a solid base.

The quantities m_S and k_S for the structure can be considered to be known, but the characteristics of the 'upper' system have to be determined, namely the machine mass m_M and the spring stiffness k_M. The 'machine mass' is not necessarily the mass m only but can also include any appropriately attached mass m_B – e.g. a concrete pedestal – required to fulfil the dynamic requirements (see Figure 16.39).

The eigenfrequencies f_1 and f_2 of the two-degree system may be directly determined through the expression (Bachmann and Ammann, 1987)

$$f_{1/2} = \left[\frac{1}{8 \cdot \pi^2} \cdot \left(\frac{k_M}{m_M} + \frac{k_M + k_S}{m_S} \pm \sqrt{\left(\frac{k_M}{m_M} + \frac{k_M + k_S}{m_S} \right)^2 - \frac{4 \cdot k_M \cdot k_S}{m_M \cdot m_S}} \right) \right]^{1/2}$$

It is generally ensured whenever possible, through appropriately selected values, that the two frequencies are clearly separate, e.g. by making one a multiple of the other. This

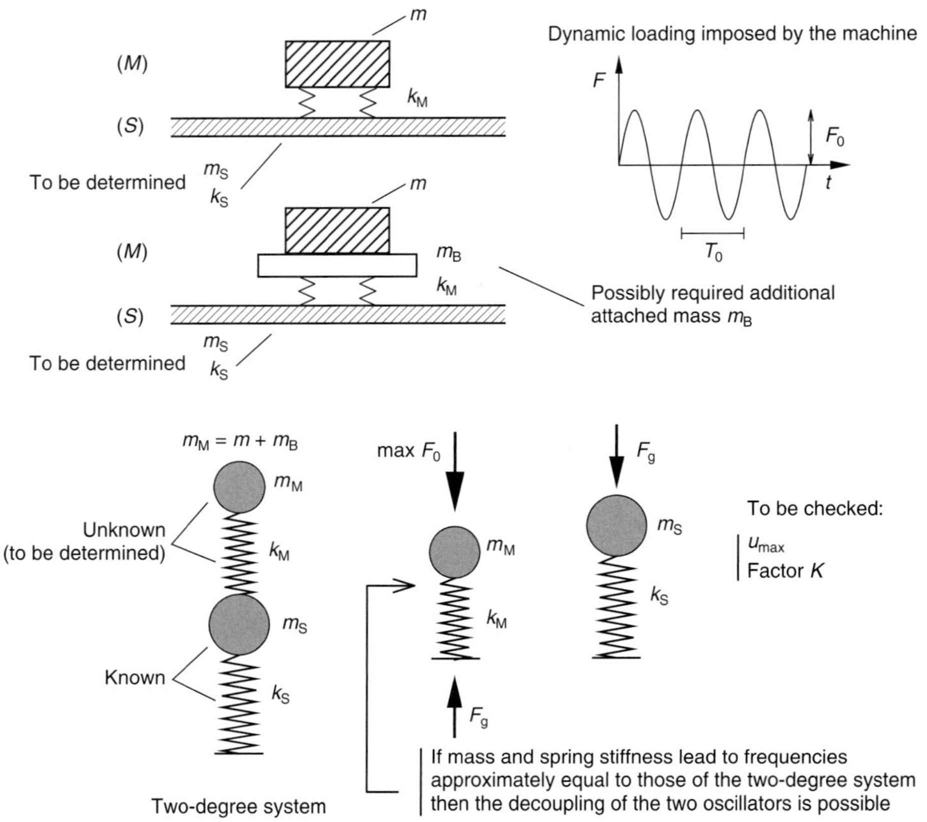

Figure 16.39 Support of a machine in a continuous system through springs

allows the two systems to be decoupled, so that each mass can be considered as connected to a fixed base through its spring. Then, the frequencies f_1 and f_2 obtained from the above expression coincide – at least approximately – with the frequencies f_M and f_S of the individual oscillators, respectively.

Thus, if the two requirements

$$f_M = \frac{1}{2 \cdot \pi} \cdot \sqrt{\frac{k_M}{m_M}} = \left[\frac{1}{8 \cdot \pi^2} \cdot \left(\frac{k_M}{m_M} + \frac{k_M + k_S}{m_S} - \sqrt{\left(\frac{k_M}{m_M} + \frac{k_M + k_S}{m_S}\right)^2 - \frac{4 \cdot k_M \cdot k_S}{m_M \cdot m_S}}\right)\right]^{1/2}$$

and

$$f_S = \frac{1}{2 \cdot \pi} \cdot \sqrt{\frac{k_S}{m_S}} = \left[\frac{1}{8 \cdot \pi^2} \cdot \left(\frac{k_M}{m_M} + \frac{k_M + k_S}{m_S} + \sqrt{\left(\frac{k_M}{m_M} + \frac{k_M + k_S}{m_S}\right)^2 - \frac{4 \cdot k_M \cdot k_S}{m_M \cdot m_S}}\right)\right]^{1/2}$$

lead – approximately – to technically plausible values for the sought-after quantities m_M and k_M, then the dynamic uncoupling of the two masses is feasible.

Then, in the system $[m_M, k_M]$, the maximum machine displacement as well as the maximum transmitted force on the base of the machine are first determined, according to Section 16.2.4, and, on the basis of this force, the system $[m_S, k_S]$ is afterwards analysed with respect to its relevant dynamic characteristics of interest, which concern the supporting structure itself.

More specifically, by considering for the machine component – according to the above – its total mass ($m_M = m + m_B$), as well as the stiffness k_M of its corresponding spring (see Figure 16.39), the corresponding eigenfrequency is obtained as

$$f_M = \frac{1}{2\pi} \cdot \sqrt{\frac{k_M}{m_M}}$$

The mass of the equivalent oscillator with a damping coefficient ζ will be equal to

$$m_S = \lambda_M \cdot M$$

The spring stiffness k_S is determined through the effective stiffness k_{eff}, according to Section 16.4, as

$$k_S = \lambda_P \cdot k_{eff}$$

The eigenfrequency f_S, representing the fundamental eigenfrequency of the supporting structure, is

$$f_S = \frac{1}{2\pi} \sqrt{\frac{k_S}{m_S}}$$

If, now, the two systems are considered to be uncoupled, as previously explained, then, for the mass m_M (machine), the maximum displacement u_{max} as well as the maximum transmitted force F_g to the supporting base can be readily determined, according to Section 16.2.4, through the following equations ($\zeta = 0$):

$$u_{max} = u_{stat} \cdot \frac{1}{1 - (\Omega/\omega)^2} = \frac{F_0}{k_M} \cdot \frac{1}{1 - (f_0/f_M)^2} \quad \text{(must be } < \delta_{max}\text{)}$$

and

$$F_g = F_0 \cdot \frac{1}{1 - (f_0/f_M)^2} = u_{max} \cdot k_M$$

Due to decoupling, the above force may be considered to be acting directly on the supporting structural system, with a cyclic frequency Ω corresponding to f_M. Thus, the resulting maximum displacement may also be expressed, according to Section 16.2.4, as

$$u_{max} = \frac{F_g}{k_S} \cdot \frac{1}{\sqrt{(1 - f_M/f_S)^2 + (2 \cdot \zeta \cdot f_M/f_S)^2}}$$

As was previously pointed out, the decisive criterion for the acceptance of the vibrational characteristics of the supporting structure is the value of the corresponding sensitivity factor K, which may be directly calculated as

$$K = u_{max} \frac{0.80 \cdot f_S^2}{\sqrt{1 + 0.032 \cdot f_S^2}}$$

References

Bachmann H., Ammann W. (1987) *Vibrations in Structures – Induced by Man and Machines*. Zurich: IABSE.

Bachmann H., Ammann W. (1995) *Vibration Problems in Structures*. Boston: Birkhäuser Verlag.

Biggs J.M. (1964) *Introduction to Structural Dynamics*. New York: McGraw-Hill.

Müller F.P., Keintzel E. (1984) *Erdbebensicherung von Hochbauten*. Berlin: Ernst.

Stavridis L. (1986) Static and dynamic analysis of multistory systems. *Technika Chronika Scientific Journal of the Technical Chamber of Greece* **6**(2), 187–219.

Tedesco J.W., McDougal W.G., Ross C.A. (1999) *Structural Dynamics – Theory and Applications*. California: Addison Wesley Longman.

17

Supporting the structure on the ground

17.1 Overview

All structural systems can be considered as free bodies in equilibrium under the loads for which they have been designed and the forces acting on them from the ground on which they are supported. These forces (i.e. the 'reactions') are, of course, accompanied by other equal and opposite forces, which are acting on the soil. The transfer of the last-mentioned forces to the ground must be accomplished in an absolutely safe manner regarding induced deformations and the soil stresses developed. For the successful transfer of such a force to the ground, an appropriately formed contact region of the structure with the soil is required. This constructional layout is called a *foundation*, and consists of a concrete structure firmly connected to the superstructure, so that by the term 'load-bearing structure' one means the whole system of 'superstructure' plus 'foundation' which is actually supported on the bearing ground. In this sense, the foundation through its contact surface with the soil has a double role: it must be designed to withstand the self-equilibrating system of soil stresses and the actions from the superstructure – equal and opposite to the developing 'reactions' – on the one hand, and to introduce such stresses in the soil mass that they can be safely taken up by it, on the other (Figure 17.1). In this respect, another point comes into question, namely the fact that the implied soil deformations – as also imposed on the structure (the foundation included) – create an additional response to the otherwise rigidly supported structural system (see Figure 17.1). These deformations are, of course, not known in advance, but they arise as a result of the *soil–structure interaction*, and in this respect the foundation design plays a decisive role.

Thus, in the design of a structure, the factor 'supporting soil' is – through the foundation – always directly involved, influencing the whole structural concept.

The following development is not intended to cover the examination of the mechanical properties of soils, nor the various constructional layouts of the foundation possibilities. The aim is simply to point out particular basic structural characteristics involved in the design of some typical forms of foundation systems, in order to assist the understanding of their load-bearing behaviour.

17.2 General mechanical characteristics of soils

There are two main categories into which the vast majority of soil types fall: the so-called *non-cohesive* and the *cohesive* soils.

Structural systems: behaviour and design
978-0-7277-4105-9
Copyright Thomas Telford Limited © 2010
All rights reserved

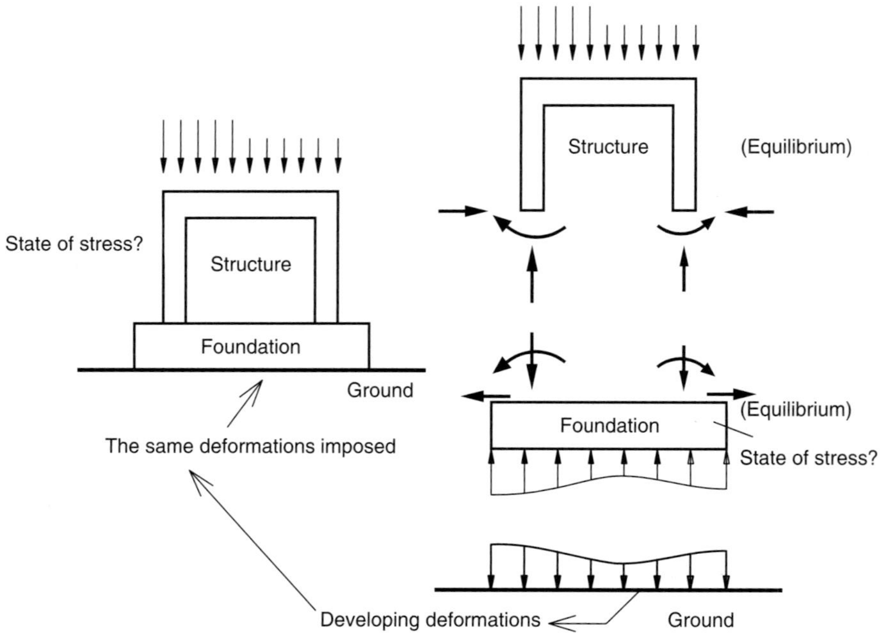

Figure 17.1 The system superstructure–foundation–soil interaction

17.2.1 Non-cohesive soils

Non-cohesive soils (sand, gravel) constitute in dry conditions a mass of loose, unconnected grains, not smaller than 0.06 mm, mutually transmitting friction forces through their small contact areas. These friction forces are proportional to the normal pressure σ acting on the contact surface, through the *angle of internal friction* φ of the soil examined, the maximum value τ being considered as its *shearing strength*. It should be pointed out that the shearing strength of soils represents the determining factor for their actual resistance, unlike solid materials, whose strength is instead represented by normal tensile or compression stresses. So, $\tau = \sigma \cdot \tan \varphi$.

Water, which can penetrate the gaps between the grains, easily escapes under pressure. This is why settlement in these soils is developed within a short period of time, once the loads are imposed. Non-cohesive soils, in a layer of medium to high density and of adequate thickness, generally constitute a good foundation ground. It should be noted, however, that in fine-grained, loose and water-saturated soil masses, the danger of loss of shearing strength exists (i.e. the loss of friction between the grains), known as *liquefaction*, particularly in the case of intense cyclic loading induced by a strong earthquake.

The specific weight γ of these soils in dry conditions is of the order of 20 kN/m^3. However, when the soil layer lies under the water level, the soil grains are under buoyancy and the above value is reduced to 20–10 = 10 kN/m^3.

Non-cohesive soils in natural free embankments exhibit a slope that tends to coincide with the angle φ of internal friction (Figure 17.2). If, however, this embankment slope is

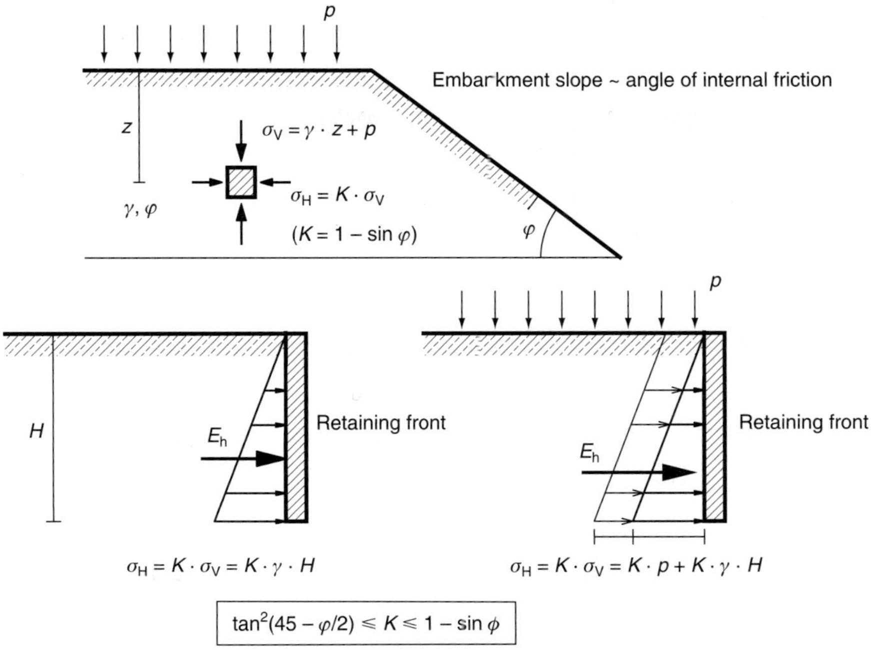

Figure 17.2 Earth pressure in non-cohesive soils

prevented through a vertical front, e.g. a retaining wall against a certain soil volume, then this front is acted on over its entire height H by soil pressures that increase as the internal angle of friction φ decreases. The horizontal component σ_H of these stresses is proportional to the prevailing vertical pressure σ_V in the considered depth h, due either to the soil self-weight ($\sigma_V = \gamma \cdot h$), or to a vertical distributed load p acting on the free soil surface ($\sigma_V = p$). Thus $\sigma_H = K \cdot \sigma_V$, where the coefficient K represents the reducing effect of the internal friction (e.g. against water, where, of course, $K = 1$), moving between a minimum value: $K_a = \tan^2 (45 - \varphi/2)$ and a maximum value $K_0 = 1 - \sin \varphi$ (see Figure 17.2). The minimum value K_a corresponds to the case where slight resilience of the vertical front is allowed, referring to the so-called *active pressure*, whereas the maximum value K_0 corresponds to an absolutely unyielding front, and refers to the so-called *pressure at rest*. The resultant horizontal force E_h acting on the vertical front due to the soil self-weight, under a triangular pressure distribution, is then $E_h = \gamma \cdot H^2 \cdot K/2$, whereas that due to a surface load p it is $E_h = p \cdot K \cdot H$ (uniform pressure distribution). For the combined action of the two loadings, the above results are accordingly superposed (see Figure 17.2). Moreover, in Figure 17.3(a), the influence of a differentiated value of K in the case of a layered soil mass is shown.

If water is present in the soil mass, the specific weight of the soil γ is reduced to $\gamma' = \gamma - \gamma_W$, due to buoyancy, and this should, together with the full hydrostatic action on the retaining structure, be taken into account (Figure 17.3(b)). It should be noted that the angle of internal friction φ remains essentially unaffected.

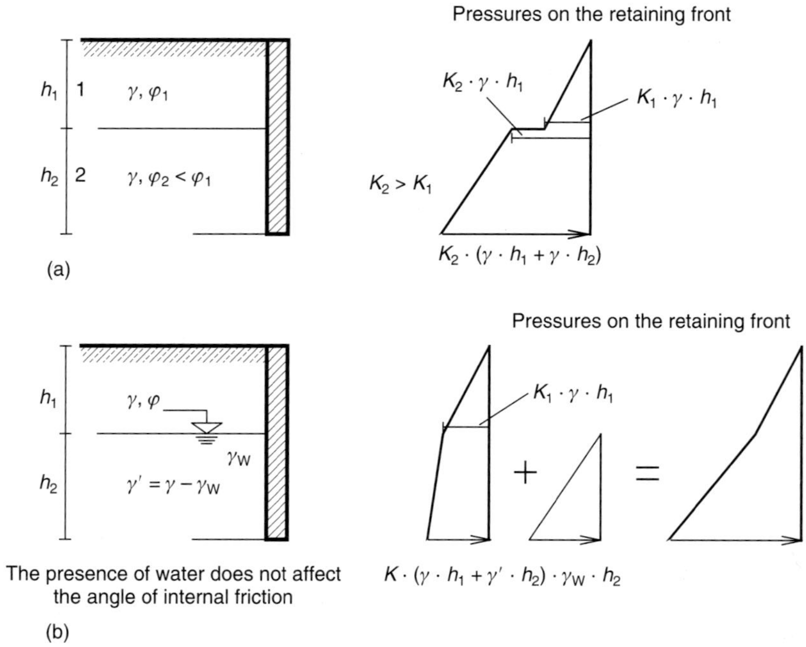

Figure 17.3 Earth pressures in a layered soil in the presence of a water table

17.2.2 Cohesive soils

Cohesive soils (clay) constitute a consistent mass even in dry condition. They consist of particles of oblong or plate form with diameters ranging from 0.0002 mm up to 0.06 mm, having many more common contact points than non-cohesive soils. Given that the voids between the particles are much larger than the dimensions of the grains themselves, any water present cannot escape as easily as in non-cohesive soils under an applied external pressure, because of the resistance offered by the many contact points of the grains. Thus, settlement develops very slowly and may continue for greater periods of time (months or even years).

A basic characteristic of cohesive soils is, apart from certain friction forces between grains, the development of cohesion forces as a result of mutually acting electrical forces between the particles. It should be pointed out that friction forces are decreased because of the existance of the water pressure u in the voids between the soil grains, called the *pore pressure*. The pore pressure, in the presence of a water table, is identical to the hydrostatic pressure at the point considered.

Thus, the shearing strength τ basically consists of *cohesion* c (kN/m^2), as well as of the friction resistance $\sigma' \cdot \tan\varphi$. The stress σ' represents the actual normal pressure between the solid grains, which, of course, results from the externally imposed normal pressure σ reduced by the existing pore pressure u. The angle of internal friction φ generally has more moderate values than in non-cohesive soils. The cohesion c in water-saturated soils has a value of the order of 10–40 kN/m^2. Thus,

$$\tau = c + (\sigma - u) \cdot \tan\varphi$$

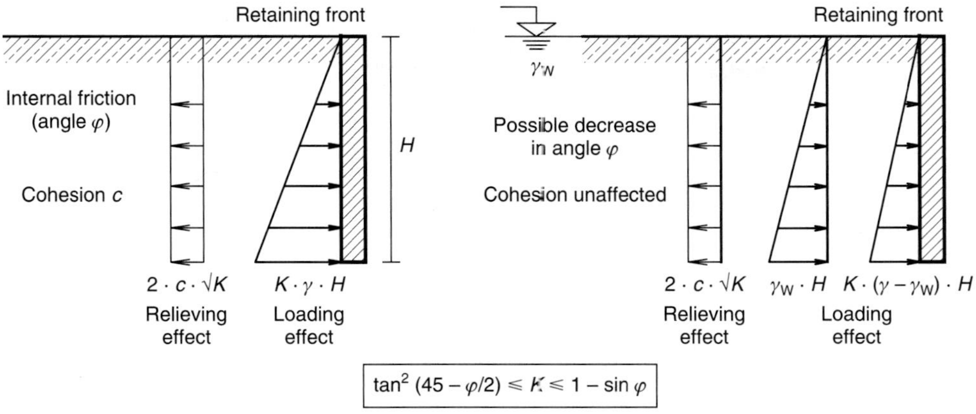

Figure 17.4 Earth pressure in cohesive soils

The stress $\sigma - u$ is called the *effective stress*. It is clear that the shear strength of the cohesive soils depends, to a significant degree, on the presence of water in the pores, as well as on the cohesion c, which plays an important role, sometimes being more decisive than the internal friction. It should be pointed out that the presence of water may cause a decrease in the angle of internal friction φ, in contrast to what happens in non-cohesive soils. Moreover, it is noted that in a cohesive saturated soil, the additionally imposed pressure on the soil mass will not imply a direct increase in the contact pressure between the grains, as the extra pressure will be completely taken up by the water mass, without thereby improving the shear strength of the soil at all.

The influx of water in cohesive soils generally affects their strength adversely, while a possible drying may give rise to pronounced deformations, accompanied by intense cracking, which will facilitate the further influx of water in the soil mass. In any case, however, under the same loading pressure the settlement of cohesive soils is greater than in the non-cohesive soils.

Regarding the pressures on vertical retaining walls, the presence of the cohesion c definitely has a relieving role (Figure 17.4). Thus, the total horizontal force on a vertical front of height H consists of the force acting under the existing internal friction and the relieving force due to cohesion:

$$E_h = \gamma \cdot H^2 \cdot K/2 - 2 \cdot c \cdot H\sqrt{K}$$

It can be seen that the contribution of self-weight to the above force can be depicted by a triangular diagram, whereas the influence of cohesion is represented by a constant value over the height. The treatment of coefficients K is the same as in the case of non-cohesive soils.

Moreover, in the above expression for E_h, the possible presence of a water table within the soil mass must be taken into account by the appropriate reduction of the specific weight γ due to buoyancy, this effect being equivalent to the fact that the pore pressure u is equal to the exerted hydrostatic pressure.

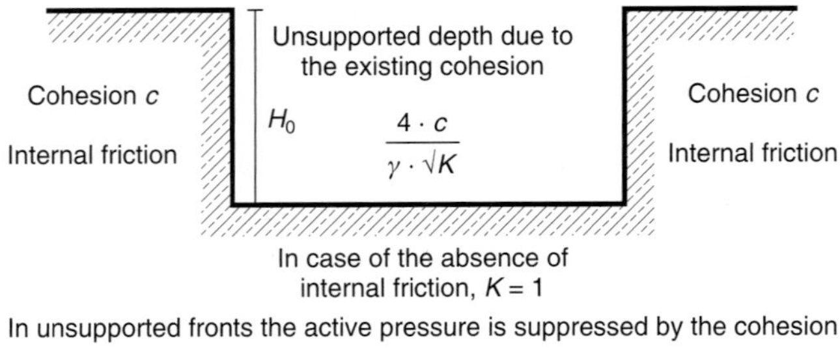

Figure 17.5 Effect of cohesion on the unbraced height of an open excavation

The above relation allows the estimation of the height H_0 of an unsupported vertical cut, where $E_h = 0$ must hold (Figure 17.5):

$$H_0 = 4 \cdot c/\gamma \cdot \sqrt{K}$$

In the absence of internal friction ($\varphi = 0$), where, of course, $K = 1$, the minimum height of an unbraced vertical cut of a cohesive soil is

$$H_0 = 4 \cdot c/\gamma$$

17.3 Shallow foundations

It is reasonable, in order to provide foundation to a structure, to always seek its safe support at the shallowest level possible, since the cost of a foundation drastically increases with its depth. This, of course, presumes that the bearing capacity of the soil at the selected foundation level is adequate to permit the uptake of all the anticipated loads on the structure. Such a foundation is called a *shallow foundation*, and comprises the majority of foundations. However, if such a foundation is not feasible, a special structural layout must be provided so that the imposed loads are transmitted to lower soil layers with satisfactorily safe bearing capacity. The foundation is then characterised as a *deep foundation*. The cost of such a foundation constitutes a significant percentage of the total cost of the structure.

What is presented below mainly concerns shallow foundations. However, some basic structural characteristics of deep foundations will be examined in the final sections of this chapter.

17.3.1 The deformational behaviour of elastic soil under vertical loads

Soil may be considered as an elastic medium – possibly consisting of intermediate layers – being acted on by vertical loads on its free surface. Under these loads, soil behaves elastically. This assumption is supported by the results of geotechnical tests performed at the proposed site. Thus, each soil layer is characterised by its *compression modulus* E_s, as well as by the corresponding Poisson ratio ν, determined by laboratory tests performed during the geotechnical exploration of the site. The compression modulus E_s is the 'elasticity modulus' of the soil material, and represents a quantity that

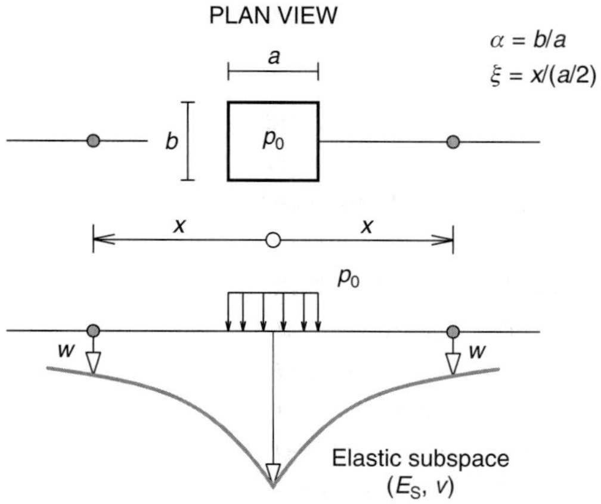

Value of λ

α	ξ				
	0.0	2.0	6.0	10.0	14.0
1.0	7.04	2.06	1.00	0.39	0.28
0.50	9.60	3.69	1.31	0.78	0.55
0.20	13.19	6.72	3.00	1.88	1.34

$$w = \frac{1-\nu^2}{E_S \cdot 2\pi} \cdot p_0 \cdot a \cdot \lambda$$

Figure 17.6 Settlement of the elastic subspace due to a uniformly loaded orthogonal area

cannot be as precisely determined as in the case of, for example, steel. Its value corresponds to the geotechnical nature of the soil examined. Thus, non-cohesive soils exhibit values from 30 000 up to 300 000 kN/m^2, depending on the grain size and the degree of consolidation (sand, 30 000–100 000 kN/m^2; gravel, 70 000–300 000 kN/m^2), whereas in cohesive clay soils the values of E_s, clearly lower, are between 10 000 and 40 000 kN/m^2. The Poisson ratio has much narrower variation margins and a significantly smaller influence on the deformation. Its order of magnitude for sand is about 0.3, while for clay it is about 0.4.

Assuming that the examined soil mass, of theoretically infinite depth, has a uniform compression modulus E_s throughout and its free surface is loaded by a uniformly distributed load p_0 over a rectangular area $(a \cdot b)$, then the settlement w developed at a distance x from its central point and in a parallel to the direction of side a, is (Figure 17.6)

$$w = \frac{1-\nu^2}{E_s \cdot 2\pi} \cdot p_0 \cdot a \cdot \lambda$$

where

$$\lambda = (1-\xi)\cdot \ln\frac{\sqrt{(1-\xi)^2+\alpha^2}+\alpha}{\sqrt{(1-\xi)^2+\alpha^2}-\alpha} + (1+\xi)\cdot \ln\frac{\sqrt{(1+\xi)^2+\alpha^2}+\alpha}{\sqrt{(1+\xi)^2+\alpha^2}-\alpha}$$

$$+ 2\alpha \cdot \ln\frac{\sqrt{(1-\xi)^2+\alpha^2}+(1-\xi)}{\sqrt{(1+\xi)^2+\alpha^2}-(1+\xi)}$$

and $\alpha = b/a$, $\zeta = 2x/a$ (Stavridis, 1997). ■

On the basis of the above result and in view of a later examination of soil–structure interaction, a strip of length L and constant width b is considered, being divided into

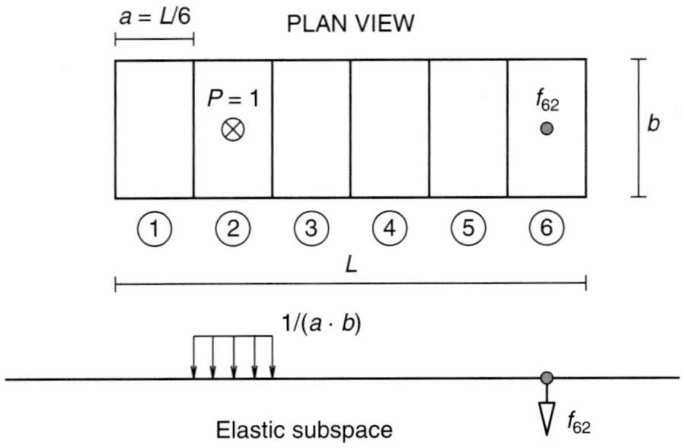

Figure 17.7 Influence of a loaded area on the settlement of a point lying on the soil surface

a number of n equal segments of length (L/n), as shown in Figure 17.7, whereby the mutual deformational influence of each of the above segments on each other is sought.

Applying a unit vertical load $P = 1$ consecutively on the midpoint of each segment and assuming it to be uniformly distributed over that segment with an intensity p_0,

$$p_0 = \frac{1}{(L/n) \cdot b}$$

The settlements at the midpoint of all the other segments can be readily determined on the basis of the last relation. If the segments are numbered in ascending order from 1 to n, a number of n^2 settlements f_{sr} may be determined. f_{sr} represents the settlement that develops in segment number s, when the segment number r is loaded with $P = 1$. Thus, with a total number of segments equal to $n = 3$ (Figure 17.8), loading, for example,

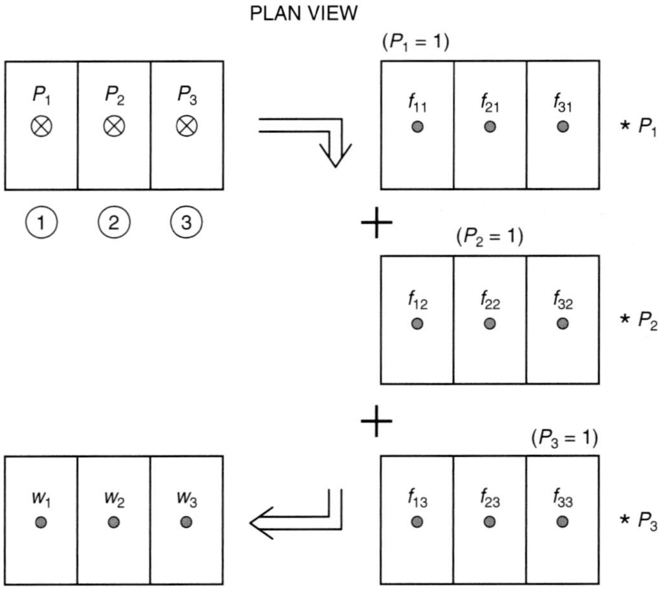

Figure 17.8 Determination of settlement due to specific concentrated loads

segment number 1, on the basis of the above relation, the quantities f_{11}, f_{21} and f_{31} may be directly determined. Similarly, by loading segment number 2, the quantities f_{12}, f_{22} and f_{32} may be calculated, and by loading segment number 3, settlements f_{13}, f_{23}, f_{33} are analogously obtained.

The above deformations allow the determination of the settlement at all midpoints, if each segment is acted on by a certain load P_r. Thus, in this example (see Figure 17.8), the settlement w of each segment is, respectively,

$$w_1 = f_{11} \cdot P_1 + f_{12} \cdot P_2 + f_{13} \cdot P_3$$

$$w_2 = f_{21} \cdot P_1 + f_{22} \cdot P_2 + f_{23} \cdot P_3$$

$$w_3 = f_{31} \cdot P_1 + f_{32} \cdot P_2 + f_{33} \cdot P_3$$

The extension to a greater number of segments is obvious:

$$w_1 = f_{11} \cdot P_1 + f_{12} \cdot P_2 + \ldots + f_{1n} \cdot P_n$$

$$w_2 = f_{21} \cdot P_1 + f_{22} \cdot P_2 + \ldots + f_{2n} \cdot P_n$$

$$\ldots$$

$$w_n = f_{n1} \cdot P_1 + f_{n2} \cdot P_2 + \ldots + f_{nn} \cdot P_n$$

By assembling settlements 'f' in a square matrix **[F]** $(n \times n)$, which may be called the *flexibility soil matrix*, and further assembling settlements 'w_i' and loads 'P_i' in the column matrices $(n \times 1)$ $\{\mathbf{W}\}$ and $\{\mathbf{P}\}$, respectively, the above expressions may be written in the compact matrix form (see Section 16.3.1):

$$\mathbf{W} = \mathbf{F} \cdot \mathbf{P}$$

The assumption of a constant compression modulus E_s over the whole depth of a certain soil mass is not always realistic. The case of consecutive vertically arranged horizontal layers, each with a compression modulus corresponding to the thickness of the layer, is not unusual. On the other hand, even in a 'homogeneous' soil mass, the compression modulus increases approximately linearly with depth. Although for preliminary design purposes the consideration of an estimated 'average' compression modulus may be sufficient, a more 'correct' determination of the settlement w is presented below for the case of a layered soil profile (Stavridis, 2002).

As a basic tool, the following result is used, regarding the settlement w developed at any corner of an orthogonal area $(a \cdot b)$ under a uniform load p_0, lying over a soil layer of thickness z, and a compression modulus E_s. This layer is assumed to rest on an absolutely rigid base (Figure 17.9):

$$w(a, b) = u(z)/Y$$

where

$$u(z) = \frac{a \cdot p_0}{4 \cdot \pi} \zeta \cdot \arctan\left(\frac{\beta}{\zeta \cdot \gamma}\right) + \ln\left(\frac{\gamma - \beta}{\gamma + \beta} \cdot \frac{\omega + \beta}{\omega - \beta}\right) + \beta \cdot \ln\left(\frac{\gamma - 1}{\gamma + 1} \cdot \frac{\omega + 1}{\omega - 1}\right)$$

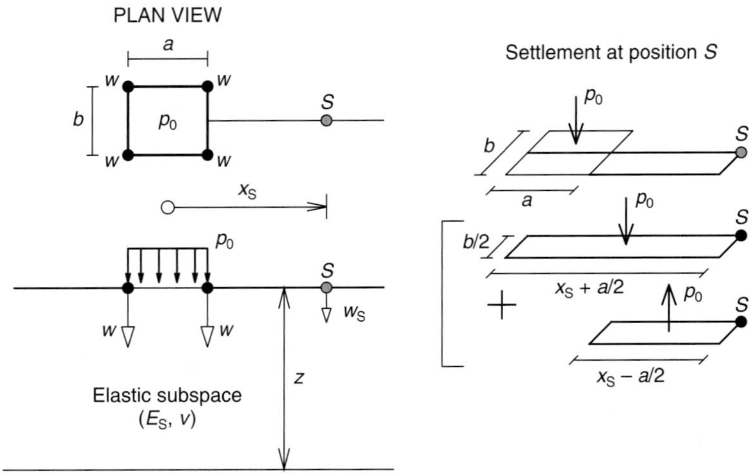

Figure 17.9 Settlement of an orthogonal area over a soil layer of finite thickness

and

$$Y = \frac{E_s}{1 - \nu^2}$$

$$\beta = b/a$$

$$\zeta = 2 \cdot z/a$$

$$\omega = \sqrt{1 + \beta^2}$$

$$\gamma = \sqrt{1 + \beta^2 + \zeta^2}$$

The settlement of a point S on the soil surface at a distance x_s from the central point of the area $(a \cdot b)$, in a parallel direction to side a, can be expressed as

$$w_s = 2 \cdot [w(x_s + a/2, b/2) - w(x_s - a/2, b/2)]$$

where the quantities $w(\ldots, \ldots)$ inside the brackets refer to the corner settlements of the orthogonal areas, $[(x_s + a/2) \cdot (b/2)]$ and $[(x_s - a/2) \cdot (b/2)]$, respectively, according to the above expression for $w(a, b)$ (see Figure 17.9).

By superposing the contribution of the compressibility of each soil layer on the development of the final settlement of point S (Figure 17.10), the following result is obtained:

$$w_s = u(z_1) \cdot (1/Y_1 - 1/Y_2) + u(z_2) \cdot (1/Y_2 - 1/Y_3) + \ldots + u(z_{k-1}) \cdot (1/Y_{k-1} - 1/Y_k) + u(z_k) \cdot 1/Y_k$$

It is clear that the formulation of the foregoing mutual interaction of the adjacent strip segments, according to Figure 17.7, remains the same in the case of a stratified soil.

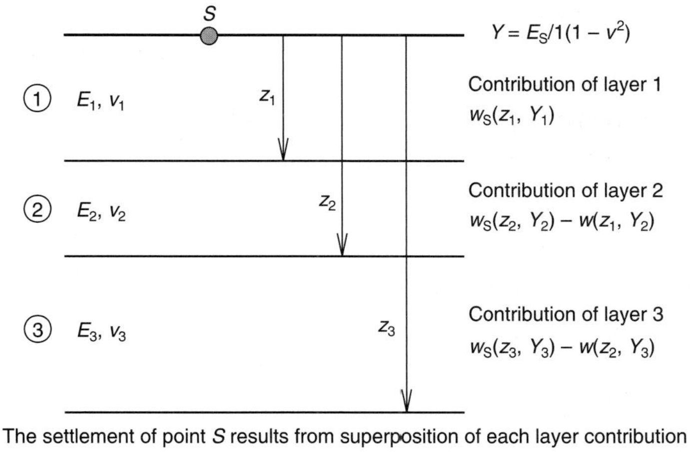

Figure 17.10 Contribution of the existing layers to the soil surface settlement

17.3.2 Rectangular spread footings

In order to provide a foundation for the base of a column element, where a specific vertical force V_1, a horizontal force H and a moment M_1 must be transmitted, an orthogonal concrete footing with a specific area and depth is usually constructed, normally having its larger base dimension parallel to the plane of action of the moment M_1 (Figure 17.11). The reason why such a footing is necessary is obvious: the direct support by the soil of a column element, developing at its base the sectional forces (V_1, H, M_1), is generally impossible because the soil cannot transmit to the element those stresses required for the 'realisation' of the above sectional forces. Thus, for the proper design of a footing the following points have to be taken into consideration:

(A) The maximum developed contact stress with the soil, which should not exceed a safe value determined by the geotechnical study of the site, characterised as the *allowable bearing pressure*.

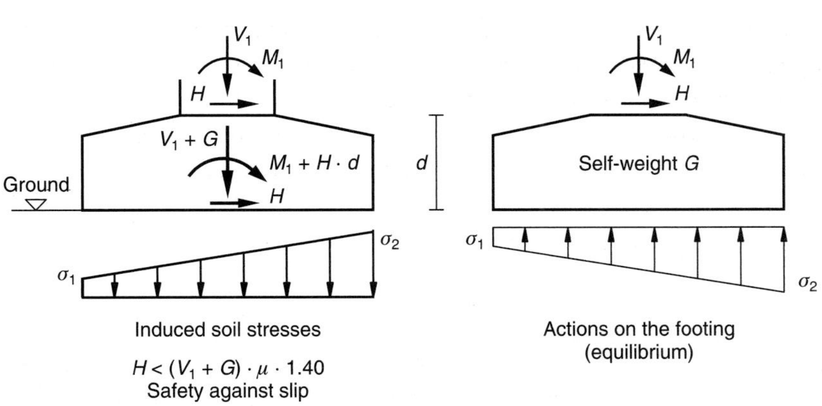

Figure 17.11 Footing equilibrium and soil pressures

(B) The absolute magnitude of the maximum soil settlement developed, as well as the angle of rotation of the footing.
(C) The state of stress of the footing, which equilibrates as a free body under the acting sectional forces V_1, H and M_1, its self-weight G, as well as the acting contact pressures from the soil. The concrete footing must be reinforced and checked appropriately so that it can take up all the above forces.

Of course, on the contact area between the footing and soil, a vertical force $V = V_1 + G$, a horizontal force H as well as a moment $M = M_1 + H \cdot d$ are transmitted. It is under these forces that the above points (A) and (B) should be checked.

The horizontal force H must be 'covered' by the 'friction capacity' $(V \cdot \mu)$ between the soil and foundation. A safe value for the friction coefficient μ is 0.50 for cohesive soils and 0.60 for non-cohesive soils. However, by adopting a safety factor γ – usually taken as 1.40 – the following relation must be satisfied: $H < \gamma \cdot (V \cdot \mu)$ (see Figure 17.11). Normally, the horizontal force H is not further involved in the design.

17.3.2.1 Soil pressures and settlements

In a first attempt to estimate the developed soil pressures, the contact area of the footing is considered as a normal orthogonal section under the forces V and M. It is assumed that force V also includes the weight of the overlying soil mass. Force V is assumed to act on the centroid of the footing area (Figure 17.12). It is, of course, clear that moment M causes a shift e of the force V along its plane of action, being equal to $e = M/V$. Thus, the pair of forces (V, M) is statically equivalent to force V alone, shifted, however, by distance e. As only compression contact pressures make sense over the foundation area, it is appropriate to check whether or not the shifted force V lies within the core of the corresponding section, i.e. whether it is $e \leq a/6$. In that case, the maximum and the minimum contact (compression) stresses are, according to the bending formula (see Section 2.2.1), respectively,

$$\sigma_{max} = \frac{V}{A} + \frac{M \cdot (a/2)}{I_f} = \frac{V}{a \cdot b} + \frac{6 \cdot M}{a^2 \cdot b}$$

$$\sigma_{min} = \frac{V}{a \cdot b} - \frac{6 \cdot M}{a^2 \cdot b}$$

Of course, if the moment M is not acting, a uniform pressure $\sigma = V/(a \cdot b)$ is developed.

In the case where the shifted force V exceeds the core limits – something that should generally be avoided – the above relations do not apply, since σ_{min} becomes tensile. Then, the force V must be equilibrated by a triangular pressure diagram, having its maximum value on the nearest side to the force V, while its zero value to which it is extending will obviously lie within the foundation area (see Figure 17.12). As the distance c of the force V to the near side is $c = (a/2 - e)$, the triangular diagram of the contact pressures extends over a length of $3 \cdot c$, and the above maximum pressure results from the equilibrium requirement:

$$\tfrac{1}{2} \cdot \sigma_{max} \cdot (3 \cdot c) \cdot b = V \quad \blacksquare$$

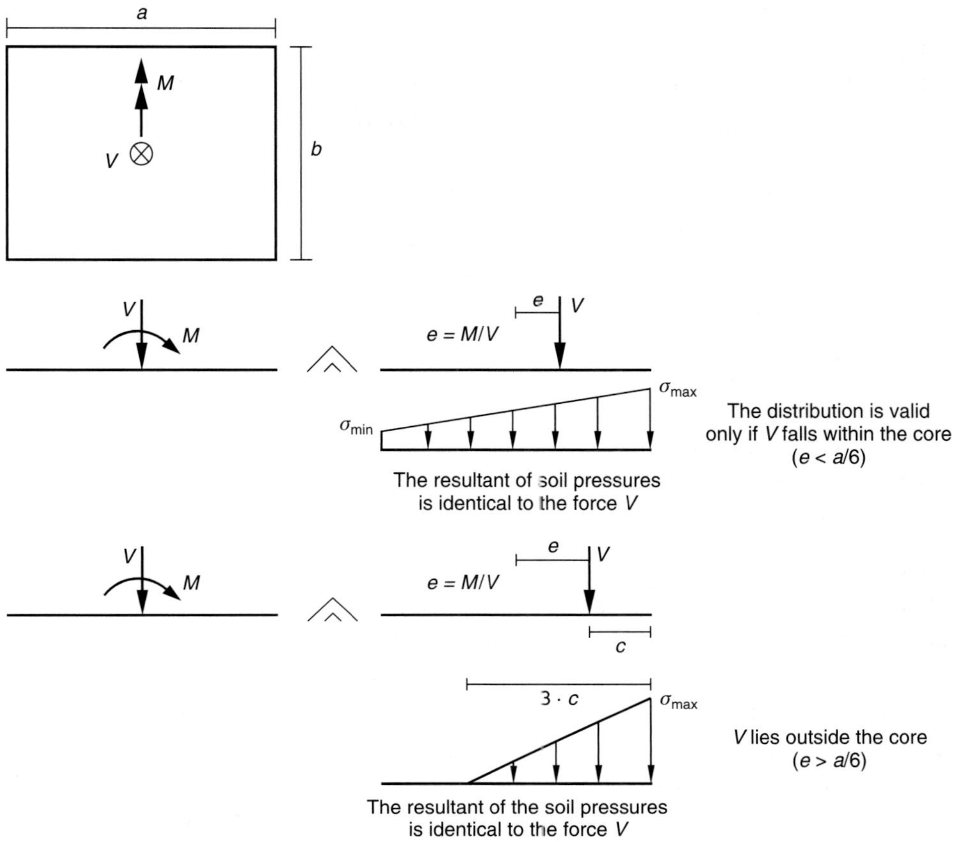

Figure 17.12 Influence of the location of the resultant force on the resulting soil pressures

Although it is clear that all the above pressure distributions in each case satisfy the corresponding equilibrium conditions for the magnitudes V and M, it is easily seen that they lead to deformations which do not correspond to reality. More specifically, as the footing can be considered as essentially undeformable, the soil settlements, being identical with those of the foundation, should lie on a single plane exhibiting a certain rotation angle, because of the presence of the moment M. But even in the 'simple' case of the central application of the force V, $M = 0$, it is clear that the resulting uniform pressure $V/(a \cdot b)$ on the soil over a rectangular base does not lead – according to Section 17.3.1 – to a uniform settlement of all the points of the soil surface.

Thus, for the case of the central action of the force V, after dividing the foundation surface into a certain number n of transverse strips of width a/n, the loads P_i for each strip have to be determined, which sum to the value V, on the one hand, and lead to the same settlement (w) for all strips, on the other hand (Figure 17.13).

On the basis of the result of Section 17.3.1, the last requirement is expressed through the following equations:

$$w = f_{11} \cdot P_1 + f_{12} \cdot P_2 + \ldots + f_{1n} \cdot P_n$$

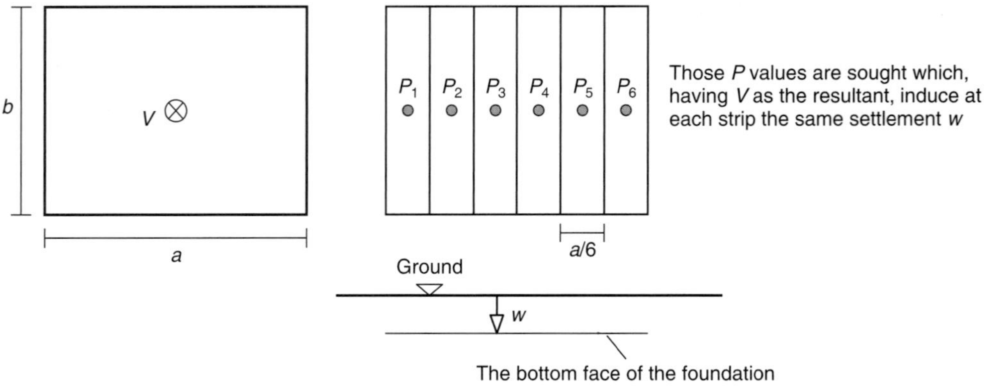

Figure 17.13 More accurate determination of soil pressures and of uniform settlement

$$w = f_{21} \cdot P_1 + f_{22} \cdot P_2 + \ldots + f_{2n} \cdot P_n$$

$$\ldots$$

$$w = f_{n1} \cdot P_1 + f_{n2} \cdot P_2 + \ldots + f_{nn} \cdot P_n$$

In addition, the following equilibrium condition must be valid:

$$V = P_1 + P_2 + P_3 + \ldots + P_n$$

The $n + 1$ unknown quantities $P_1, P_2, P_3, \ldots, P_n$ and w, can be determined through the above linear system of equations.

By selecting a relatively small number of strips, e.g. $n = 6$, it can be ascertained that in an orthogonal footing under a central compression force V, forces P, instead of being equal, clearly increase towards the edges of the footing (Figure 17.14).

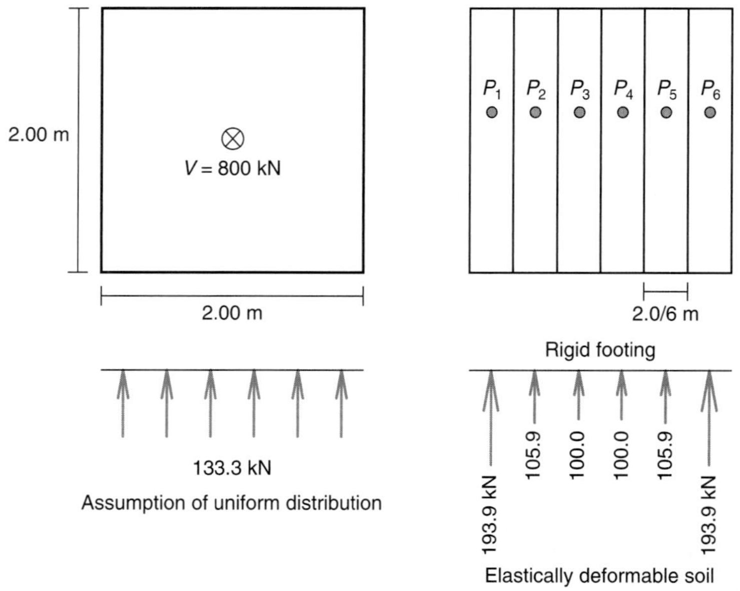

Figure 17.14 Determination of soil pressures under a uniform settlement

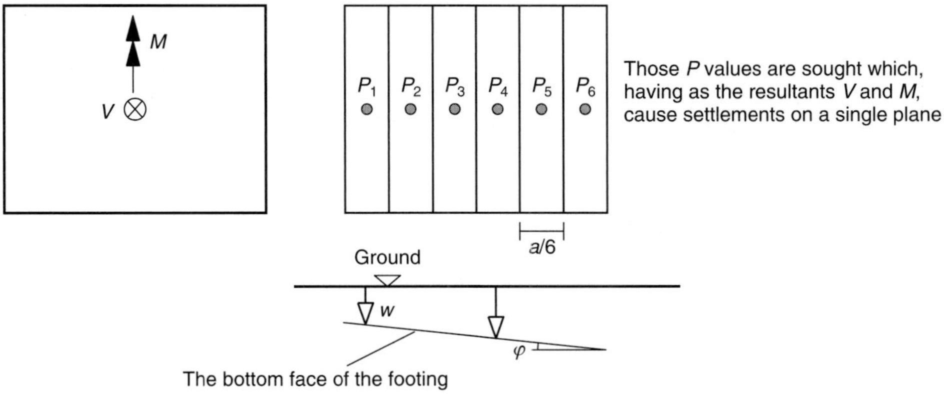

Figure 17.15 More strict determination of soil pressures and deformation for an eccentric loading

The presence of the moment M introduces one more unknown, namely the inclination φ of the resulting deformation plane of the soil surface (Figure 17.15). Thus, the above equations are modified as follows:

$$w + 0 \cdot \varphi \cdot (a/n) = f_{11} \cdot P_1 + f_{12} \cdot P_2 + \ldots + f_{1n} \cdot P_n$$

$$w + 1 \cdot \varphi \cdot (a/n) = f_{21} \cdot P_1 + f_{22} \cdot P_2 + \ldots + f_{2n} \cdot P_n$$

$$\ldots$$

$$w + (n-1) \cdot \varphi \cdot (a/n) = f_{n1} \cdot P_1 + f_{n2} \cdot P_2 + \ldots + f_{nn} \cdot P_n$$

Moreover, the static equivalence of magnitudes V and M with the forces $P_1, P_2, P_3, \ldots, P_n$ is required. Using the location of the force P_1 as a reference point (see Figure 17.15),

$$V = P_1 + P_2 + P_3 + \ldots + P_n$$

$$V \cdot (a/2 - a/2n + M/V) = P_1 \cdot (1-1) \cdot (a/n) + P_2 \cdot (2-1) \cdot (a/n) + P_3 \cdot (3-1) \cdot (a/n) + \ldots + P_n \cdot (n-1) \cdot (a/n)$$

It is clear that the $n+2$ unknowns $P_1, P_2, P_3, \ldots, P_n$, w and φ, can be determined through the above $n+2$ linear equations. Obviously, for practical setting out and solution of both the above linear systems, the use of appropriate computer software is essential. Moreover, it can be seen that the above equations are acceptable only for positive (i.e. compressive) values of the forces 'P'. Otherwise, an iterative process must be followed. The developed soil pressures σ are obtained at each point by evaluating $n \cdot P/a$, whereas the maximum settlement is equal to $w + (n-1) \cdot \varphi \cdot (a/n)$ (Stavridis, 2009).

17.3.2.2 Support on an elastic base

A less reliable but clearly more practical alternative to describing the deformational behaviour of the soil is offered by the concept of the *modulus of subgrade reaction*, represented by the coefficient k. It is based on the assumption that the settlement of

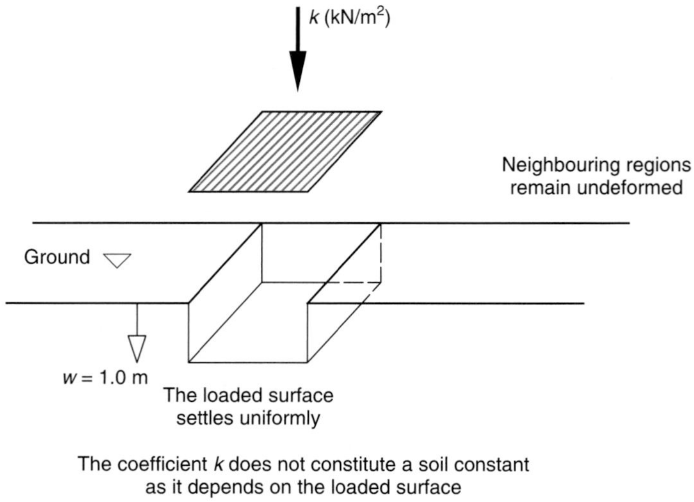

Figure 17.16 Concept of subgrade modulus

any region on the soil surface under a distributed load p is uniform over its entire extent and is equal to p/k. The coefficient k (kN/m^3) represents the pressure (kN/m^2) required to be applied to a soil surface in order to produce a settlement of 1 m, thus expressing the stiffness of the soil surface (Figure 17.16).

Although the concept of the subgrade modulus k is very convenient, being consistent with the concept of elastic support (see Section 2.3.7), it does not constitute a measurable soil property. More specifically, according to the above, the settlement of, for example, a rectangular surface under a pressure p is equal to p/k, i.e. it is independent of the surface itself. This is clearly not correct – as may easily be deduced from Figure 17.6 – since, under the same pressure p, an increase in the loaded area always leads to an increase in the corresponding settlement too (see Section 17.3.1). Thus, the determination of a subgrade modulus k has no meaning as a characteristic parameter of a given soil layer, as it depends directly on the specific loaded area. In addition, the assumption that all the points in the loaded area undergo the same settlement, while the surrounding region remains undeformed, is not realistic.

However, the computational convenience, together with the opportunity for the qualitative and quantitative estimation of the structural behaviour offered by the use of the parameter k, often overweigh, in practice, its serious 'natural disadvantages'.

Certainly, a logical value of k for a specific orthogonal area may be obtained if the settlement w, at its central point, under a pressure equal to 1 kN/m^2 is calculated according to Section 17.3.1, even by using the table in Figure 17.6. Of course, $k = 1/w$. However, for a square surface with a side equal to a, it is approximately

$$k \approx E_s/a \quad (\text{kN/m}^3)$$

In this way, the soil under each single footing having an area A, may be simulated by a spring with a stiffness k_s equal to $k_s = k \cdot A$, since, according to the above, a centrally

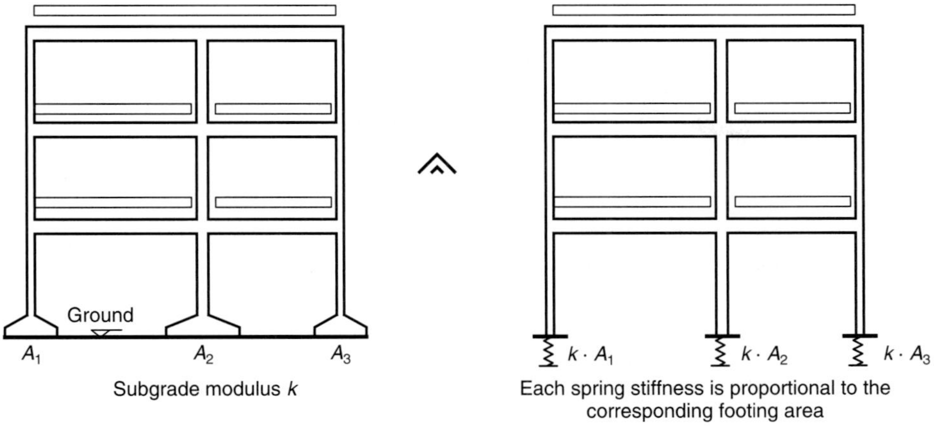

Figure 17.17 Simulation of the soil deformability through translational springs

acting load P on the footing causes a settlement $w = (P/A)/k$. Then, the whole structure may be considered as including all the springs corresponding to its footings, being itself naturally supported on a rigid base (Figure 17.17). ■

It can be seen that the induced rotation of a footing under the action of a moment M determined – as usual – by assuming a fixed column base implies a reduction in this moment, according to what has been discussed in Sections 3.2.9 and 3.3.6 regarding indeterminate structures. This footing rotation and the ensuing moment reduction needs to be assessed, since it will lead to more favourable soil stresses.

Thus, whether the resilience of the footing can be described by a rotational spring with a specific elastic rotation flexibility f_φ, as defined in Section 2.3.7.2, i.e. as the developed footing rotation under the action of a unit moment, will be examined. Obviously, such a moment should be accompanied by a vertical load V, in order to ensure the development of contact pressures over the entire footing (Figure 17.18).

In the case of a rigid footing, transmitting the sectional forces V and M over its bearing area A and, provided that only compressive pressures are developed over the entire foundation surface – meaning that the shifted resultant V falls within the core of the rectangular area – then, on the basis of a subgrade modulus k and according to Section 17.3.2.1, the settlements s_1 and s_2 at its two edges will be

$$s_1 = \frac{\sigma_{max}}{k} = \frac{V}{A \cdot k} + \frac{M \cdot (a/2)}{I_f \cdot k}$$

$$s_2 = \frac{\sigma_{min}}{k} = \frac{V}{A \cdot k} - \frac{M \cdot (a/2)}{I_f \cdot k} \quad (I_f = b \cdot a^3/12)$$

Consequently, the rotation φ of the footing will be

$$\varphi = \frac{s_1 - s_2}{a} = \frac{M}{I_f \cdot k}$$

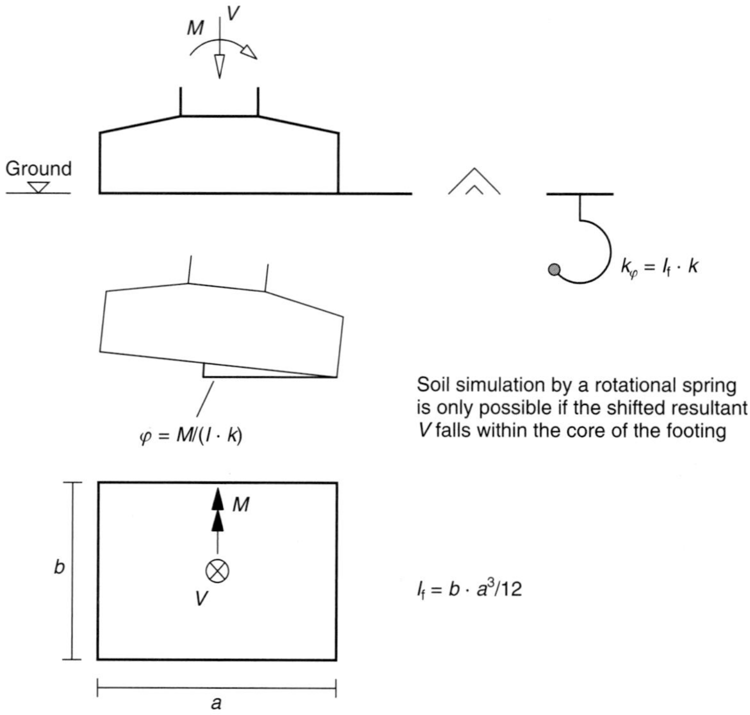

Figure 17.18 Conditions for the soil simulation through a rotational spring

This expression for φ allows the determination of the coefficient of elastic rotation f_φ – according to Section 2.3.7.2 – or, equivalently, of the stiffness coefficient with respect to rotation, $k_\varphi = 1/f_\varphi$, simulating in this way the soil through a rotational spring (see Figure 17.18):

$$f_\varphi = \frac{1}{I_f \cdot k}$$

$$k_\varphi = I_f \cdot k$$

However, it should be pointed out that, while, on the basis of the previous more exact approach to soil deformability (see Section 16.3.2.1) through the compression modulus E_s, the load eccentricity may exceed the core limits of the foundation surface and yet maintain the compression contact stresses everywhere, the adoption of a subgrade modulus k does not lead to such a consequence, as can be ascertained by means of the above expressions.

When the resultant force V acts outside the core, the settlement s_1, according to the above, is

$$s_1 = \frac{\sigma_{max}}{k} = \frac{V}{3 \cdot \left(a/2 - \frac{M}{V}\right) \cdot b \cdot k}$$

and the corresponding rotation φ of footing will be

$$\varphi = \frac{s_1}{3 \cdot c} = \frac{V}{9 \cdot \left(\frac{a}{2} - \frac{M}{V}\right)^2 \cdot b \cdot k}$$

It is clear that rotation φ is not proportional to the acting moment – depending also on the value of the vertical force V – and, therefore, in this case the soil cannot be represented by a rotational spring. ■

On the basis of the resistance k_φ offered by the rotational spring against the imposition of a unit rotation, the reduction of the moment M previously mentioned may be estimated as follows.

The rotational spring introduces the developed moment to the column, according to the rotation of their common joint, which obviously rotates together with the spring. If an external moment is considered to act on this node so that its rotation vanishes, then this moment M must clearly be equal to that corresponding to a clamped column base (Figure 17.19). Thus, in order to make this moment disappear, since it does not actually

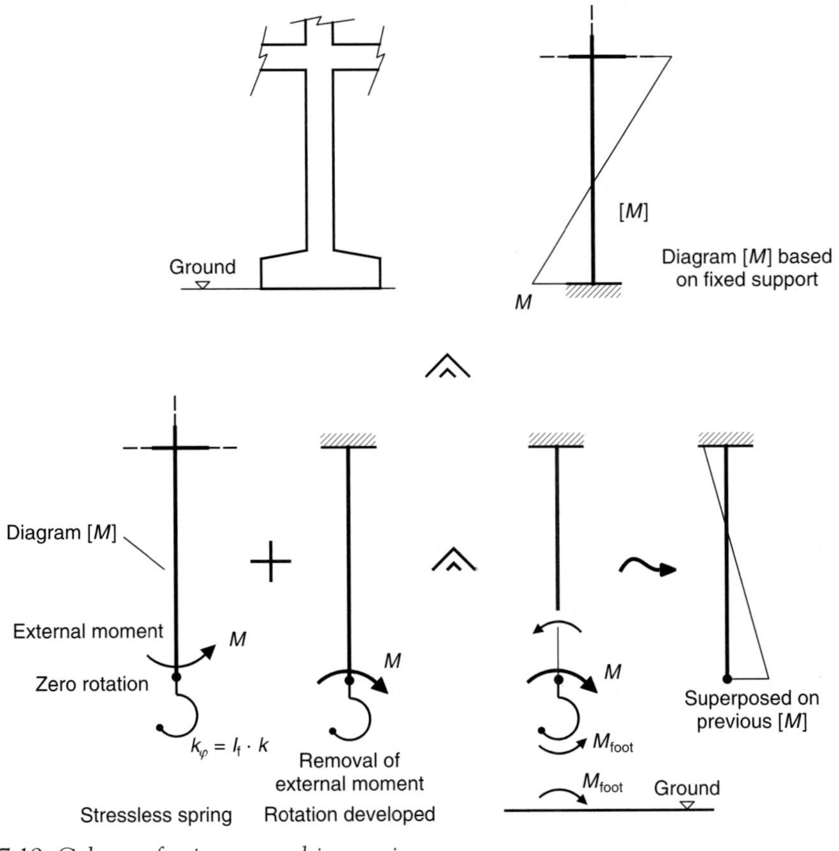

Figure 17.19 Column–footing–ground interaction

exist, an equal and opposite moment M must be superposed, according to the process used in Section 3.3.1.

The above moment M causes a rotation φ to both the rotational spring and the column base, which may be determined from the equilibrium requirement of their common joint. Assuming the upper end of the column to be fixed (see Section 3.3.3.2):

$$\frac{\varphi}{f_\varphi} + \frac{4 \cdot EI}{H} \cdot \varphi = M$$

Thus, the moment developed in the spring, i.e. the moment M_{foot} taken up by the footing and transmitted to the ground, is

$$M_{foot} = \frac{\varphi}{f_\varphi} = \frac{M}{1 + \frac{4 \cdot EI}{H} \cdot f_\varphi}$$

and is clearly reduced compared with M (see Figure 17.19).

As can be seen, this result is identical to that based on Sections 3.2.10 and 3.3.7, namely that the moment M is distributed in proportion to the 'adjacent' stiffnesses:

$$M_{foot} = M \cdot \frac{1/f_\varphi}{\frac{1}{f_\varphi} + \frac{4 \cdot EI}{H}}$$

The moment M_{foot} may be further reduced if beams are inserted that connect the lower end of the examined column with the neighbouring ones (Figure 17.20).

In this case, too, the equilibrium of the joint of the rotational spring, which connects the column and the two connecting beams – with their far ends fixed – leads to the distribution of the moment M proportionally to the stiffness of the three adjacent members (see Figure 17.20). The corresponding portion M_{foot} assigned to the footing surface is

$$M_{foot} = M \cdot \frac{1/f_\varphi}{\frac{1}{f_\varphi} + \sum \frac{4 \cdot EI}{L}} = M \cdot \frac{1}{1 + f_\varphi \cdot \sum \frac{4 \cdot EI}{L}}$$

Generally it is useful to provide such 'connecting ribs' between the lower ends of the columns. However, these are indispensable in cases where an eccentric layout of the footing with respect to the column is constructionally imposed, e.g. along property lines where each column load V necessarily exhibits a certain eccentricity e (Figure 17.21).

Of course, the isolated footings normally follow a symmetric layout with respect to the column section, in view of the possible shift of the column load V to either side, due to the possible occurrence of an alternating moment action M, e.g. during an earthquake.

Returning now to the constructional eccentricity e of a footing, it can be seen that the column moment M deduced by assuming its base to be fixed must be – in the most adverse case – increased by $V \cdot e$ and, consequently, the last expression for the

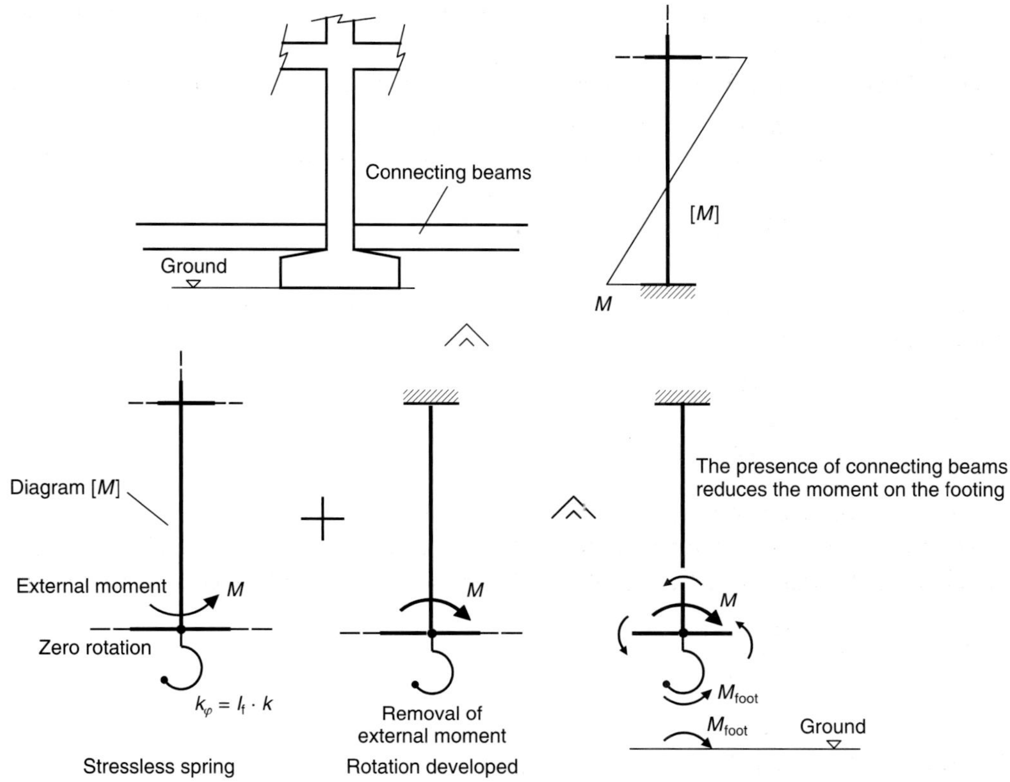

Figure 17.20 The structural action of the connecting ribs

moment taken up by the foundation surface becomes

$$M_{\text{foot}} = \frac{M + V \cdot e}{1 + f_\varphi \cdot \sum \frac{4 \cdot EI}{L}}$$

where, in the denominator, the column member and the connecting rib (i.e. two members) have to be taken into account. It is clear that in evaluating the contact

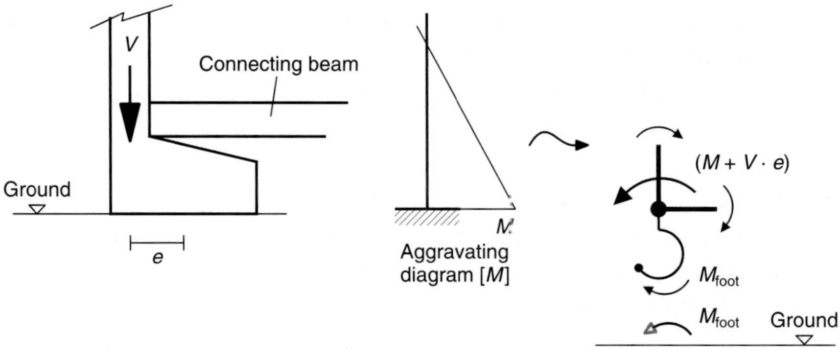

Figure 17.21 Treatment of an eccentric footing

pressures according to the pair (M_{foot}, V), the force V is assumed to act centrally (see Figure 17.21). ■

It is noted that the coefficient f_φ may be determined either through its previous expression as $f_\varphi = 1/(I_f \cdot k)$, which in the case of a square footing with side a, is $f_\varphi = a/(I_f \cdot E_s)$, or, more 'reliably', through the consideration of the foundation soil as an elastic medium, as shown in Section 17.3.2.1.

It should be pointed out here that the more resilient the foundation soil, the smaller is the moment M_{foot} acting on the ground. This is structurally convenient, because a smaller E_s generally means a lower allowable soil pressure but also a greater reduction in the 'effective moment' acting on the foundation surface. However, in the case of a large E_s the coefficient f_φ has a small value, and the reduction of the moment $M + V \cdot e$ is also small. On the other hand, the capacity of the soil to safely take up higher compression stresses is increased.

17.3.2.3 Dimensioning

The footing is subjected as a free body to the soil pressures and to the transmitted sectional forces of the column base (equal and opposite to the corresponding reactions). Under the above forces, the footing acts like a slab loaded 'upwards'. This loading causes bending, with the bottom face of the footing being in tension, whereas the compressed region is limited in practice by the corresponding width of the column. The tensile force, clearly equal to the compressive force, is provided by reinforcement distributed over the whole corresponding width of the footing (Figure 17.22). The dimensioning of the

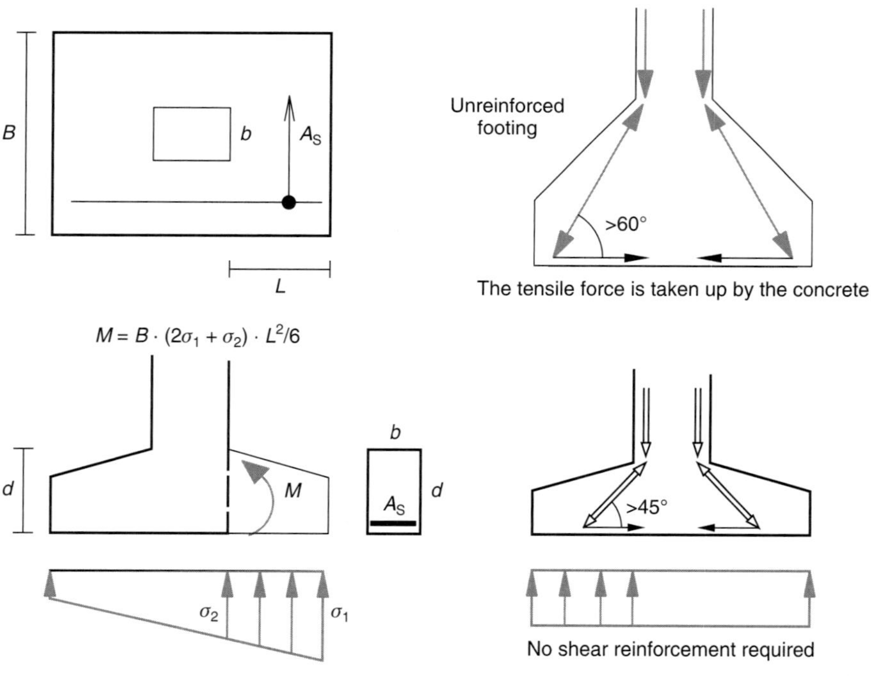

Figure 17.22 Design of a concrete footing

reinforcement refers to a section having as the width b the width of the column and as the depth d the depth of the footing itself (see Figure 17.22). It is clear that the maximum compressive force in the footing due to bending, as determined by the width b, must be kept below the corresponding allowable value.

However, if a tensile force required by the soil pressures, for their transference to the column through the truss mechanism shown in Figure 17.22, can be provided by the concrete itself, then the footing may be constructed without reinforcement. This is the case if the inclination angle of the corresponding strut in the above truss model is at least 60° (see Figure 17.22).

On the other hand, if the reinforced footing has been formed so that the resultant of the soil pressures outside the outline of the column can be directed to its edges under an angle greater than 45°, then the established truss mechanism, as shown in Figure 17.22, transfers the soil pressures to the column without needing shearing reinforcement (Menn, 1990).

If the above conditions are not met, then the footing must be checked not only in bending but also in punching shear, following the procedure in Section 11.5.2.

17.3.3 Foundation beams

As was pointed out in Section 17.3.2, the foundation of columns on single footings should take into account not only the maximum developed soil pressures but also the settlements and rotations that are developed.

It is clear that the settlement of each single footing is generally also affected by its neighbouring ones (see Section 17.3.1), except when the separation between them is more than about triple their dimensions. In this case, the footings settle independently of each other. However, it should be kept in mind that the founded structure is – almost always – a statically indeterminate one, and, as such, it is not sensitive to differential settlements of its supports. Of course, in the case of a relatively low allowable bearing pressure, the resulting footing dimensions could lead to footings lying either very near to each other or even overlapping (Figure 17.23).

For all the above reasons, when designing the foundation of columns belonging to the same vertical plane as in, for example, a frame, it is frequently deemed useful to construct a beam with an appropriate width to serve as a common foundation for the

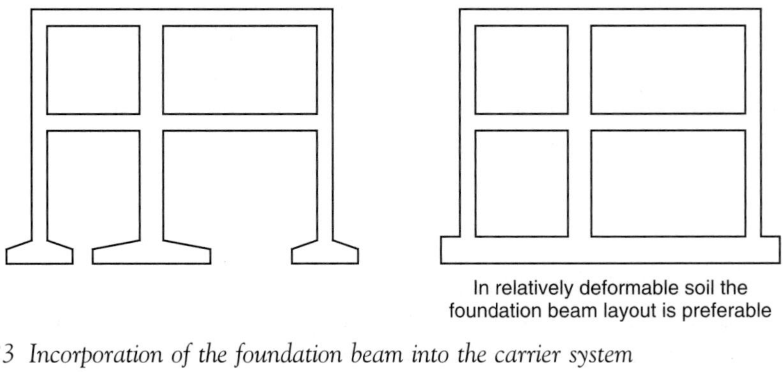

Figure 17.23 Incorporation of the foundation beam into the carrier system

group of columns under consideration. This beam, called the *foundation beam*, is loaded by the upward soil pressures and supported on its columns. Strictly speaking, the corresponding – multi-storey – frame containing the columns and the foundation beam is supported as a complete 'closed' structure on the ground (see Figure 17.23).

In such a frame, the tendency of a column to settle down as a single element is prevented by the bending stiffness of the foundation beam, which shares this settlement between the other columns, clearly limiting in this way the differential settlements. Thus, a foundation beam with a depth adequate to ensure the necessary bending stiffness is clearly superior, as a foundation system, to that consisting of single spread footings. However, while the settlement behaviour of single footings is treated simply through appropriate spring supports – translational and/or rotational ones – supported on a fixed base, the treatment of a foundation beam is more complicated, as the problem consists of the determination of the soil pressures along the foundation beam itself, which implies examination of the interaction of the entire superstructure with the soil. It is clear that this interaction affects not only the response of the foundation beam but also that of the whole structure.

The basic concepts and procedures which allow the investigation of the interaction between the soil and the frame with its foundation beam will be examined below.

17.3.3.1 Soil simulation according to the Winkler model

The behaviour of a foundation beam resting on an elastic base may be approached through the concept of the subgrade modulus k, as defined in Section 17.3.2.2. According to this concept, assuming that a segment of a foundation beam undergoes a settlement w implies that the soil is acted on by a pressure p_{soil} equal to $p_{soil} = k \cdot w$, which, representing the resistance offered by the soil to the imposition of the above settlement, is obviously acting on the beam segment itself too. Expressing this pressure through the distributed load p (kN/m) along the foundation beam of width b, $p_{soil} = p/b = k \cdot w$ and, consequently, $p = K \cdot w$, where $K = b \cdot k$ – expressed now in kN/m^2 – represents the subgrade coefficient in the longitudinal direction of the beam (Figure 17.24).

Thus, the soil is simulated by a continuous distribution of springs, having the property of exhibiting for a uniform load p (kN/m) over some length a uniform settlement equal to $w = p/K$, whereas outside this region no deformation occurs.

The foundation beam having rigidity EI, under the distributed load $q(x)$ is deflected by $w(x)$. The resistance per unit length offered by the beam to this deformation equals $EI \cdot (\mathrm{d}w^4/\mathrm{d}x^4)$, while that offered by the soil equals $K \cdot w$, and, obviously, both resistances balance the externally applied load $q(x)$. Hence (see Figure 17.24),

$$EI \cdot \frac{\mathrm{d}w^4}{\mathrm{d}x^4} + K \cdot w = q(x)$$

This differential equation, known as the *Winkler equation*, permits, through its closed solution, the determination of analytic expressions for the response of a beam segment embedded in an 'elastic soil' of the type examined, thus making possible the evaluation

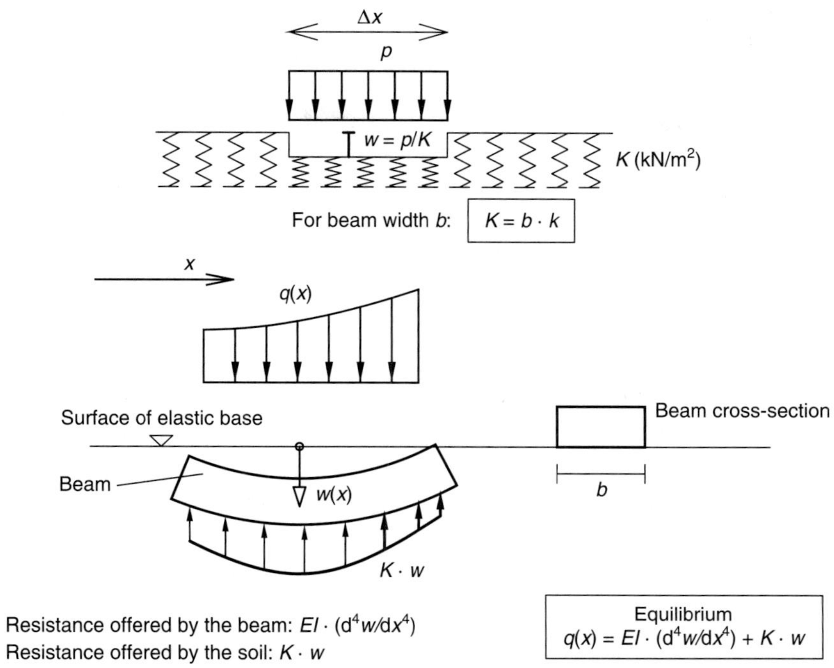

Figure 17.24 Foundation beam on a Winkler foundation

of the response of the whole frame, including, of course, the foundation beam itself. This analytic facility related to the adoption of the subgrade coefficient K makes this method of evaluating the soil–structure interaction very convenient, particularly with the use of an appropriate computer program.

The bending response of the foundation beam of a three-storey frame is shown in Figure 17.25, under a uniform girder loading of 80 kN/m. The outer and inner columns have cross-sectional dimensions of 40/40 and 60/40 cm, respectively. The girders have a uniform section of 40/80 cm, whereas the section of the foundation beam is 60/120 cm. The soil is assumed to have a compression modulus $E_s = 10\,000$ kN/m^2, while its subgrade coefficient K, according to the foregoing, is considered equal to (see Section 17.3.2.2)

$$K = 0.60 \cdot k = 0.60 \cdot (10\,000/0.60) = 10\,000 \text{ kN/m}^2$$

The results are obtained by using appropriate computer software. ■

The following points are again emphasised. The subgrade modulus k does not constitute in any case a soil constant. Ensuring consistency with its definition requires reference to a specific surface, which may be obvious for a footing of specific dimensions, but in the case of the foundation beam is rather unclear. Moreover, loading a single foundation beam with a uniform load causes a uniform settlement over the whole length of the beam, meaning the absence of any bending along the beam, which obviously cannot be valid. However, the concept of the subgrade coefficient k, although physically not

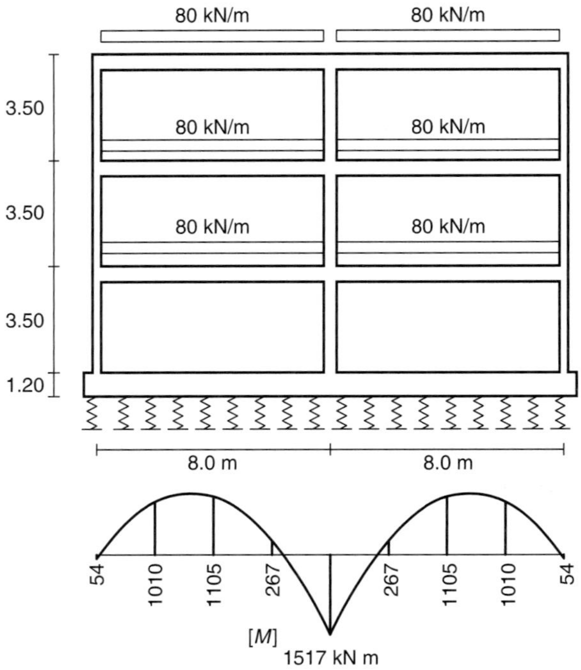

Figure 17.25 Bending response of the foundation beam of a frame on an elastic base

consistent – as pointed out in Section 17.3.2.2 – nevertheless allows an approach to the determination of the soil pressures and to the response of the whole frame. In addition, it should also be recognised that the development of a settlement at some location on the soil surface implies soil deformation outside the considered region, because of the stiffness of the loaded beam, compensating in some way for the weakness of the soil model.

The adoption of a subgrade modulus k leads to the important concept of the so-called *elastic length* L_s of the beam (see Sections 9.7 and 12.3.1):

$$L_s = \sqrt[4]{\frac{4 \cdot EI}{b \cdot k}}$$

The concept of the elastic length applies directly in the case where the foundation beam – theoretically of an infinite length – is loaded by a concentrated load F, which may represent a column load. It denotes that length over which the bending response of the beam extends from either side of the load F (Figure 17.26). Considering the soil pressure p as constant over the total double elastic length, i.e. $p \cdot b = F/(2 \cdot L_s)$, and given that the bending moment at both its ends vanishes, it is found that the bending moment under the concentrated load resulting from either of the corresponding cantilevers represents a very good approximation to the bending moment of the elastically supported long foundation beam developed in practice at this location:

$$M = (F/2L_s) \cdot L_s^2 = 0.25 \cdot F \cdot L_s$$

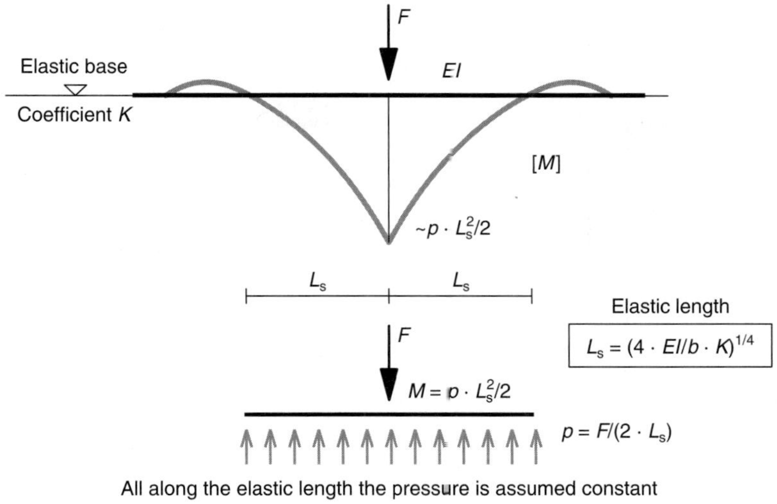

Figure 17.26 Physical meaning of the elastic length

Using the example given in Figure 17.25, the concentrated force F acting on the midpoint of the foundation beam is

$$F \approx 3 \cdot (5 \cdot 80.0 \cdot 8.0/8) \cdot 2 = 2400.0\,\text{kN}$$

and its elastic length is equal to

$$L_s = \sqrt[4]{4 \cdot 2.1 \cdot 10^7 \cdot 0.086/10\,000} = 5.18\,\text{m}$$

Then, according to the above, the bending moment at the location of the concentrated load is $M = 0.25 \cdot 2400.0 \cdot 5.18 = 3108$ kN m. The discrepancy – on the safe side – from the previously calculated value of 1517 kN m is mainly due to the fact that the foundation beam is considered here to be free, i.e. not being prevented from deforming by the stiffness of the superstructure.

17.3.3.2 Soil simulation as an elastic medium

According to the soil model examined in Section 17.3.1, the structural frame is supported with its foundation beam on an infinitely extended elastic subspace, either having a constant compression modulus E_s with a Poisson ratio ν, or consisting of a certain number of horizontal layers, with a corresponding E_s for each layer.

According to precedents, a foundation beam of width b immediately poses the problem of the determination of the distribution of soil pressures $p_{soil}(x)$ along its length. These pressures expressed as a distributed load, $p(x) = p_{soil}(x) \cdot b$ (kN/m), must fulfil the following two conditions:

(1) they must be in equilibrium with the vertical loads of the structure, the foundation beam included
(2) the deformation of the foundation beam must match exactly the surface deformation of the soil due to the load $p(x)$.

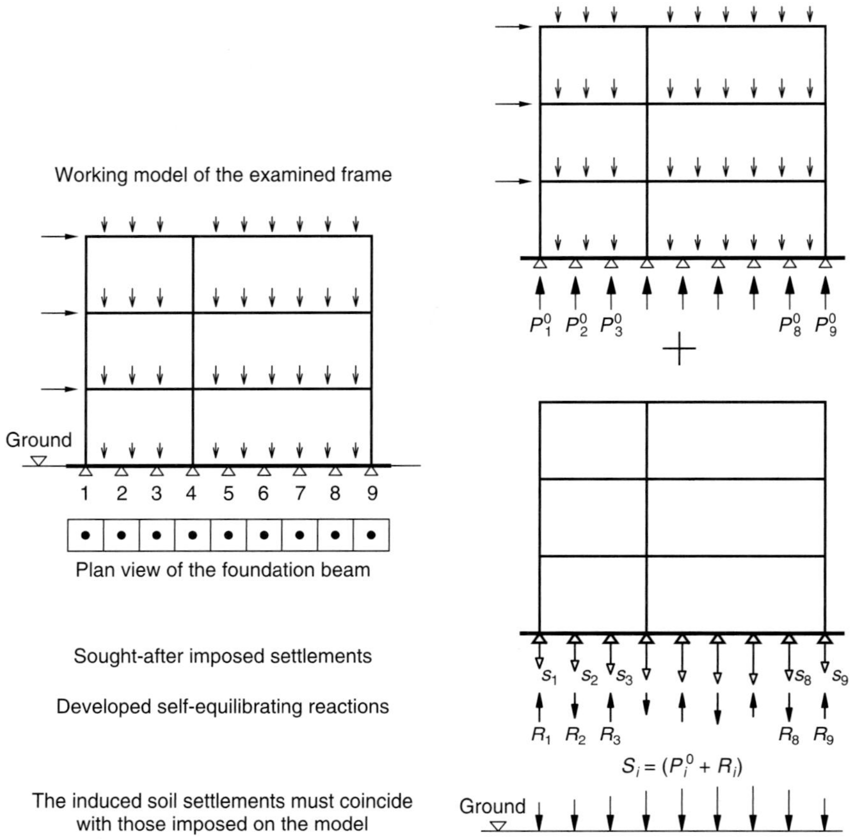

Figure 17.27 Frame foundation with a foundation beam on an elastic subspace

It should be clear that both of these conditions are satisfied by the Winkler model, with the only remark that, for the reasons mentioned earlier, the description of the soil surface deformation through the insertion of springs is not absolutely reliable. Thus, in order to take into account the soil behaviour as an elastic medium in a more consistent manner, the model shown in Figure 17.27 is considered.

The oblong orthogonal contact surface of a foundation beam of length L is divided into a number n of strips of equal width (L/n), at the midpoint of which the resultant of the corresponding soil pressures is acting. The whole structure is assumed to be simply supported at the above n midpoints. These rigid supports under the loads of the superstructure develop the reactions P_i^0, which may be assembled into the column matrix $\boldsymbol{P}_0$ $(n \times 1)$.

It can now be seen that, for a certain group of imposed settlements $s_1, s_2, \ldots, s_n$, the developed reactions of the continuously supported structure $R_1, R_2, \ldots, R_n$, will constitute a self-equilibrating force system summing to zero. The fully developed reactions due to the structure loads and the settlements will be

$$V_1 = R_1 + P_1^0, V_2 = R_2 + P_2^0, \ldots, V_n = R_n + P_n^0$$

The above forces may be considered to be acting in the opposite sense also on the ground, and, thus, the solution of the problem consists of finding the imposed settlements $s_1, s_2, \ldots, s_n$ on the fictitiously supported frame such that the induced soil settlements under the loads $V_1, V_2, \ldots, V_n$ are identical to them (see Figure 17.27). This is achieved by the following procedure (Figure 17.28) (Stavridis, 2002).

Imposing on the first support 1 a unit settlement $s_1 = 1$, a set of self-equilibrating reactions ($k_{11}, k_{21}, k_{31}, \ldots, k_{n1}$) is developed. Next, a unit settlement $s_2 = 1$ is imposed on support 2, leading to another set of self-equilibrating reactions ($k_{12}, k_{22}, k_{32}, \ldots, k_{n2}$). Proceeding consecutively in the same way up to the last support n, a unit settlement $s_n = 1$ is imposed equally, resulting in the self-equilibrating reactions $k_{1n}, k_{2n}, k_{3n}, \ldots, k_{nn}$.

If the set of settlements $s_1, s_2, \ldots, s_n$ is now imposed on the continuously supported frame, the developed reactions $R_1, R_2, \ldots, R_n$ can clearly be expressed through the following relations:

$$R_1 = s_1 \cdot k_{11} + s_2 \cdot k_{12} + s_3 \cdot k_{13} + \ldots + s_n \cdot k_{1n}$$

$$R_2 = s_1 \cdot k_{21} + s_2 \cdot k_{22} + s_3 \cdot k_{23} + \ldots + s_n \cdot k_{2n}$$

$$R_3 = s_1 \cdot k_{31} + s_2 \cdot k_{32} + s_3 \cdot k_{33} + \ldots + s_n \cdot k_{3n}$$

$$R_n = s_1 \cdot k_{n1} + s_2 \cdot k_{n2} + s_3 \cdot k_{n3} + \ldots + s_n \cdot k_{nn}$$

Assembling the forces 'k_{ij}' into a square matrix $\mathbf{K}_{\text{sup}}$ ($n \times n$) and further assembling settlements 's_i' and reactions 'R_i' in the ($n \times 1$) column matrices $\boldsymbol{s}$ and $\boldsymbol{R}$, respectively, the above expressions may be written in matrix form (see Section 16.3.1):

$$\boldsymbol{R} = \mathbf{K}_{\text{sup}} \cdot \boldsymbol{s}$$

Also taking into account the reactions $\boldsymbol{P}_0$ due to the structure loading, as previously mentioned, then the above expressions for the totally developed reactions 'V_i' in the fictitiously supported frame being assembled in the ($n \times 1$) column matrix $\boldsymbol{V}$ can be expressed in matrix form as

$$\boldsymbol{V} = \boldsymbol{P}_0 + \boldsymbol{R} = \boldsymbol{P}_0 + \mathbf{K}_{\text{sup}} \cdot \boldsymbol{s}$$

According to the above, what remains is the requirement from the forces $V_1, V_2, \ldots, V_n$, if applied in the opposite sense to the ground, to produce the soil settlements $s_1, s_2, \ldots, s_n$ (see Figure 17.28). For this purpose, the relation describing the deformational behaviour of the soil surface is recalled from Section 17.3.1:

$$\boldsymbol{W} = \mathbf{F} \cdot \boldsymbol{P}$$

where, now, $\boldsymbol{W} = \boldsymbol{s}$ and $\boldsymbol{P} = -\boldsymbol{V}$. This relation, after multiplying both sides by $\mathbf{F}^{-1}$, allows the following expression for the forces $\boldsymbol{V}$:

$$\boldsymbol{V} = -\mathbf{F}^{-1} \cdot \boldsymbol{s}$$

By substituting the above relation, the following equation is finally obtained:

$$(\mathbf{F}^{-1} + \mathbf{K}_{\text{sup}}) \cdot \boldsymbol{s} = -\boldsymbol{P}_0$$

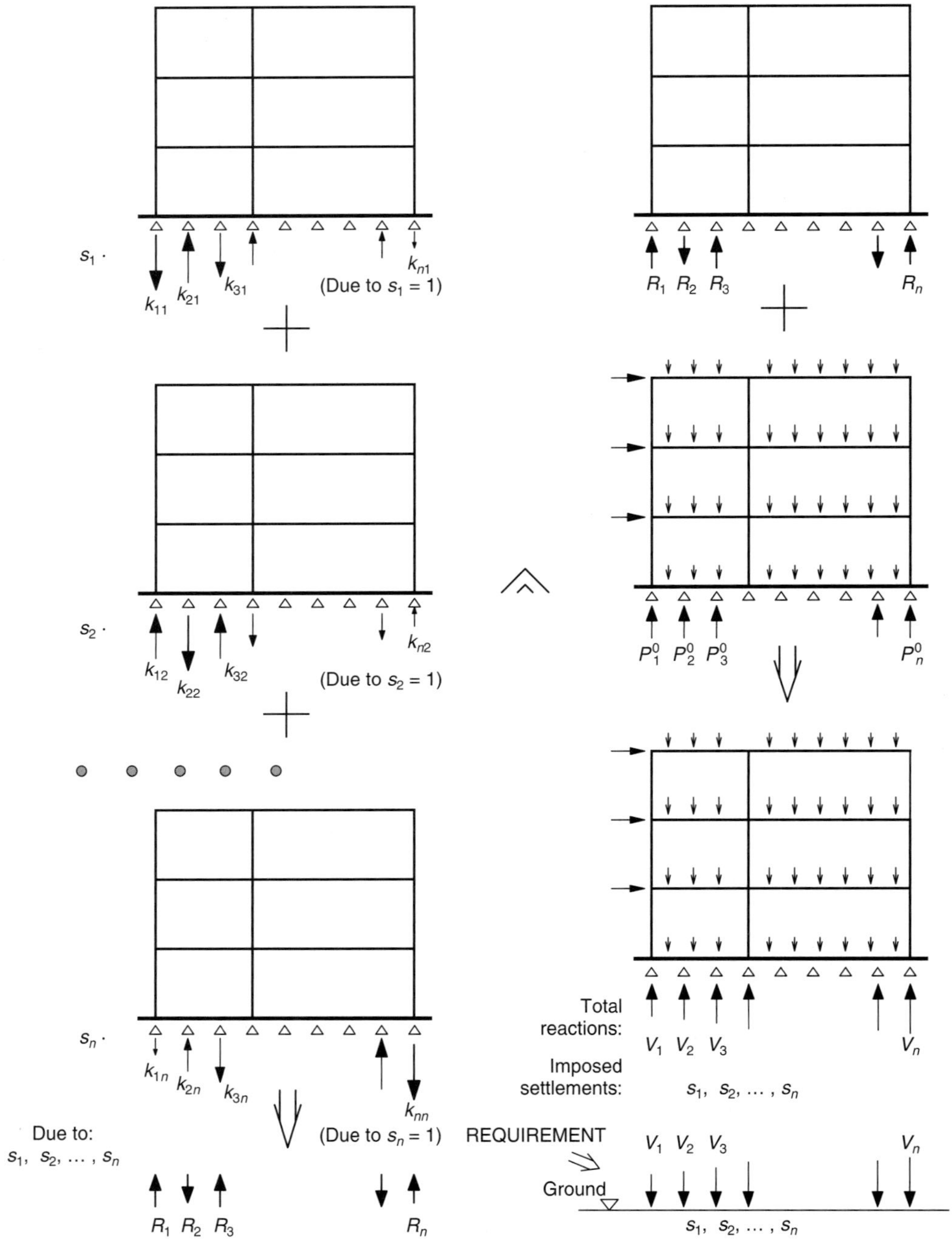

Figure 17.28 Practical analysis of the soil–frame interaction

By determining the settlements 's_i' from this linear system, the forces 'V_i' acting on the ground are readily obtained from the penultimate relation.

It is obvious that the whole procedure is feasible only by using an appropriate computer program for plane frame analysis.

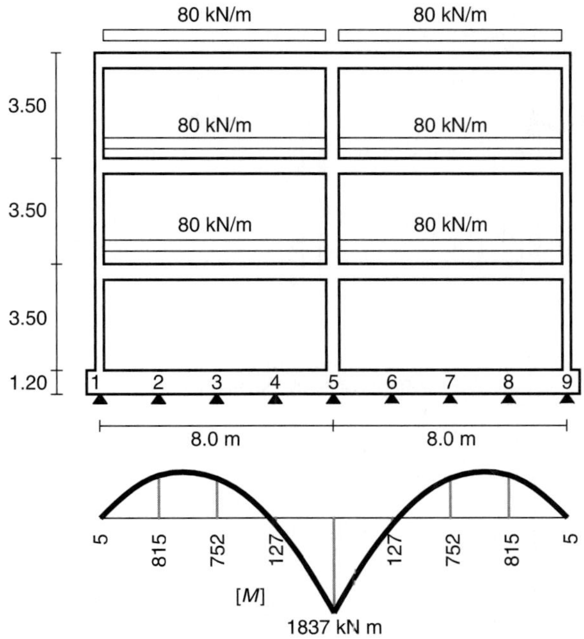

Figure 17.29 Example of frame analysis, with the frame resting on an elastic subspace

Numerical example

For reasons of direct comparison, the frame examined in Section 17.3.3.1 will again be considered, under the same loading and soil characteristics (Figure 17.29).

The model under consideration is supported at nine intermediate equidistant points of the foundation beam, as shown in Figure 17.29. According to the procedure described above, the following settlements are obtained:

$s_1 = 0.0571$ m, $s_2 = 0.0562$ m, $s_3 = 0.0565$ m, $s_4 = 0.0577$ m, $s_5 = 0.0587$ m,
$s_6 = 0.0577$ m, $s_7 = 0.0565$ m, $s_8 = 0.0562$ m, $s_9 = 0.0571$ m

This means that the state of stress and the deformation of the frame examined are identical to those of the fictitiously supported model, under the existing loading and the above support settlements.

The bending moment diagram of the foundation beam is shown in Figure 17.29. By comparing the results with those based on the adoption of the Winkler soil model, it is observed that in the present case where the soil is considered as an elastic half-space, the span moments have lower values, whereas the values of the bending response in the region of the central column are greater.

17.4 Pile foundations

17.4.1 Overview

As mentioned in Section 17.3, deep foundations transfer the loads of the superstructure, through weak unsuitable soils, to a lower stratum with sufficient bearing strength. The

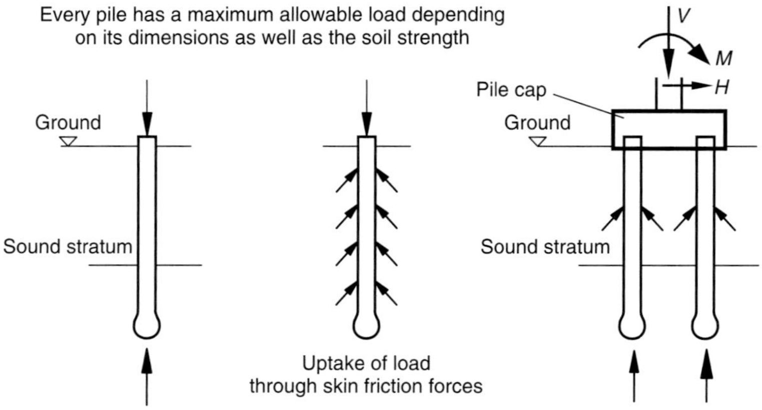

Figure 17.30 Layout and load-bearing action of piles

usual way to accomplish this transfer is through the use of piles. Obviously, piles are used when the subsurface conditions are not suitable for a shallow foundation. Piles are vertical or slightly inclined structural members made of concrete, up to 30 m long, having a circular cross-section and a diameter ranging from about 0.40 m up to 1.50 m. The piles are arranged in a group of relatively dense layout, being connected with each other at their upper end through a thick slab called a *pile cap*, which is further connected to the superstructure. At their lower end, the piles have a slightly enlarged base, resting on a soil layer of satisfactory bearing strength. In relatively rare cases, where this layer lies so deep that the length required from the piles would result in a prohibitive total cost, the equilibrium of the loads transmitted to the piles from the pile cap is mainly accomplished through skin friction forces – developed between the so-called *friction pile* – and the adjoining soil, through the whole length of the piles (Figure 17.30).

Thus, the foundation system consisting of the pile cap and the piles is in equilibrium under the action of the superstructure applied to the pile cap and the forces acting on the lower end of piles from the sound stratum. It is clear that, in the case of friction piles, the role of the acting soil force at the lower pile end is undertaken by the total force of the lateral friction. Usually, there is a coexistence of friction and bearing forces in a rather uncertain proportion that does not affect the total axial force, in practice, that is required to be taken up by the pile in order to fulfil the global equilibrium conditions.

The pile cap, as shown in Figure 17.30, is connected at its top to the supported element, which transmits the actions V, M_x, M_y (i.e. a vertical force and a moment with an arbitrary horizontal vector), as well as a horizontal force H (Figure 17.31).

The piles are usually arranged in a symmetrical layout around the base of the founded element, and they are generally placed in a vertical position. With regard to the horizontal force H, this is taken up either by the vertical piles themselves, or by *batter piles* specially constructed for that purpose, with an angle of inclination not greater than 15°.

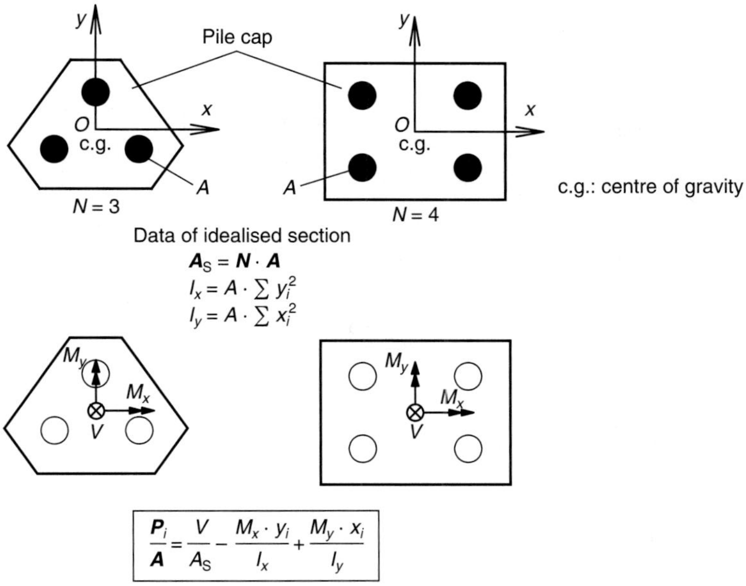

Figure 17.31 Loading of a pile system through a pile cap

17.4.2 Vertical loads

The symmetrical layout of N piles around the point of application O of the load V implies the existence of the orthogonal axes of symmetry Ox and Oy. The components M_x and M_y are referred to these two axes (see Figure 17.31).

For preliminary design purposes, it is sufficient to consider the magnitudes V, M_x, M_y as being applied to a rigid section, consisting of the total layout of circular section areas of the piles. Given the coordinates x_i, y_i of the centres of the circular pile sections, with respect to the centroid O of the section and assuming that all piles have a common section area A, the following sectional data are derived:

- Total area of section: $A_s = N \cdot A$.
- Moment of inertia about the Ox axis: $I_x = A \cdot \sum y_i^2$.
- Moment of inertia about the Oy axis: $I_y = A \cdot \sum x_i^2$.

Thus, the consideration of biaxial bending, to which the total section is subjected, leads to the determination of the axial force P_i for each pile separately (see Figure 17.31):

$$\frac{P_i}{A} = \frac{V}{A_s} - \frac{M_x \cdot y_i}{I_x} + \frac{M_y \cdot x_i}{I_y}$$

It is clear that the pile cap is in equilibrium under the loads V, M_x, M_y and the pile forces P_i, and must, therefore, be designed to safely take up the above actions. For preliminary design purposes, the eventual bending of the piles is ignored. Moreover, selecting a large enough thickness for the pile cap, i.e. equal to at least half of the clear distance between the piles, the bending action of the cap is excluded and the load introduced from the column is transferred to the piles through a strut-and-tie model, normally a

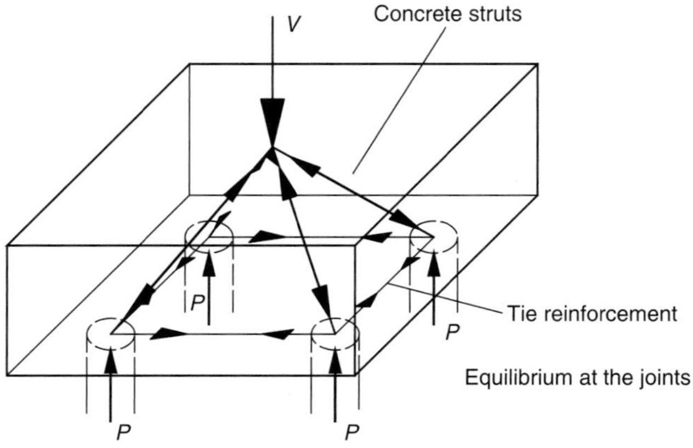

Figure 17.32 Load-bearing action of a pile cap

three-dimensional one (Figure 17.32). In this model, which depicts the load transfer through the most 'stiff' – i.e. the shortest – paths, the struts represent the compressed concrete between the column and the pile heads, whereas the ties are realised through reinforcement bars, arranged within the narrow region of the pile circular section with good anchor conditions, as shown in Figure 17.32, thus ensuring the equilibrium of the formed joints.

Designing in this way, that is, by taking up the design loads (i.e. service loads magnified by the corresponding safety factors) on the basis of a constructionally feasible equilibrium, ensures, according to the static theorem of the plastic analysis (see Section 6.6.2), that the collapse load is greater than that which the substitute strut-and-tie model can carry.

17.4.3 Horizontal loads

The uptake of a horizontal load by a vertical pile clearly implies a bending response of the pile. The pile acts as if laterally supported by an elastic medium described by a subgrade soil coefficient k_h and substituted accordingly through a continuous distribution of lateral springs, as in the case of the foundation beam (Figure 17.33).

In non-cohesive soils, the coefficient k_h can be considered to increase linearly to a depth t, according to the relation $k_h = n_h \cdot t/D$, where D is the pile diameter and n_h is a quantity representing the lateral stiffness of the soil, ranging from 2 to 18 MN/m^3, depending on the layer density.

In cohesive soils, the coefficient k_h is considered to be constant, and is $k_h = n_h/D$, where n_h is expressed in terms of cohesion c as $n_h = 160 \cdot c$.

Clearly, the uptake of a horizontal force, or even a moment acting on top of a pile (transmitted through the pile cap), is accomplished through an appropriate distribution of lateral soil reactions along the length of the pile, implying either tension or

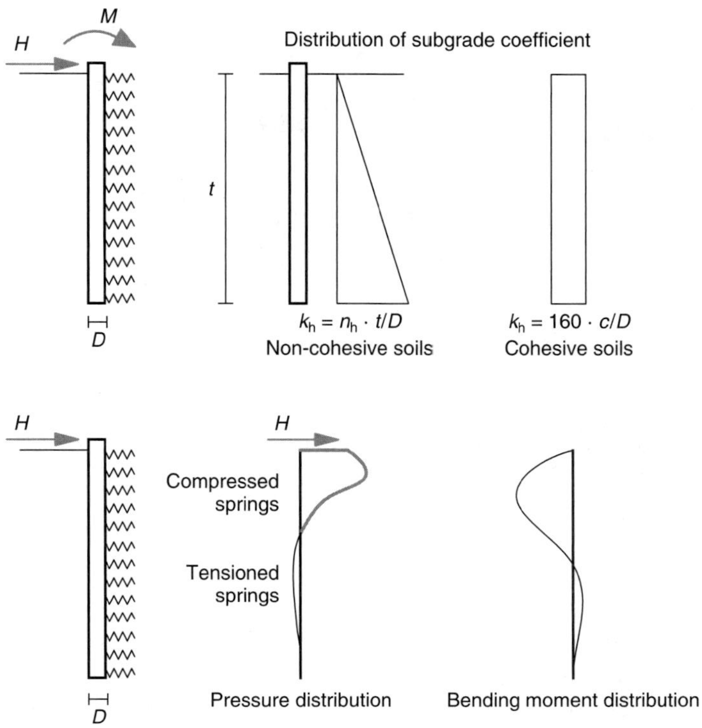

Figure 17.33 Lateral response of piles

compression of the corresponding supporting springs (see Figure 17.33). In contrast to the foundation beam, where the development of tensile spring forces is not normally allowed, in the present case this is obviously meaningful, as any resulting spring force represents compressive contact pressure between the soil and the pile.

It is clear that the above response may be determined using the same software as used for the analysis of foundation beams.

References

Menn C. (1990) *Prestressed Concrete Bridges*. Basel: Birkhäuser Verlag.

Stavridis L. (1997) Tragweke auf elastischem Boden. *Bauimgenieur* **72**(12) 565–569.

Stavridis L. (2002) Simplified analysis of layered soil–structure interaction. *Journal of Structural Engineering* (ASCE) **168**(2), 224–230.

Stavridis L. (2009) Rigid foundations resting on an elastic layered soil. *Geotechnical and Geological Engineering* **27**, 407–417.

Bibliography

Bachmann H., Ammann W. (1987) *Vibrations in Structures – Induced by Man and Machines*. Zurich: IABSE.

Bachmann H., Ammann W. *et al.* (1995) *Vibration Problems in Structures*. Boston: Birkhäuser.

Beles A.A., Soare M.V. (1972) *Berechnung von Schalentragwerken*. Wiesbaden: Bauverlag.

Bieger K.W., Lierse J., Roth J. (1993) *Stahlbeton – und Spannbetontragwerke*. Berlin: Springer-Verlag.

Biggs J.M. (1964) *Introduction to Structural Dynamics*. New York: McGraw-Hill.

Billington D. (1965) *Thin Shell Concrete Structures*. New York: McGraw-Hill.

Ciesielski R., Mitzel A., Stachurrski W., Suwalski J., Zmudzinski Z. (1970) *Behälter, Bunker, Silos, Schornsteine, Fernsehtürme und Freileitungsmaste*. Berlin: Wilhelm Ernst.

Duddeck H. (1984) *Traglasttheorie der Stabwerke – Beton Kalender*, Teil II. Berlin: Ernst.

Flügge W. (1960) *Stresses in Shells*. Berlin: Springer-Verlag.

Franz G., Schäfer K. (1988) *Konstruktionslehre des Stahlbetons*, Band II, A. Berlin: Springer-Verlag.

Franz G., Hampe E., Schäfer K. (1991) *Konstruktionslehre des Stahlbetons*, Band II, B. Berlin: Springer-Verlag.

Gaylord E. H., Gaylord C. N. (1957) *Design of Steel Structures*. New York: McGraw-Hill.

Herzog M. (1996) *Kurze baupraktische Festigkeitslehre*. Düsseldorf: Werner Verlag.

Kollar L. (1984) *Schalenkonstruktionen – Beton – Kalender 1984*, Teil II. Berlin: Wilhelm Ernst.

Leonhardt F., Zellner W. (1980) Cable stayed bridges. *IABSE Surveys*, S-13/80.

Mann W. (1976) Kippnachweis und Kippaussteifung von schlanken Stahlbeton – und Spannbetonträgernn. *Beton – und Stahlbetonbau* **2**, 37–42.

Menn C. (1990) *Prestressed Concrete Bridges*. Basel: Birkhäuser Verlag.

Müller F.P., Keintzel E. (1984) *Erdbebensicherung von Hochbauten*. Berlin: Wilhelm Ernst.

Nervi P.L. (1956) *Structures*. New York: McGraw-Hill.

O'Connor C. (1971) *Design of Bridge Superstructure*. New York: John Wiley.

Pflüger A. (1963) Zur praktischen Berechnung der axial gedrückten Kreiszylinderschale. *Stahlbau* **32**, 161–164.

Pflüger A. (1966) Zur praktischen Berechnung der axial gedrückten Kreiszylinderschale unter Manteldruck. *Stahlbau* **35**, 249–252.

Roik K. (1983) *Vorlesungen über Stahlbau. Grundlagen*. Berlin: Wilhelm Ernst.

Rosman R. (1983) *Erdbeben–widerstandsfähiges Bauen*. Berlin: Wilhelm Ernst.

Salvadori M. and Levy M. (1967) *Structural Design in Architecture*. Englewood Cliffs, NJ: Prentice-Hall.

Schlaich J. and Scheef H. (1982) *Concrete Box-girder Bridges*. Zurich: IABSE.

Seide P. (1981) Stability of cylindrical reinforced concrete shells. *American Concrete Institute ACI* SP **67**, 43–62.

Stavridis L.T. (1986) Static and dynamic analysis of multistory systems. *Technika Chronika Scientific Journal of the Technical Chamber of Greece* **6**(2), 187–219.

Stavridis L.T. (1993) Statische und dynamische Analyse von orthotropen rechteckigen Platten. *Stahlbau* **62**, 73–80.

Stavridis L.T. (1994) Beanspruchung von mehrstöckigen Bauten infolge Temperaturänderung. *Bauingenieur* **69**(3), 117–122.

Structural systems: behaviour and design
978-0-7277-4105-9

Copyright Thomas Telford Limited © 2010
All rights reserved

Stavridis L.T. (1997) Tragweke auf elastischem Boden. *Bauimgenieur* **72**(12), 565–569.

Stavridis L.T. (2001) Alternative layout for the prestressing of slab bridges. *Journal of Bridge Engineering of the ASCE*, **6**(5), 324–332.

Stavridis L.T. (2002) Simplified analysis of layered soil–structure interaction. *Journal of Structural Engineering of the ASCE* **168**(2), 224–230.

Stavridis L.T. (2008) A simplified analysis of the behavior of suspension bridges under live load, *Structural Engineering and Mechanics* **30**(5), 559–576.

Stavridis L.T. (2009) Rigid foundations resting on an elastic layered soil. *Geotechnical and Geological Engineering* **27**, 407–417.

Stavridis L.T. (2010) Evaluation of static response in stress-ribbon concrete pedestrian bridges, *Structural Engineering and Mechanics* **34**(2), 213–229.

Stavridis L.T., Armenakas A.E. (1988) Analysis of shallow shells with rectangular projection: applications. *Journal of Engineering Mechanics of the ASCE* **114**(6), 943–952.

Tedesko J.W., McDougal W.G., Ross C. Allen (1999) *Structural Dynamics – Theory and Applications*. California: Addison Wesley Longman.

Timoshenko S. (1956) *Strength of Materials*, Part II. Princeton, NJ: Van Norstrand.

Timoshenko S., Young D. (1965) *Theory of Structures*. New York: McGraw-Hill.

Tveit P. (1966) Design of network arches. *Structural Engineer* **44**(7), 247–259.

Vlasov V.Z. (1961) *Thin Walled Elastic Beams*. London: Oldbourne Press.

VSL (1984) *VSL Stay Cables for Cable-stayed Bridges*. Berne, VSL.

Walther R., Miehlbradt M. (1990) *Dimensionnement des Structures en Béton*. Lausanne: Presses polytechniques et universitaires romandes.

Zbirohowski–Koscia K. (1967) *Thin Walled Beams – From Theory to Practice*. London: Crosby Lockwood.

Index

Page numbers in *italics* refer to figures separate from the corresponding text.

Structural systems: behaviour and design
978-0-7277-4105-9

Copyright Thomas Telford Limited © 2010
All rights reserved